SUGARCANE PRODUCTION

Environmental Impacts and Socioeconomic Issues

SUGARCANE PRODUCTION

Environmental Impacts and Socioeconomic Issues

Krishan K. Verma, PhD
Xiu-Peng Song, PhD
Lin Xu, PhD
Rajan Bhatt, PhD
Yang-Rui Li, PhD
Editors

AAP | APPLE ACADEMIC PRESS

First edition published 2026

Apple Academic Press Inc.
1265 Goldenrod Circle, NE,
Palm Bay, FL 32905 USA

760 Laurentian Drive, Unit 19,
Burlington, ON L7N 0A4, Canada

CRC Press
2385 NW Executive Center Drive,
Suite 320, Boca Raton FL 33431

4 Park Square, Milton Park,
Abingdon, Oxon, OX14 4RN, UK

Library and Archives Canada Cataloguing in Publication

CIP data on file with Canada Library and Archives

Library of Congress Cataloging-in-Publication Data

CIP data on file with US Library of Congress

ISBN: 978-1-779641-12-0 (hbk)
ISBN: 978-1-779641-13-7 (ebk)
DOI: 10.1201/9781779641137

About the Editors

Krishan K. Verma, PhD
Sugarcane Research Institute, Guangxi Academy of Agricultural Sciences, Nanning, Guangxi, China

Krishan K. Verma PhD, is working as a Principal Scientist at the Sugarcane Research Institute, Guangxi Academy of Agricultural Sciences/Guangxi Key Laboratory of Sugarcane Genetic Improvement/Key Laboratory of Sugarcane Biotechnology and Genetic Improvement (Guangxi) of the Ministry of Agriculture and Rural Affairs in Nanning, Guangxi, China. He has 14 years of research experience in the field of biotic and abiotic stresses, such as environmental toxicology, plant physiology, bioengineering, and nanobioscience. His research focuses on environmental toxicology, bioengineering modeling, plant physiology, and molecular biology, and their impacts on growth and physiological and molecular adaptation strategies of plants to stressful environments and how they affect plant structure. He is working with Prof. Yang-Rui Li, a world-renowned plant scientist. He received an "Award of Excellence-2021" for his contribution to agriculture science from the Sugarcane Society of the Guangxi Academy of Agricultural Sciences (GXAAS), Nanning, Guangxi, China. He has also been elected as a Fellow of the Society for Sugarcane Research & Promotion (FSSRP-2024) for his outstanding contribution to sugarcane crop production. He has been included in the list of 2% world's top scientists by Elsevier BV and Stanford University. He is a Regional Editor of *Fluoride* journal, published by ISFRI, Dunedin, New Zealand. He has been serving as an editorial board member for various peer-reviewed journals.

Xiu-Peng Song, PhD
Sugarcane Research Institute, Guangxi Academy of Agricultural Sciences, Nanning, Guangxi, China

Xiu-Peng Song is working as Associate Professor at the Sugarcane Research Institute, Guangxi Academy of Agricultural Sciences/Guangxi Key Laboratory of Sugarcane Genetic Improvement/Key Laboratory of Sugarcane Biotechnology and Genetic Improvement (Guangxi) of Ministry of Agriculture and Rural Affairs, Nanning, Guangxi, China. His ongoing research focuses on plant physiology and biochemistry of stress resistance, occurrence mechanisms, and comprehensive control of diseases and insect pests in sugarcane. He has presided over four funded projects, such as by the National Natural Science Foundation, China, and participated in 16 projects that involved national and international cooperation. He has won six awards, including a Guangxi Science and Technology Progress Award. He has published over 65 scientific publications, three book chapters, and four books (Springer & Apple Academic Press, a group of Taylor & Francis). He received his master's and PhD degrees (Biochemistry and Molecular Biology) from Guangxi University, Guangxi, China.

Lin Xu, PhD
Sugarcane Research Institute, Guangxi Academy of Agricultural Sciences, Nanning, Guangxi, China

Lin Xu is an Associate Researcher working at the Sugarcane Research Institute, Guangxi Academy of Agricultural Sciences, Nanning, Guangxi, China. She is mainly engaged in conventional breeding and cultivation of sugarcane, innovation of sugarcane germplasm resources, and key technology innovation of the sugarcane seed industry. She has participated in and presided over several provincial, ministerial, and prefectural level projects. She has been granted several invention and utility model patents and has participated in the grant of other invention patents. She has also participated in the grant of 26 patents (as first inventor) and has obtained three computer software copyrights. She has participated in the compilation of two local standards in Guangxi.

Her awards include the Guangxi Science and Technology Award for Scientific and Technological Progress (first prize), third prize of Guangxi Science and Technology Award for Technological Invention, and the first prize from the Guangxi Academy of Agricultural Sciences for Scientific and Technological Progress. She has published over 40 scientific publications in high repute journals.

Rajan Bhatt, PhD
Krishi Vigyan Kendra, Amritsar, Punjab, Punjab Agricultural University (PAU), Punjab, India

Rajan Bhatt is currently working as an Associate Professor of Soil Science at Krishi Vigyan Kendra, Amritsar, Punjab Agricultural University (PAU), India. He formerly worked at the Regional Research Station at Punjab Agricultural University, Ludhiana, India. He has more than 13 years of research experience in enhancing farmers' livelihoods and more than five years in enhancing sugarcane crop growth and yield, insect-pest management [early shoot borer (*Chilo infuscatellus*) and stalk borer (*Chilo auricilius*)] under differential regimes of irrigation and potash under stressed conditions. Dr. Bhatt also worked on foliar application of potassium salt of active phosphorus (PSAP) to mitigate insect pests and improve yield along with sugarcane quality in response to agroclimatic conditions of Punjab for the first time. He assessed the changes in soil organic pool and carbon preservation capacity of macro- and micro-aggregates in response to land-use change in northwestern India and tested polyhalite ($K_2Ca_2Mg(SO_4)_4 \cdot 2H_2O$) in Punjab, India, which positively influences sugarcane's growth, yield, and quality in potassium and calcium-deficient soils in the semiarid tropics. He actively improves farmer livelihoods by reducing water, fertilizer, and carbon footprints in their fields.

Yang-Rui Li, PhD
Sugarcane Research Center, Chinese Academy of Agricultural Sciences; Chief-Expert, National Joint Research Program for Elite Sugarcane Variety Development, China

Yang-Rui Li has been serving as Professor (Life-time and Director of the Sugarcane Research Center of Chinese Academy of Agricultural Sciences since September 2007. He has also been serving as Chief-Expert of the National Joint Research Program for Elite Sugarcane Variety Development in China since 2018. He is President of the International Association for Professionals of Sugar and Integrated Technologies (IAPSIT), Vice-President of the Society for Sugar Research and Promotion (SSRP), and President of the Chinese Sugarcane Industry Association for Technological Innovation (CSIATI). His research interests include sugarcane physiology, biochemistry and molecular biology, sugarcane genetics, breeding and cultivation, and chemical regulation of growth and development of sugarcane. He has published 16 books and more than 1000 papers and has received 23 scientific research achievement awards from the Chinese government to date as well as 14 awards from international organizations, including a Lifetime Achievement Award—SSRP (2011), Lifetime Achievement Award—IAPSIT (2014), and Leadership Excellence Award from the Thailand Society of Sugarcane and Sugar Technologists (2018). He has been affiliated with Guangxi Agricultural University in many roles, including lecturer, professor, university vice president as well as President of the Guangxi Academy of Agricultural Sciences.

Contents

Contributors

Noé Aguilar-Rivera
Universidad Veracruzana, Facultad de Ciencias Biológicas y Agropecuarias, Córdoba, Veracruz, México

Musa Opeyemi Ahmed
Nigeria Sugar Institute, Ilorin, Kwara State, Nigeria and Department of Agricultural and Environmental Engineering, University of Ibadan, Oyo State, Nigeria

Mohammad Ashfaque
Department of Biosciences, Integral University, Lucknow, Uttar Pradesh, India

Rajan Bhatt
PAU-Krishi Vigyan Kendra, Amritsar, Punjab, India

Mandapelli Sharath Chandra
AICRP on Integrated Farming System,
Professor Jayashankar Telangana State Agricultural University, Telangana, India

Faith Eweluegim Enahoro-Ofagbe
Nigeria Sugar Institute, Ilorin, Kwara State, Nigeria and Department of Agronomy, University of Ilorin, Kwara State, Nigeria

Alvina Farooqui
Department of Bioengineering, Integral University, Lucknow, Uttar Pradesh, India

Anshu Gangwar
Department of Agricultural Engineering, Krishi Vigyan Kendra, (DRPCAU, Pusa), Bihar, India

Gael Francisco García-Merino
Universidad Veracruzana, Facultad de Ciencias Biológicas y Agropecuarias, Córdoba, Veracruz, México

Akhtar Hussain
Department of Biosciences, Integral University, Lucknow, Uttar Pradesh, India

Adepoju Moronkola Idris
Nigeria Sugar Institute, Ilorin, Kwara State, Nigeria and Department of Agronomy, University of Ilorin, Kwara State, Nigeria

Irum
Department of Bioengineering, Integral University, Lucknow, Uttar Pradesh, India

Kingsley O. Iwuozor
Nigeria Sugar Institute, Ilorin, Kwara State, Nigeria and Department of Pure and Industrial Chemistry, Nnamdi Azikiwe University, Akwa, Anambra State, Nigeria

V. P. Jaiswal
ICAR-Indian Institute of Sugarcane Research, Lucknow, Uttar Pradesh, India

Jaisuriyan K.
Department of Agronomy, School of Agriculture, Lovely Professional University, Phagwara, Punjab, India

Praveen V. Kadam
Agronomy, KSNUAHS, Shivamogga, Karnataka, India

Rayapati Karthik
College of Agriculture, Professor Jayashankar Telangana State Agricultural University, Hyderabad, Telangana, India

Prasann Kumar
Department of Agronomy, School of Agriculture, Lovely Professional University, Phagwara, Punjab, India

Navnit Kumar
Department of Agronomy, Sugarcane Research Institute,
Dr. Rajendra Prasad Central Agricultural University, Pusa, Samastipur, Bihar, India

Nallagatla Vinod Kumar
Malla Reddy University, Maisammaguda, Hyderabad, Telangana, India

Ch. Pragathi Kumari
AICRP on Integrated Farming System,
Professor Jayashankar Telangana State Agricultural University, Telangana, India

Yang-Rui Li
Sugarcane Research Institute, Guangxi Academy of Agricultural Sciences/Key Laboratory of Sugarcane Biotechnology and Genetic Improvement (Guangxi), Ministry of Agriculture and Rural Affairs/Guangxi Key Laboratory of Sugarcane Genetic Improvement, Nanning, Guangxi, China

Qiang Liang
Sugarcane Research Institute, Guangxi Academy of Agricultural Sciences/Key Laboratory of Sugarcane Biotechnology and Genetic Improvement (Guangxi), Ministry of Agriculture and Rural Affairs/Guangxi Key Laboratory of Sugarcane Genetic Improvement, Nanning, Guangxi, China

Jyotirmay Mahapatra
ICAR-Central Institute of Agricultural Engineering, Bhopal, Madhya Pradesh, India

A. K. Mall
ICAR-Indian Institute of Sugarcane Research, Lucknow, Uttar Pradesh, India

Bhupendra Mathpal
Department of Agronomy, School of Agriculture, Lovely Professional University, Phagwara, Punjab, India

Kamlesh K. Mishra
Department of Botany, University of Lucknow, Lucknow, Uttar Pradesh, India

Varucha Misra
ICAR-Indian Institute of Sugarcane Research, Lucknow, Uttar Pradesh, India

Vikas Pagare
Central Institute of Agricultural Engineering, Bhopal, Madhya Pradesh, India

L. Peace Raising
Department of Agricultural Sciences, Sharda University, Greater Noida, Uttar Pradesh, India

Lalita Rana
Department of Agronomy, Sugarcane Research Institute, Dr. Rajendra Prasad Central Agricultural University, Pusa, Samastipur, Bihar, India

Ramesh K. Sahni
Central Institute of Agricultural Engineering, Bhopal, Madhya Pradesh, India; Center for Precision and Automated Agricultural Systems, Washington State University, Pullman, WA, USA

Lalan Sharma
ICAR-Indian Institute of Sugarcane Research, Lucknow, Uttar Pradesh, India

S.K. Shukla
ICAR-Indian Institute of Sugarcane Research, Lucknow, Uttar Pradesh, India

Bhaskar Pratap Singh
Department of Agricultural Engineering, ANDUAT-Krishi Vigyan Kendra, Kathaura, Amethi, Uttar Pradesh, India

Dinesh Singh
ICAR-Indian Institute of Sugarcane Research, Lucknow, Uttar Pradesh, India

Jagjeet Singh
Central Institute of Agricultural Engineering, Bhopal, Madhya Pradesh, India

Sujeet Pratap Singh
Division of Plant Pathology, Sugarcane Research Institute, U.P. Council of Sugarcane Research, Shahjahanpur, Uttar Pradesh, India

S. P. Singh
Division of Plant Pathology, Sugarcane Research Institute, U.P. Council of Sugarcane Research, Shahjahanpur, Uttar Pradesh, India

Trisha Sinha
Department of Plant Physiology, Narayan Institute of Agricultural Sciences, Gopal Narayan Singh University, Rohtas, Bihar, India

Xiu-Peng Song
Sugarcane Research Institute, Guangxi Academy of Agricultural Sciences/Key Laboratory of Sugarcane Biotechnology and Genetic Improvement (Guangxi), Ministry of Agriculture and Rural Affairs/Guangxi Key Laboratory of Sugarcane Genetic Improvement, Nanning, Guangxi, China

Sumit Sow
Department of Agronomy, PGCA, Dr. Rajendra Prasad Central Agricultural University, Pusa, Samastipur, Bihar, India

Gyanendra Tripathi
Department of Bioengineering, Integral University, Lucknow, Uttar Pradesh, India

Krishan K. Verma
Sugarcane Research Institute, Guangxi Academy of Agricultural Sciences/Key Laboratory of Sugarcane Biotechnology and Genetic Improvement (Guangxi), Ministry of Agriculture and Rural Affairs/Guangxi Key Laboratory of Sugarcane Genetic Improvement, Nanning, Guangxi, China

S.K. Vishwakarma
Division of Plant Pathology, Sugarcane Research Institute, U.P. Council of Sugarcane Research, Shahjahanpur, Uttar Pradesh, India

Lin Xu
Sugarcane Research Institute, Guangxi Academy of Agricultural Sciences/Key Laboratory of Sugarcane Biotechnology and Genetic Improvement (Guangxi), Ministry of Agriculture and Rural Affairs/Guangxi Key Laboratory of Sugarcane Genetic Improvement, Nanning, Guangxi, China

Abbreviations

ABA	abscisic acid
AICRP	All India Coordinated Research Project
CAT	catalase
CAX	cation/H+ exchanger
CRISPR	clustered regularly interspaced short palindromic repeats
CSTs	climate-smart technologies
CWPs	cell wall proteins
DEGs	differentially expressed genes
DL	deep learning
DRIS	Diagnosis and Recommendation Integrated System
EC	electrical conductivity
EST	expressed sequence tag
ET	evapotranspiration
FC	field capacity
FIRB	furrow irrigated raised bed
FYM	farm yard manure
GB	glycine betaine
GHCs	greenhouse gas emissions
GHG	greenhouse gas
GIS	geographic information systems
HTL	hydrothermal liquefaction
ICAR	Indian Institute of Sugarcane Research
IISR	Indian Institute of Spices Research
INM	Integrated Nutrient Management
IRT	iron-regulated transporter
JA	jasmonic acid
LAI	leaf area index
LCA	life cycle analysis
LCB	lignocellulosic biomass
LIBS	laser-induced breakdown spectroscopy
LT	leaf temperature
LTG	low-temperature gasification
MAPK	mitogen-activated protein kinase
MDA	malondialdehyde
ML	machine learning
MREs	metal stress-responsive elements

MTFs	metal-responsive transcription factors
PM	particulate matter
PSII	photosystem II
PTO	power take-off
PW	permanent wilting
ROS	reactive oxygen species
RWC	relative water content
SA	salicylic acid
SAM	S-adenosylmethionine
SDGs	sustainable development goals
SOD	superoxide dismutase
SSI	sustainable sugarcane initiative
SSNM	site-specific nutrient management
STCR	soil test crop response
STD	sett treatment device
TPP	trehalose-6-phosphate phosphatase
Tre6P	trehalose 6-phosphate
TSase	trehalose synthase
UAVs	unmanned aerial vehicles
VRA	variable rate application
VRI	variable rate irrigation

Acknowledgment

The authors would like to thank the Sugarcane Research Institute, Guangxi Academy of Agricultural Sciences, Nanning, Guangxi, China, for providing the necessary facilities. This research was financially supported by the Guangxi Key R & D Program (Guike AB24010140), Guangxi Natural Science Foundation (2021GXNSFAA220022; 2023GXNSFAA026460), Guangxi Innovation Teams of Modern Agriculture Technology (nycytxgxcxtd-2021-03), Guangxi Characteristic Crop Experimental Station (GTS2022022), National Key Research and Development Project (2022YFD2301102-07), The National Natural Science Foundation of China (31760415; 32060468), Fund of Guangxi Academy of Agricultural Sciences (2021YT011), Science and Technology Major Project of Guangxi (Guike AA22117002-1), Fundamental Research Fund of Guangxi Academy of Agriculture Sciences (2023YM55), Key Research and Development Program of Nanning (20232060) and Guangxi Key Laboratory of Sugarcane Genetic Improvement Project (21-238-16-K-03-09). The book content expresses the authors' opinions and future research directions for sugarcane improvement.

Preface

Sugarcane (*Saccharum* spp. hybrid) is a bioenergy crop globally. It is cultivated extensively in arid and semiarid regions for the production of sucrose, which is extracted from the mature stalks of the sugarcane plants. Sugarcane cultivation requires a warm climate with a sufficient water irrigation supply. The cultivation cycle of sugarcane varies depending on the agroclimatic factors and the specific variety of the plant.

Sugarcane is a key ingredient in the production of bioethanol, a renewable fuel that is blended with gasoline to reduce carbon emissions. Sugarcane production plays a crucial role in the global economy, providing sugar and biofuels, making it essential to address its environmental and socioeconomic impacts. Sustainable practices, such as improved water management, minimize organic fertilizers, and other related practices. Additionally, policies that promote crop diversification and support for affected communities can help ensure the long-term sustainability of sugarcane production.

The use of chemical pesticides and fertilizers in sugarcane cultivation can lead to contamination of soil and water. In areas, where agricultural land is limited, the prioritization of sugarcane for biofuel or sugar production over food crops can impact food security and safety. The biomass residues (bagasse) from sugarcane processing are used in power generation, providing a renewable energy source for sugar mills and surrounding communities. Molasses, a by-product of sugar production, is used in food products as well as in the production of alcoholic beverages. Advances in agricultural technology, such as precision farming, genetically modified crops for higher yields and disease resistance, and improved water management systems, are crucial for the sustainability of the sugarcane industry. However, balancing economic benefits with environmental sustainability and social responsibility is essential for the long-term viability of the cane agro-industries.

Hence, this book summarizes the production, management, and socioeconomic impacts, and its related sustainability issues. The distinguished editors and experts have tried their level best to present the agri-techniques that can be significant for sugarcane crop production to sustain the supply of sugar and bioenergy around the globe.

CHAPTER 1

Sugarcane, Environment, and Sustainable Development: An Overview

GAEL FRANCISCO GARCÍA-MERINO and NOÉ AGUILAR-RIVERA*

Universidad Veracruzana, Facultad de Ciencias Biológicas y Agropecuarias, Córdoba, Veracruz, México

*Corresponding author

ABSTRACT

Sugarcane (*Saccharum* spp.) is a major bioenergy crop grown in over 100 tropical and subtropical regions worldwide. Its relevance lies in the fact that it is the raw material for the sugar agroindustries, panela industry bioethanol, and bioelectricity. The by-products (molasses, filter mud, vinasse, ashes, bagasse, and trash) are minimally processed or confined as waste in landfills or the environment. Although there are many technological proposals and sustainable development research for their conversion to bio-products, this agribusiness maintains that the cane sugar monoproduction is highly unsustainable in most regions. Besides, the sugarcane agroindustry in recent years has presented growth in the energy sector but with several socioeconomic and environmental limitations derived from climate change, abiotic and biotic stress conditions, and inadequate management practices such as harvest burning, resulting in exponential environmental impacts in soils, water, air, and health of the population. This chapter systematically reviews environmental factors, crop management, processing, productive diversification, markets, competing bio-products, and public policies. It analyzed the challenges of this agroindustry and biorefineries to transition to sustainable development by incorporating frameworks such as Agriculture 4.0, Industry 4.0, and circular economy.

Sugarcane Production: Environmental Impacts and Socioeconomic Issues.
Krishan K. Verma, Xiu-Peng Song, Lin Xu, Rajan Bhatt, and Yang-Rui Li (Eds.)

DOI: 10.1201/9781779641137-1

1.1 INTRODUCTION

Sugarcane (*Saccharum* spp.) is one of the most efficient crops at converting sunlight, water, and CO_2 into energy in the form of simple (sucrose) and complex (holocellulose) carbohydrates. In recent years, the crop has faced major global and regional challenges due to environmental problems arising from climate change (Verma et al., 2020a, 2020b, 2022a; Huang et al., 2024). In addition, the sugar agroindustries have also been affected by various factors that negatively impact production, having repercussions on yield, productivity, and the quality of final products, thus directly reducing the competitiveness of the agroindustrial activity and facing a great challenge at the global level (Bhatt et al., 2023a, 2023b; Deng et al., 2023; Chen et al., 2023). It is urgent to apply innovative approaches for the development of sustainable and diversified production that generates well-being and contributes significantly to the objectives of the 2030 Agenda and the Paris Agreement.

Table and industrial sugar are the target output of this agroindustry, as food meets the basic dietary needs of the population and generates synergies in various production chains. Sugarcane, the main raw material, is a perennial plant with sucrose (10–18%). Of the plant material, one-third are leaves, inflorescence, and buds, and two-thirds stalk, of which one-third fiber, and one-third juice. Sugar is produced mainly and marginally noncentrifuged sugar in some Asian and Latin American countries, and ethanol, electric energy in some countries as Brazil, without considering the rest the potentially more than 100 bio-products (Lee et al., 2023).

Tayade et al. (2023) concluded that this crop requires the latest generation of elite cultivars. The current ones in operation in sugarcane plantations require significant amounts of agrochemicals and are highly susceptible to the climate, and management practices generate important and significant environmental impacts on soil, air, and water bodies and the health of workers and neighboring populations. Sugarcane planted in the traditional slash-and-burn system, deforesting rainforests and pastures (Guo et al., 2023; Joshi et al., 2023; Verma et al., 2023a, 2023b; Zheng et al., 2023). These varieties or cultivars were designed for other productive environments, severely impacted by biotic and abiotic stresses and generalized harvesting with the burning of cane fields (preburning, burning, and reburning of residues) that were practiced in all producing countries. Directly influencing the stagnation of sugarcane yield productivity and sucrose in some regions of less than 40 t/ha (Pérez Medina et al., 2023; Li et al., 2023; Singh et al., 2023).

Therefore, sugarcane producers with different production systems in most producing countries are based on land tenure (public and private) and farm size from half a hectare and upward. Sugarcane growers search for options to reduce costs, increase productivity, and develop a competitive value chain prospectively (Verma et al., 2019a, 2019b; Guo et al., 2022; Huang et al., 2022). However, the innovations that would transform this agroindustry have been conditioned in part by the monoculture. As a result, sugarcane has a high socioeconomic impact worldwide. It is used to produce food products (sucrose and noncentrifuged sugar), bioenergy (ethanol and cogeneration), agricultural products (boiler ash, filter sludge, and vinasse), and various molasses derivatives (biotechnological products). However, conventional cultivation practices, mainly fertilization, phytosanitary management, and harvesting by burning, jeopardize the sustainability of this agroindustry (El Chami et al., 2020; Verma et al., 2022b, 2022c).

1.2 CONVENTIONAL MANAGEMENT FOR SUGARCANE PRODUCTION AND PROCESSING

The activities carried out in conventional management for the plant cycle (new crop) are related to planting, seedlings or buds. They are those required for land preparation, planting, cultural practices, and inputs to be used. Only cultural work is performed for ratoon (regrowth cane), and inputs are incorporated, so production costs decrease from planting to ratoon. It is worth mentioning that production costs and management practices vary according to the production zone and among growers. Although they have the same type of management, they do not apply the same inputs; they prefer to omit some activity or carry it out differently (Oliveira et al., 2022; Solanki et al., 2023; Verma et al., 2024a). Furthermore, the cost of organic or agroecological management is lower due to activities that are omitted and that are carried out in conventional management, such as land clearing and burning residues, in addition to buying the fertilizers to be used, because in organic management, the producers manufacture their biofertilizers, thus reducing their production costs. However, due to the soil damage caused by monoculture and the nontechnical use of agrochemicals, the potential of the soil is severely reduced; therefore, agroecological and organic systems are long-term productive options (Grandis et al., 2023; Verma et al., 2021a, 2021b, 2022d, 2022e).

Moreover, as the objective of the harvest for growers is not to consume the sugarcane itself but to sell it to industrial facilities (sugar mills, distilleries, or biorefineries) to extract sugar and other derivatives, the industrial yield projections are also required, where the crop reaches its maximum plant development, to determine in advance the area and production of sugarcane. This simulation cannot based on static or average situations (Song et al., 2019, 2021; Singh et al., 2021; Xu et al., 2022). These studies often provide an incomplete scenario of what happens in sugarcane plantations due to different environmental conditions and cultivation practices. As a fundamental activity, this agricultural planning increases profitability without additional costs (Wu et al., 2022; Verma et al., 2021c, 2021d, 2022f, 2022g). Nevertheless, to avoid depending on subjective information, make immediate decisions, avoid speculative action, and intervene favorably in managing the controllable factors of sugarcane production, it is necessary to collect as much information as possible systematically and orderly. This information would allow crop management planning by knowing the cause–effect relationship that a certain agricultural practice has caused in the yield, productivity, and quality of the crop (Verma et al., 2020c, 2020d, 2020e; Singh et al., 2020; Chen et al., 2020; Jagadesh et al., 2024).

Concerning the above, the environmental factors that have the greatest impact on agricultural activities are abiotic factors (climate soil), biotic factors (weeds, pests, and diseases), and conventional management practices that directly affect field yields (Krishna et al., 2022; Joshi et al., 2022; Kumari et al., 2022; Verma et al., 2024b). However, due to its physiological mechanism and remarkable photosynthetic efficiency, sugarcane becomes a crop with highly demanding essential nutrients, which obtained from the soil. The depletion of these nutrients causes soil deterioration (Anas et al., 2021a, 2021b; Guo et al., 2021; Verma et al., 2022h, 2022i; Martini et al., 2020, 2024). Moreover, the need of the crop to extract more nutrients for its development requires the use of chemical or organic fertilizers that provide essential nutrients such as nitrogen, phosphorus, and potassium, which are supplied according to the need of the phenological cycle of the sugarcane crop, to promote its optimal physiological development (Verma et al., 2021e; Sharma et al., 2021, 2022; Malviya et al., 2021, 2022).

Moore et al. (2017) concluded that sugarcane production has a higher demand for chemical inputs to correct the lack of soil nutrients and the intensive use of agricultural machinery, which generates degradation and infertility of agricultural soil, causing soil organic matter deficiency, which consequently increases erosion, infertility and decrease of beneficial microbial populations for the soil and its cultivation (Song et al., 2022; Verma et al., 2022j, 2022k; Kumari et al., 2023a, 2023b). Therefore, according to Bridhikitti et al. (2023), byproducts from the production of sucrose, bioethanol, and bio-products for biofertilizers and their return to sugarcane soils are key factors for sustainability (Tian et al., 2023; Chaudhary et al., 2023).

On the other hand, several highly harmful practices, such as burning and re-burning harvest residues, seriously affect the soil microbiota and severely impact the health of ecosystems and neighboring populations (Mugica-Alvarez et al., 2018; Verma et al., 2023c). Consequently, the territory has been exploited for decades as a sugarcane plantation. Hence, the environmental impact is irremediable in the short term, and value creation is present. It does not consider the repair of the damage to the environment generated by the farmers and entrepreneurs who consume the sugarcane stalks.

For this reason, the basic factors that affect the sustainability of sugarcane cultivation are mainly the following:

- Physical: including regional soil and climatic conditions, climate change, pests, and diseases.
- Structural: corresponding to the farming systems and management practices adopted, road infrastructure, public services, and competitive environment.
- Institutional: involving efficient public policies and government actions affecting prices, credit, marketing, incentives, research, development, technology transfer, management practices, and sustainable crop management.
- These are related to innovations to increase production and solve the physical and environmental problems that restrict activities related to sustainable sugarcane agriculture.
- There is a low level of diversification of sugarcane farms with intercropping, no linkage with livestock farmers for sugarcane feed, and no recovery of manure to recover soils with composting techniques.

- Most sugarcane plantations worldwide have heterogeneous soil and climate conditions that impact productivity according to the environment. This indicates that it is necessary to characterize the environments or agroecological zones with homogeneous conditions to obtain maximum productivity of sugarcane cultivation as the first step toward sustainability and reduction of environmental impacts.
- Information technology, data management, precision farming, and other management technologies are not well known to most growers.

1.3 ENVIRONMENTAL IMPACTS OF SUGARCANE AND SUGARCANE DERIVATIVES PRODUCTION

The sugarcane agroindustry affects different environmental components, such as biological, physical, social, and cultural. However, the most affected is the physical component, especially the air component, due to the volume of gases from the burning harvest and the nontechnological handling of agrochemicals that emit various pollutants into the atmosphere (Verma et al., 2023d). Although no globally applicable research is available on the impact of this agroindustry on the environment, it is evident that the atmospheric component at the local level receives the greatest effects.

1.3.1 SUGARCANE HARVESTING

Harvesting is one of the most important stages in sugarcane production. Its objective to collect the raw material available in the field with minimum loss and high efficiency, guaranteeing the supply of raw material in sufficient quantity and quality to the biorefinery, sugar mills, and distillery. The shortest time between harvesting and milling, with low levels of foreign matter (especially leaves, tips, flowers, and soil), and with the lowest transportation and environmental costs, to obtain high quality of by-products, that is., sugars, ethanol, noncentrifuged sugar, electric energy, bioplastics, organic fertilizers, and so on at competitive prices.

Notwithstanding, sugarcane is harvested in most countries in a social and political environment with few environmental restrictions. Before manual or mechanized cane cutting, the plantation is set on fire to eliminate most dry foliage, flowers, and tops. Once the cane field burned, cutters enter to cut the stalks with cutting tools from the lowest part, separating the foliage not burned (green leaves and tops). Stacks of cut stalks formed and oriented perpendicularly in the direction of the furrows. Once the cane has been cut and lined up in bundles, the mechanical or manual loader enters and deposits them in a transport unit for transfer to the factory. All the remaining foliage left on the ground in a similar orientation to that of the stalks for 2–6 days to dry and then finally eliminated in a second burning or reburning. When a burn is carried out, a rain of burned particulate material and ashes observed over the surrounding areas. However, the potential uses of trash are ignored or not considered (Powar et al., 2022).

The aim is to ensure the burned cane does not remain in the field for more than 24 h. If it happens, the cane loses weight and industrial quality (Rasche and Sos Del Diego, 2020). In comparison to the consequences as mentioned earlier caused by burning, some benefits of traditional harvesting are as listed below:

- Increase the cutting capacity of the workers.
- Reduces transportation costs due to water evaporation and the reduction of foreign matter, which serves as a sponge during milling, capturing the juice extracted from the cane and reducing production efficiency.
- Facilitates the production process since precooked material is processed.

Worldwide, multiple multi-, inter-, and transdisciplinary approaches have been carried out about sugarcane burning and the efficiency of the sugar manufacturing process, its influence on the cutting process, and postharvest deterioration. Nonetheless, the impacts on the soil, biotics of the sugarcane plantation (interaction, soil, roots, and plant), and atmosphere are only disciplinary, without considering social, environmental, and governance aspects among producers and society in general. However, harvesting by burning, mainly manually with workers, prevails for an eminently economic purpose without considering the mechanizable and nonmechanizable zones. Therefore, semimechanized sugarcane harvesting is another alternative due to its low cost, low soil compaction, low damage to roots, low soil compaction, and low damage to roots and sugarcane plants.

Although burning sugarcane and postharvest residues not only pollutes the environment and aggravates the situation of extreme heat during the harvest season, it also seriously affects fertility by reducing the organic matter and nitrogen content of the soil. It is the main cause of forest fires and the generation of highly harmful emissions and contributes to the phenomenon of global warming of the atmosphere (Hartner et al., 2024; Geldenhuys et al., 2023). Green harvesting is a better option because harvesting by burning reduces the agro-industrial quality of sugarcane, the percentage of sucrose and increases the amount of noncrystallizable sugars.

Among the main pollutants derived from the burning of sugarcane fields in the atmosphere are greenhouse gases and suspended particles particulate matter (PM) 10 μm (in diameter) are those particles suspended in the air that are characterized by having an aerodynamic diameter equal to or less than 10 μm, which cause respiratory diseases. The size of these particles PM 10 μm allows them to reach beyond the throat, up to the respiratory tract, causing respiratory diseases. In contrast, particles finer than 2.5 μm known as fine particulate matter contribute to forming atmospheric aerosols (Stem et al., 2024; Ruths et al., 2023; Moitinho et al., 2021). Besides, Linnenluecke et al. (2018) reviewed the environmental effects on sugarcane field burning, climate change, and crop productivity. Green cane harvesting leaves approximately 20–25% of the total biomass as residue, whereas burnt cane harvesting results in around 16% residue remaining in the field. The most sustainable uses for these residues are as follows:

- Incorporate them into sugarcane soils to improve their organic matter content (efficient technologies).
- Packing them for livestock feed (preparation of rations and silage).
- Extracting from the field for energy generation (use of biomass).
- Use as feedstock in lignocellulosic biomass biorefineries.

Therefore, both manual and mechanical harvesting of green sugarcane offers the opportunity to develop new technologies to significantly increase productivity and profitability. There are also valuable challenges and opportunities to develop new technologies to exploit the sugarcane plant to its full potential, such as alternative uses for sugarcane and byproducts.

1.3.2 USE OF WATER AND POLLUTION

Sugarcane and water are the fundamental inputs in the agro-industrial production chain in the field and the sugar mill or biorefinery. The high volumes of water used and the enormous discharges of pollutants into natural water reservoirs imply a strong pressure on groundwater and land springs. Unlike sugarcane, which has a monetary cost for its production, water, despite being an essential raw material in producing sugarcane derivatives, represents a very low cost for entrepreneurs, given its abundance, availability, and scarce legislation. However, the environmental cost derived from the irrational use of water has been little visible. The above is derived from the fact that public environmental policies have not improved the rational use of water in sugarcane plantations and industrial facilities that process sugarcane since its consumption has not been substantially reduced, causing the reduction of aquifer reserves and wastewater discharges are still very high because water from production processes has not been technologically reused.

In addition, in recent years, water management and availability have become one of the most relevant topics on the global environmental agenda, particularly in the case of the 2030 agenda for sustainable development. The issues surrounding this resource have worsened due to its overexploitation, the high contamination of aquifers, and the erroneous environmental policies. Even the water alert in most countries has begun to affect the areas with the greatest availability, including the regions close to Ecuador. This environmental problem results from agroindustry, sugar mills, distilleries, and biorefineries maintaining the same production patterns for several decades. This is reflected in a scarce innovation of the productive processes and technological patterns, although in most of the phases of the productive process, the use of water is indispensable.

It is highly recommended that the water used in industrial facilities be returned clean to the water system and reused. With this strategy, there is no pressure on the water supply sources, which requires the availability of land for treatment in specialized plants and large investments in cooling towers. The excess cooled condensed water and fall water can be fully utilized in washing instead of fresh water, thus reducing both factory water consumption and wastewater pollution. It is imperative to invest in research to develop water-saving technologies (Salazar et al., 2023).

1.3.4 AIR POLLUTION

The main solid waste discharges come from raw material preparation (soil and plant remains), steam generators (ash) and extract purification (filter mud), lignocellulosic materials (bagasse, trash), process sludge (diluted mud), calcium hydroxide "lime" (dumps), sugar dust, grease, and oils used in the mill's engines and to lubricate machinery. Toxic gas and air emissions come from the boiler system (flue gases from the combustion of solid, liquid, and gaseous materials), volatile substances (soot and ash), raw material preparation, extraction, juice purification, and evaporation, and biochemical reactions of the organic components of the wastewater in the stratified ponds (ammonia and sulfuric acid), which generate excess SO_2 and SO_3. An important indicator to consider in the environmental sustainability of industrial facilities that process sugarcane stalks is oil consumption per ton of sugar produced. This saving is due to the substitution of petroleum for bagasse as fuel used in their boilers, allowing bioenergy to be cogenerated.

However, the use of bagasse represents an important saving for companies and has improved their environmental image by promoting the use of alternative energies to fossil fuels. It is not exempted from having an impact on the environment since the combustion of bagasse in the boilers produces a great deal of pollution due to all the black carbon and CO_x emitted, since it is burned with high humidity, contributing to global warming. Carbon particles and ashes also generate important damage, such as contamination of water bodies, soil contamination, and intoxication of flora and fauna. They are also responsible for respiratory and eye diseases suffered by workers and inhabitants living near the mills, distilleries, and biorefineries. On the other hand, the deforestation of the forest biomass cut to establish sugarcane plantations never replenished, resulting in deforestation of the areas surrounding the industrial facilities, sugar mills, artisanal sugar mills (trapiches for production of noncentrifuged sugar), distilleries and/or biorefineries.

1.4 SUSTAINABILITY OF SUGARCANE AGROINDUSTRY AND BIOREFINERIES

The environmental approach in sugarcane plantations, sugar mills, and distilleries, but mainly in biorefineries related to the sugarcane agroindustries.

Highly relevant issues in the context of sustainability and climate change mitigation for the different nations directly or indirectly involved in this agroindustry. This environmental approach involves the management, evaluation, and optimization of all aspects of the process, from the raw material collection, processing, and conversion to the use of the final products or byproducts (Aguilar-Rivera and Michel-Cuello, 2020; Karp et al., 2022). Some key issues to consider include energy efficiency, reduction of greenhouse gas emissions, and sustainable waste management (Solomon et al., 2019; de Souza et al., 2021; Cherubin et al., 2021). Due to the growing scientific information in the last decade on analyzing the performance of sugarcane processing biorefineries, a data-driven approach known as bibliometric mapping was selected.

Bibliometrics, a quantitative research methodology, allows using statistical techniques to identify trends based on scientific data in various forms of publications, including scientific articles, book chapters, reviews, conference papers, and other documents (Donthu et al., 2021). This methodology uses data analysis to provide a comprehensive overview of the most relevant results found in the bibliographic material analyzed, making it possible to move the variables desired by the author, such as words of concurrence, year of publication, and type of document from which the information obtained. Through this type of analysis, it is possible to examine the patterns of the scientific literature at regional, national, and world levels (Al-Jamimi et al., 2022).

The application of bibliometrics has made it possible to study scientific activities in fields such as environmental and agricultural sciences and the behavior of innovations in the sugarcane sector (Li et al., 2021; Malanski et al., 2021; Santoyo-Castelazo et al., 2023). The main results are presented through visual representations of the field generated using VOSviewer software to perform a bibliometric analysis. These visualizations allow the researcher to examine the relationships between key terms, countries, and even authors involved in the research. The data come from the title, abstract, and keywords of scientific publications. These relationships are structured using a clustering algorithm, which makes it possible to identify meaningful sets when analyzing the literature (Sahni and Kaurav, 2023). Despite the relevance of sugarcane agroindustry biomass processing and transformation, no comprehensive review covers an environmental approach, sustainability issues, climate change, and current innovations using bibliometric tools. Although previous research has used scientometric analysis on sugarcane, the approaches taken in those studies differ from the present analysis.

The present analysis is divided into two parts. The first part details and discusses the analysis of the countries involved in providing information over time, implementing the keywords sugarcane, environment, and biorefineries. The second part involves the results and discussion of the concurrent words related to the abovementioned keywords.

1.5 NETWORK OF COAUTHORSHIP COUNTRIES INVOLVED IN ENVIRONMENTAL FOCUS IN SUGARCANE BIOREFINERIES

The bibliographic networks based on distance are maps where links are generated, implying that a smaller distance denotes a more intense connection (Donthu et al., 2021). The analysis shows a network of co-authorship nations using VOSviewer for the total count of published papers and citations in the last five years (Figure 1.1). A *frame* symbolizes a country, and its dimension indicates its contribution to research on environmental focus in sugarcane agroindustry biorefineries. In addition, the width of the lines and their coloration indicate the close collaboration between countries. Researchers from 24 countries participated in the analysis and made significant contributions through the abovementioned publications. A minimum threshold of 25 papers per country was applied as an inclusion criterion to ensure equitable representation in the bibliographic network. As a result, eight nations were identified and retained, which were grouped into six distinct categories. The first bibliographic network reveals that Brazil tops the list with 18 documents, followed by India with nine, the United States with four, and China, Thailand, Pakistan, Saudi Arabia, the United Kingdom, and South Korea with three documents each (Figure 1.1A).

The main collaboration network was observed between different nations of the Asian continent, where Saudi Arabia is among the union of countries such as South Korea, where it interacts with the United States and China, which has a close connection with Pakistan. In addition, India interacts with the United Kingdom, although its ties are somewhat distant compared to those of other Asian nations. Among the nations mentioned above, China and India stand out since they are the two most populated countries in the world and share several similarities (Han et al., 2022).

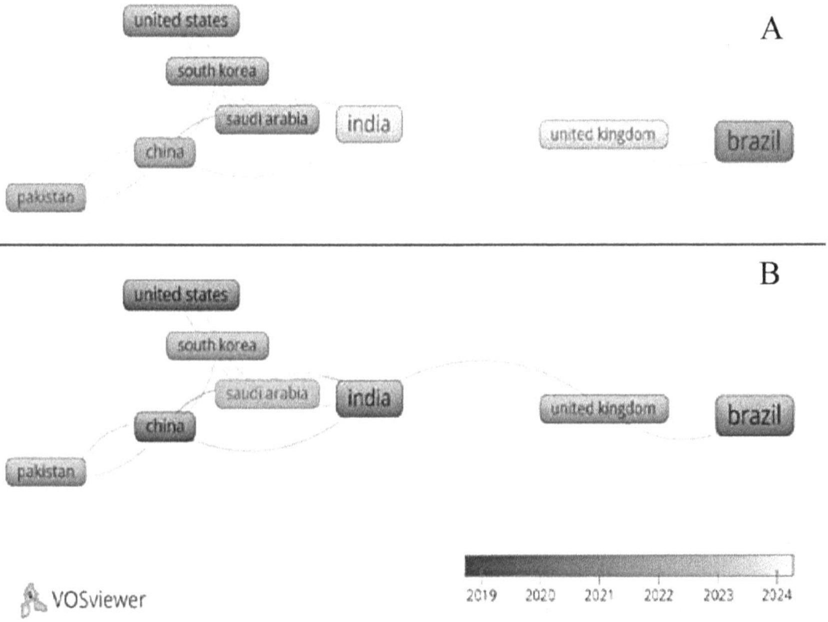

FIGURE 1.1 International co-authorship network of countries with research-related publications emphasizing the environmental approach in sugarcane biorefineries: (A) by a number of papers and (B) papers published in the last five years.

Both in the last 10 years have had an exponential increase in the environmental focus of biorefineries, are some of the main producers of sugarcane, and consumers of derivatives such as sugar, noncentrifuged sugar, and so on. They motivated by a combination of factors that are worrying for the wellbeing of their societies (Tam et al., 2023). Both countries highly depend on imported fossil fuels to meet their energy needs. Therefore, sugarcane biorefineries based on sugarcane juice, bagasse, and trash, molasses offer more sustainable alternative by providing biofuels, and other biomass-derived products, which can help reduce dependence on fossil fuels and improve energy security (Velidandi et al., 2023).

Since the beginning of the new millennium, China and India have faced serious air pollution and climate change problems due to high emissions of greenhouse gases and air pollutants (Farajzadeh et al., 2023). Biorefineries can produce biofuels for transportation that emit fewer greenhouse gases and air pollutants than fossil fuels, contributing to emission reductions, improving air quality, and promoting greater energy resilience (Ebrahimian et al., 2023). Diversification of the energy matrix is crucial to ensure long-term stability and sustainability (De Rosa et al., 2022; Gitelman et al., 2023). Likewise, both countries have large rural populations and a significant agricultural base. Recently, biorefineries have offered an opportunity to utilize agricultural and forestry residue as dedicated energy crops, which can boost rural development, create jobs in rural areas, and diversify agricultural incomes (Gonzalez et al., 2023).

China and India have signed international agreements to reduce greenhouse gas emissions and promote clean and sustainable energy, such as the United Nations Framework Convention on Climate Change (UNFCCC), the Stockholm Convention on Persistent Organic Pollutants and the Paris Agreement. The development of approaches that promote the environment as a factor in development through the participation of agribusinesses will offer options for meeting these commitments and improving their image on the international stage in terms of environmental sustainability.

Brazil shares a stronger interaction with the United Kingdom on the other side of the network. Brazil is one of the world's leading biomass producers, with a large amount of land dedicated to agriculture and a vast amount of agricultural and forestry residues (Chandel et al., 2021). In the case of sugarcane, it is the world's leading producer of sugarcane, sugar, ethanol, and its derivatives. On the other hand, the United Kingdom possesses expertise and advanced technology in areas such as process engineering, biotechnology, and environmental management (Jaworski et al., 2024). Brazil can benefit from UK experience and know-how in areas such as process optimization and byproduct valorization. In contrast, the United Kingdom can benefit from access to natural resources and practical experiences in the field of biomass from Brazil. This complementarity of resources and knowledge has created opportunities for collaboration in developing and implementing biorefinery projects (Liu et al., 2022; Hung, 2022).

Brazil and the United Kingdom have a shared interest in addressing environmental challenges and developing sustainable technologies, which has led both to have in common several international conventions such as the UNFCCC, the Kyoto Protocol, and the Paris Agreement. Research institutions and universities in Brazil and the United Kingdom have established collaborations to address environmental and sustainable energy issues (Nicolletti et al., 2020). These collaborations may include researcher exchanges, joint research projects, and joint funding programs. Collaboration in scientific research related to biorefineries and their environmental focus can benefit both countries.

The study has revealed an uneven distribution in the amount of scientific information on environmental approaches in biorefineries among different countries over the last five years (Figure 1.1B). At the most up-to-date part of the list, we find Saudi Arabia, followed by South Korea, Pakistan, and the United Kingdom. On the other hand, India and Brazil are lagging behind in terms of the recent production of scientific information in this field. This pattern can be attributed to several factors. On the one hand, leading research countries, such as Saudi Arabia and South Korea, have demonstrated a strong commitment to R&D investment in the biorefinery sector, possibly driven by the need to diversify their economies and reduce their dependence on fossil fuels (Waqas et al., 2023; Lee et al., 2023). In addition, these countries have established strong international collaborations and developed advanced research infrastructures (Ali et al., 2023).

On the other hand, countries such as India and Brazil have faced challenges in terms of funding constraints, less focused government policies on environmental research, and the need to strengthen their research infrastructures; however, this is likely to change with the recent formation of the BRICS group (Brazil, Russia, India, China and South Africa) (Zeraibi et al., 2023). Leading research countries have a significant advantage in terms of access to resources and knowledge and the ability to implement sustainable technologies in the biorefinery sector (Karp et al., 2022). For example, the United Kingdom has pioneered research on second-generation biorefineries and has established clear policies to encourage investment in this field (Gaffey et al., 2023). In contrast, India and Brazil have faced challenges translating their vast natural resource potential into significant scientific advances due to a lack of coordination between government, industry, rural producers, and academia (Hassan et al., 2024).

The inequitable distribution of research on this topic has important implications for sustainable development at the global level. It is important to remember that the present study is based on the amount of scientific information available and may not fully reflect the research efforts in each country. In addition, the quality and impact of research has not been considered in this ranking.

1.6 NETWORK OF CONCURRENCE WORDS INVOLVED IN AN ENVIRONMENTAL APPROACH IN SUGARCANE BIOREFINERIES

For this bibliometric analysis, the criterion applied was that each term had to be repeated at least three times. A set of 93 terms was categorized into three distinct groups. It was decided to exclude the term cane sugar (Figure 1.2). In the blue group of the map, terms related to biomass processing were identified. The red group comprises terms associated with sustainable development, life cycle, and climate change. In contrast, the green group encompasses aspects related to the biochemistry of sugarcane, bio-products, byproducts, and their derivatives.

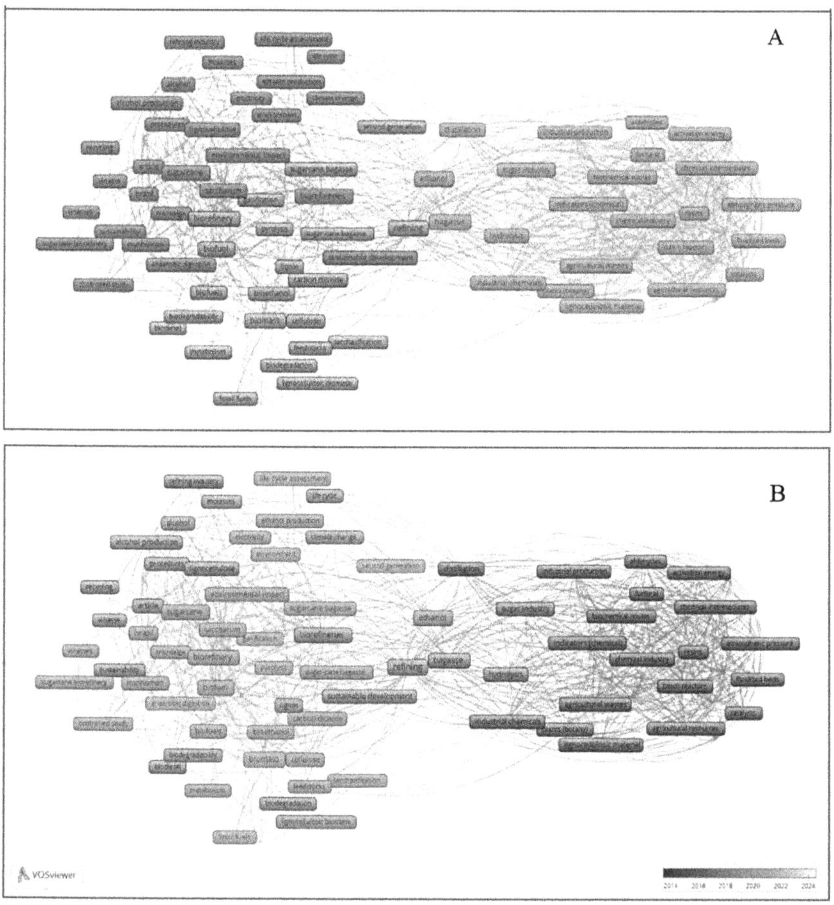

FIGURE 1.2 Network that shows the co-occurrence of terms in research with the environmental approach in the sugarcane biorefineries with three co-occurrences.

In the blue cluster, a number of key terms were obtained that point toward a comprehensive approach to research on technologies related to lignocellulosic biomass, with a particular focus on sugarcane bagasse as a potentially rich source of biomass resources for the production of biofuels and other products. Biofuel sourcing emerges as a central theme in this grouping of terms, reflecting the growing interest in finding sustainable alternatives to conventional fossil fuels (Vandenberghe et al., 2022). In this context, sugarcane bagasse presented as a promising feedstock due to its abundance, renewability, and potential to mitigate greenhouse gas emissions (Ajala et al., 2021). Another word found in this group is pyrolysis, which is a thermal conversion process that decomposes biomass in the absence of oxygen and emerges as an important area of research (Wang et al., 2020).

This process can play a key role in the production of solid, liquid, and gaseous biofuels, as well as in the generation of high-value-added products such as biochar, which can be used as a soil amendment or for carbon sequestration (Tripathi et al., 2016). The reference to biodegradation indicates an interest in understanding and improving the natural processes of biomass decomposition, which can influence the efficiency and sustainability of biofuel production. Optimization of these processes can help increase biofuel yields and reduce dependence on fossil resources (Ambaye et al., 2021; Malode et al., 2021). The mention of biorefineries suggests an integrated approach to biomass conversion that goes beyond simple biofuel production to generate various value-added products and byproducts (Rodionova et al., 2022). This cascade approach, which maximizes the utilization of each biomass component, is essential to optimize the efficiency and cost-effectiveness of production processes and can significantly contribute to the transition to a more circular and sustainable economy.

In the red group are terms that highlight the relevance of environmental impact, sustainable development, and the sugarcane life cycle optimization, including diversification into byproducts such as ethanol, vinasse, and lignocellulosic derivatives. This finding reflects an evolution in sugarcane industry research and highlights the emergence of a comprehensive approach that addresses environmental and economic concerns in this area. Biorefineries represent an innovative way to address contemporary challenges related to biomass production, offering an integrated approach to converting renewable feedstocks into a wide variety of products and byproducts (Freitas et al., 2021).

In the case of the current sugarcane agroindustry, which produces exclusively sucrose, ethanol, and bioenergy, with increased use of agrochemicals, water and fossil fuels, and highly polluting agribusiness, biorefineries can play a fundamental role in maximizing the added value of the raw material, while minimizing the environmental impact of the production processes (Vandenberghe et al., 2022; Lee et al., 2023). The key to biorefineries lies in their ability to efficiently harness every component of sugarcane, from the sucrose used for sugar production to the lignocellulose that can be converted into biofuels or high-value chemicals. This cascade approach, which prioritizes the sequential and complete utilization of resources, not only increases the profitability of the industry but also reduces the amount of waste and unwanted byproducts, thus contributing to environmental sustainability. In addition, biorefineries offer the flexibility to adapt to changing market demands and fluctuations in feedstock availability (Duque et al., 2021). Their ability to diversify the product range and adjust production processes according to market needs gives them a significant competitive advantage in a dynamic and constantly evolving economic context (Solarte-Toro and Alzate 2021).

For the last green group, the network was formed with key terms that highlight crucial aspects related to industrial chemistry, hydrolysis, ethanol production, resins, and the sugar industry. These terms reflect a convergence of research areas essential for the development and optimization of biorefineries derived from the sugar industry. This sector needs to diversify its products and improve its operational efficiency in a context of growing demand and environmental pressures (Korasapati et al., 2023).

Industrial chemistry covers a broad spectrum of disciplines and techniques transforming raw materials into chemical and value-added materials. Interaction on the previous term refers to hydrolysis technology that emerges as a key process in producing ethanol from lignocellulosic biomass, such as bagasse and trash. This process implies the breakdown of the bonds of cellulose and hemicellulose molecules to release fermentable sugars, which can converted into ethanol through fermentation (Brar et al., 2020). Research in hydrolysis seeks to improve the efficiency and profitability of this process and develop more effective and sustainable technologies (Gomes et al., 2020; Morais Junior et al., 2021). Ethanol production is a crucial component in searching for sustainable alternatives to fossil fuels in the transport sector and creating new biomolecules. Sugarcane agroindustry, with an abundance of renewable raw material (juice, molasses, bagasse, and trash), plays an important role in ethanol production since sugar extracted from cane can be fermented to produce this biofuel (Hernandes et al., 2021).

Research in this area focuses on improving production processes, increasing energy efficiency, and reducing the environmental impact of ethanol production (Karp et al., 2021). Including resins suggests an interest in diversifying products derived from the sugar industry, which goes beyond traditional products such as sugar and ethanol. Resins can have a wide range of applications in construction, automotive, and chemical industries. This represents an opportunity to add value to sugarcane by-products and diversify industry income (Patil et al., 2023). In the sugar industry, the application of industrial chemistry may include the optimization of extraction, purification, and transformation of sugarcane and its byproducts into various products, from biofuels to specialized chemicals. An integrated approach that takes advantage of the state-of-the-art industrial chemistry, hydrolysis, ethanol production, and resin development can help maximize the value of sugarcane and its byproducts while contributing to the environmental and economic sustainability of the sector. These findings offer valuable perspectives for innovation and development in the sugarcane industry. They can input to develop public policies and practices aimed at improving the competitiveness and resilience of this important industrial sector.

The results in Figure 1.2 reveal a dynamic picture in the field of scientific research related to the environmental focus of sugarcane agroindustry biorefineries in the last five years. Identifying a lag in research output linked to the chemical industry, industrial waste, distillate production, and catalysis suggests that these areas may have received less attention than the other groups obtained. This finding raises important questions about the need to direct more resources and efforts toward these neglected areas, given their potential to significantly influence the efficiency and sustainability of biorefineries.

Notwithstanding, the abundance of recent scientific information on the generation and management of biomass of diverse origin and composition, sustainable production of biofuels and bio-products and byproducts of these processes reveals a promising trend toward the adoption of more sustainable and environmentally friendly practices in the sugarcane agroindustry. This evolution reflects the growing awareness of the importance of reducing the carbon footprint and mitigating the negative environmental impacts associated with traditional sugar and sugar byproduct production (Linnenluecke et al., 2020; Cheesman et al., 2023). The emphasis on environmental impact, climate change, and sustainable production indicates a paradigm shift toward a more holistic and responsible view of sugarcane agribusiness (Kumar et al., 2022). Research in these areas seeks to develop more efficient and cleaner technologies and practices and address global challenges related to environmental conservation and the fight against climate change.

Nonetheless, it is essential to recognize that the path to sustainability in the sugarcane industry presents multiple challenges of a transdisciplinary nature. The transition to more sustainable production models and the adoption of innovative technologies can face significant barriers, such as technical, economic, and regulatory constraints (Aguilar-Rivera and Michel-Cuello, 2020). In addition, the complexity of the supply chain and the interactions between the actors involved can further complicate the change process. In this sense, it is crucial to foster collaboration between industry, academia, government, and other stakeholders to effectively address these challenges and promote innovation and the adoption of best practices throughout the sugarcane industry value chain. Interdisciplinary research and global collaboration can play a key role in identifying comprehensive and sustainable solutions that will drive the future sustainable development of the sugarcane industry.

1.7 CONCLUSION

Sugarcane (*Saccharum officinarum*) is the raw material used in sugar agroindustry, distilleries, and biorefineries. It is a productive activity with a high social impact due to its production. The employment it creates in plantations and because sugar and other derivatives are deeply rooted in the economy and culture and a staple food in the world diet, making it a potentially conflictive activity. In recent years has suffered the impact of numerous factors that threaten its viability as an economic activity. However, although it could be a wealth of industrial raw materials, for bioproducts its primary producers have relatively low profits and depend on a vulnerable market because its main product, sugar, is a commodity. In addition, the cultivation and processing of sugarcane generate strong, complex environmental impacts that require commitments from stakeholders and decision-makers to reduce externalities and increase the resilience of the crop and reconvert the agroindustry into biorefineries with the support of available technology, management, and administrative options.

KEYWORDS

- **sustainability**
- **profitability**
- **environmental challenges**
- **sugarcane growth**
- **development**

REFERENCES

Aguilar-Rivera N, Michel-Cuello C. (2020) Challenges and prospects for sugarcane biorefineries in mitigating climate change. *Handbook of Climate Change Management: Research, Leadership, Transformation*, pp. 1–17. https://doi.org/10.1007/978-3-030-57281-5_245

Ajala EO, Ighalo JO, Ajala MA, Adeniyi AG, Ayanshola AM. (2021) Sugarcane bagasse: A biomass sufficiently applied for improving global energy, environment and economic sustainability. *Bioresour. Bioprocess*. 8, 1–25. https://doi.org/10.1186/s40643-021-00440-z

Ali U, Guo Q, Nurgazina Z, Sharif A, Kartal MT, Depren SK, Khan A. (2023) Heterogeneous impact of industrialization, foreign direct investments, and technological innovation on carbon emissions intensity: Evidence from Kingdom of Saudi Arabia. *Appl. Energy* 336, 120804. https://doi.org/10.1016/j.apenergy.2023.120804

Al-Jamimi HA, BinMakhashen GM, Bornmann L. (2022) Use of bibliometrics for research evaluation in emerging markets economies: A review and discussion of bibliometric indicators. *Scientometrics* 127(10), 5879–5930. https://doi.org/10.1007/s11192-022-04490-8

Ambaye TG, Vaccari M, Bonilla-Petriciolet A, Prasad S, van Hullebusch ED, Rtimi S. (2021) Emerging technologies for biofuel production: A critical review on recent progress, challenges and perspectives. *J. Environ. Manag*. 290, 112627. https://doi.org/10.1016/j.jenvman.2021.112627

Anas M, Verma KK, Haq IU, Naeem M, Li Q, Liao F, Xiang YX, Li YR. (2021a) Characterization of exotic, native and wild-type genotypes of sugarcane (*Saccharum* spp. Hybrids) for internal nitrogen use efficiency under different nitrogen levels. *Sugar Tech*. 23, 1258–1267. https://doi.org/10.1007/s12355-021-01014-1

Anas M, Verma KK, Riaz M, Qiang L, Liao F, Liu Y, Li YR. (2021b) Physio-morphological and biochemical mechanism of nitrogen use efficiency in sugarcane (*Saccharum* spp.) genotypes under different growth stages and nitrogen levels. *J. Plant Interact.* 16, 332–343.

Bhatt R, Verma KK, Jamwal NS. (2023a) Can paddy-compost eliminate N fertilization for climate-smart sugarcane? *Sugar Tech.* 25, 1036–1046. https://doi.org/10.1007/s12355-023-01270-3

Bhatt R, Verma KK, Kumar R, Sanghera GS. (2023b) Foliar application of potassium salt of active phosphorus (PSAP) mitigates insect pests and improves yield along with sugarcane quality in response to agroclimatic conditions of Punjab. *Sugar Tech.* 25(3), 660–669.

Brar KK, Santo MCE, Pellegrini VO, deAzevedo ER, Guimaraes FE, Polikarpov I, Chadha BS. (2020) Enhanced hydrolysis of hydrothermally and autohydrolytically treated sugarcane bagasse and understanding the structural changes leading to improved saccharification. *Biomass Bioenergy* 139, 105639. https://doi.org/10.1016/j.biombioe.2020.105639

Bridhikitti A, Kaewsuk J, Karaket N, Friend R, Sallach B, Chong JP, Redeker KR. (2023) Balancing agriculture and industry through waste utilization for sugarcane Sustainability. *Sustainability* 15(20), 14711. https://doi.org/10.3390/su152014711

Chandel AK, Forte MB, Gonçalves IS, Milessi TS, Arruda PV, Carvalho W, Mussatto SI. (2021) Brazilian biorefineries from second generation biomass: Critical insights from industry and future perspectives. *Biofuels, Bioprod. Biorefining* 15(4), 1190–1208. https://doi.org/10.1002/bbb.2234

Chaudhary P, Xu M, Ahamad L, Chaudhary A, Kumar G, Adeleke BS, Verma KK, Hu D-M, Siric I, Kumar P, Popescu SM, Fayssal SA. (2023) Application of synthetic consortia for improvement of soil fertility, pollution remediation, and agricultural productivity: A review. *Agronomy* 13, 643. https://doi.org/10.3390/agronomy13030643.

Cheesman AW, Brown F, Farha MN, Rosan TM, Folberth GA, Hayes F, Sitch S. (2023) Impacts of ground-level ozone on sugarcane production. *Sci. Total Environ.* 904, 166817. https://doi.org/10.1016/j.scitotenv.2023.166817

Chen G-L, Zheng FJ, Lin B, Lao SB, He J, Huang Z, Zeng Y, Sun J, Verma KK. (2020) Phenolic and volatile compounds in the production of sugarcane vinegar. *ACS Omega* 5, 30587–30595.

Chen G-L, Zheng F-J, Lin B, Yang Y-X, Fang X-C, Verma KK, Yang L-F. (2023) Vinegar: A potential source of healthy and functional food with special reference to sugarcane vinegar. *Front. Nutr.* 10, 1145862. https://doi.org/10.3389/fnut.2023.1145862

Cherubin MR, Carvalho JLN, Cerri CEP, Nogueira LAH, Souza GM, Cantarella H. (2021) Land use and management effects on sustainable sugarcane-derived bioenergy. *Land* 10(1), 72. https://doi.org/10.3390/land10010072

De Rosa M, Gainsford K, Pallonetto F, Finn DP. (2022) Diversification, concentration and renewability of the energy supply in the European Union. *Energy* 253, 124097. https://doi.org/10.1016/j.energy.2022.124097

de Souza NRD, Duft DG, Bruno KMB, de Souza Henzler D, Junqueira TL, Cavalett O, Hernandes TAD. (2021) Unraveling the potential of sugarcane electricity for climate change mitigation in Brazil. *Resour. Conserv. Recycl.* 175, 105878. https://doi.org/10.1016/j.resconrec.2021.105878

Deng ZN, Huang HR, Huang CM, Verma KK, Luo HB, Jiang SL, Li YJ, Cao HQ, Wu KC, Xu L. (2023) Short-term limited water irrigation influences the photosynthetic pigments, enzymatic and non-enzymatic activities in *Saccharum spontaneum* L. at vegetative stage. *Sugar Tech.* 25, 110–118. https://doi.org/10.1007/s12355-022-01175-7

Donthu N, Kumar S, Mukherjee D, Pandey N, Lim WM. (2021) How to conduct a bibliometric analysis: An overview and guidelines. *J. Business Res.* 133, 285–296. https://doi.org/10.1016/j.jbusres.2021.04.070

Duque A, Álvarez C, Doménech P, Manzanares P, Moreno AD. (2021) Advanced bioethanol production: From novel raw materials to integrated biorefineries. *Processes* 9(2), 206. https://doi.org/10.3390/pr9020206

Ebrahimian F, Khoshnevisan B, Mohammadi A, Karimi K, Birkved M. (2023) A biorefinery platform to valorize organic fraction of municipal solid waste to biofuels: An early environmental sustainability guidance based on life cycle assessment. *Energy Conversion Manag.* 283, 116905. https://doi.org/10.1016/j.enconman.2023.116905

El Chami D, Daccache A, El Moujabber M. (2020) What are the impacts of sugarcane production on ecosystem services and human well-being? A review. *Ann. Agric. Sci.* 65, 188–199 https://doi.org/10.1016/j.aoas.2020.10.001

Farajzadeh Z, Ghorbanian E, Tarazkar MH. (2023) The impact of climate change on economic growth: Evidence from a panel of Asian countries. *Environ. Develop.* 47, 100898. https://doi.org/10.1016/j.envdev.2023.100898

Freitas JV, Bilatto S, Squinca P, Pinto AS, Brondi MG, Bondancia TJ, Farinas CS. (2021) Sugarcane biorefineries: Potential opportunities towards shifting from wastes to products. *Indust. Crops Products* 172, 114057. https://doi.org/10.1016/j.indcrop.2021.114057

Gaffey J, Rajauria G, McMahon H, Ravindran R, Dominguez C, Ambye-Jensen M, Sanders JP. (2023) Green Biorefinery systems for the production of climate-smart sustainable products from grasses, legumes and green crop residues. *Biotechnol. Adv.* 66, 108168. https://doi.org/10.1016/j.biotechadv.2023.108168

Geldenhuys G, Orasche J, Jakobi G, Zimmermann R, Forbes PB. (2023) Characterization of gaseous and particulate phase polycyclic aromatic hydrocarbons emitted during preharvest burning of sugar cane in different regions of Kwa-Zulu Natal, South Africa. *Environ. Toxicol. Chem.* 42(4), 778–792. https://doi.org/10.1002/etc.5579

Gitelman L, Kozhevnikov M, Visotskaya Y. (2023) Diversification as a method of ensuring the sustainability of energy supply within the energy transition. *Resources* 12(2), 19. https://doi.org/10.3390/resources12020019

Gomes MG, Gurgel LVA, Baffi MA, Pasquini D. (2020) Pretreatment of sugarcane bagasse using citric acid and its use in enzymatic hydrolysis. *Renew. Energy* 157, 332–341. https://doi.org/10.1016/j.renene.2020.05.002

Gonzalez JM, Tomlinson JE, Martínez Ceseña EA, Basheer M, Obuobie E, Padi PT, Harou JJ. (2023) Designing diversified renewable energy systems to balance multisector performance. *Nat. Sustain.* 6(4), 415–427. https://doi.org/10.1038/s41893-022-01033-0

Grandis A, Fortirer JS, Navarro BV, de Oliveira LP, Buckeridge MS. (2023) Biotechnologies to improve sugarcane productivity in a climate change scenario. *BioEnergy Res.* 17, 1–26. https://doi.org/10.1007/s12155-023-10649-9

Guo D-J, Li D-P, Chen Z-D, Yang B, Verma KK, Singh RK, Singh P, Khan Q, Sharma A, Qin Y, Zhang B-Q, Song X-P, Li Y-R. (2023) Effect of endophytic diazotroph *Enterobacter roggenkampii* ED5 on nitrogen-metabolism-related microecology in the sugarcane rhizosphere at different nitrogen levels. *Front. Microbiol.* 14, 1132016. https://doi.org/10.3389/fmicb.2023.1132016

Guo D-J, Li D-P, Singh RK, Singh P, Sharma A, Verma KK, Qin Y, Khan Q, Lu Z, Malviya MK, Song X-P, Xing Y-X and Li Y-R. (2021) Differential protein expression analysis of two sugarcane varieties in response to diazotrophic plant growth-promoting endophyte *Enterobacter roggenkampii* ED5. *Front. Plant Sci.* 12, 727741. https://doi.org/10.3389/fpls.2021.727741

Guo DJ, Li DP, Singh RK, Singh P, Verma KK, Sharma A, Qin Y, Khan Q, Song XP, Malviya MK, Xing YX, Li YR. (2022) Comparative transcriptome analysis of two sugarcane varieties in response to diazitrophic plant growth promoting endophyte *Enterobacter roggenkampii* ED5. J. Plant Interactions, 17(1), 75–84.

Han G, Zhou T, Sun Y, Zhu S. (2022) The relationship between night-time light and socioeconomic factors in China and India. *PLos One* 17(1), e0262503. https://doi.org/10.1371/journal.pone.0262503

Hartner E, Gawlitta N, Gröger T, Orasche J, Czech H, Geldenhuys GL, Zimmermann R. (2024) Chemical fingerprinting of biomass burning organic aerosols from sugar cane combustion: Complementary findings from field and laboratory studies. *ACS Earth Space Chem.* 8(3), 533–546. https://doi.org/10.1021/acsearthspacechem.3c00301

Hassan Q, Viktor P, Al-Musawi TJ, Ali BM, Algburi S, Alzoubi HM, Jaszczur M. (2024) The renewable energy role in the global energy Transformations. *Renew. Energy Focus* 48, 100545. https://doi.org/10.1016/j.ref.2024.100545

Hernandes TAD, Duft DG, dos Santos Luciano AC, Leal MRLV, Cavalett O. (2021) Identifying suitable areas for expanding sugarcane ethanol production in Brazil under conservation of environmentally relevant habitats. *J. Cleaner Prod.* 292, 125318. https://doi.org/10.1016/j.jclepro.2020.125318

Huang CM, Wu KC, Verma KK, Luo HB, Deng ZN, Shi GY, Cao HQ, Wu XJ, Ye LP, Wei YW, Xu L. (2024) Isolation and expression analysis of cellulose synthase 3 (*Ces3*) genes from sugarcane (*Saccharum officinarum* L.). *Chem. Biol. Technol. Agric.* 11, 71. https://doi.org/10.1186/s40538-024-00597-7

Huang HR, Gao YJ, Malviya MK, Verma KK, Solanki MK, Huang YX, Li X, Deng Y, Yan J, Tang SY, Wang LW, Xu L. (2022) Genetic diversity analysis of sugarcane (*Saccharum* spp. Hybrids) among high-sucrose clones of GT series and commonly used parents by using microsatellite markers in Guangxi, China. *Sugar Tech.* 24, 397–407. https://doi.org/10.1007/s12355-021-01033-y

Hung NT. (2022) Biomass energy consumption and economic growth: Insights from BRICS and developed countries. *Environ. Sci. Poll. Res.* 29(20), 30055–30072. https://doi.org/10.1007/s11356-021-17721-x

Jagadesh M, Dash M, Kumari A, Singh SK, Verma KK, Kumar P, Bhatt R, Sharma SK. (2024) Revealing the hidden world of soil microbes: Metagenomic insights into plant, bacteria, and fungi interactions for sustainable agriculture and ecosystem restoration. *Microbiol. Res.* 285, 127764. https://doi.org/10.1016/j.micres.2024.127764

Jaworski CC, Krzywoszynska A, Leake JR, Dicks LV. (2024) Sustainable soil management in the United Kingdom: A survey of current practices and how they relate to the principles of regenerative agriculture. *Soil Use Manag.* 40(1), e12908. https://doi.org/10.1111/sum.12908

Joshi A, Verma KK, Rajput VD, Arora J. (2023) Metabolic engineering of microorganisms in advancing biofuel production. In: *Advances in Lignocellulosic Biofuel Production Systems*, pp. 181–202. https://doi.org/10.1016/B978-0-323-91192-4.00016-X

Joshi A, Verma KK, Rajput VD, Minkina T, Arora J. (2022) Recent advances in metabolic engineering of microorganisms for advancing lignocellulose-derived biofuels. *Bioengineered* 13(4), 8135–8163. https://doi.org/10.1080/21655979.2022.2051856.

Karp SG, Medina JD, Letti LA, Woiciechowski AL, de Carvalho JC, Schmitt CC, Soccol CR. (2021) Bioeconomy and biofuels: The case of sugarcane ethanol in Brazil. *Biofuels, Bioprod. Biorefining* 15(3), 899–912. https://doi.org/10.1002/bbb.2195

Karp SG, Schmitt CC, Moreira R, de Oliveira Penha R, De Mello AFM, Herrmann LW, Soccol CR. (2022) Sugarcane biorefineries: Status and perspectives in bioeconomy. *BioEnergy Res.* 15(4), 1842–1853. https://doi.org/10.1007/s12155-022-10406-4

Korasapati R, Basthipati SB, Rosangzuala K, Reddy PR, Gajjala R, Babu PR, Banoth L. (2023) Value added products generation from sugarcane bagasse and its impact on economizing biorefinery and sustainability of sugarcane industry. In: *Sugarcane—Its Products and Sustainability*. IntechOpen. https://doi.org/10.5772/intechopen.107472

Krishna D, Sachan HK, Verma KK. (2022) Abiotic stresses in plants: An overview. In: *Understanding Abiotic Stresses*. Nova Science Publishers, Inc.: New York.

Kumar A, Singh UP, Kakran S. (2022) Social accounting for sugarcane industry: A bibliometric overview. *J. Positive School Psychol.* 1010–1025.

Kumari A, Dash M, Singh SK, Jagadesh M, Mathpal B, Mishra PK, Pandey SK, Verma KK. (2023a) Soil microbes: A natural solution for mitigating the impact of climate change. *Environ. Monit. Assess.* 195, 1436. https://doi.org/10.1007/s10661-023-11988-y

Kumari A, Lakshmi GA, Krishna GK, Patni B, Prakash S, Bhattacharyya M, Singh SK, Verma KK. (2022) Climate change and its impact on crops: A comprehensive investigation for sustainable agriculture. *Agronomy* 12, 3008. https://doi.org/10.3390/agronomy12123008

Kumari A, Singh SK, Mathpal B, Verma KK, Garg VK, Bhattacharyya M, Bhatt R. (2023b) The biosynthesis, mechanism of action, and physiological functions of melatonin in horticultural plants: A review. *Horticulturae* 9, 913. https://doi.org/10.3390/horticulturae9080913

Lee H, Sohn YJ, Jeon S, Yang H, Son J, Kim YJ, Park SJ. (2023) Sugarcane wastes as microbial feedstocks: A review of the biorefinery framework from resource recovery to production of value-added products. *Bioresour. Technol.* 376, 128879. https://doi.org/10.1016/j.biortech.2023.128879

Li A-M, Liao F, Wang M, Chen Z-L, Qin C-X, Huang R-Q, Verma KK, Li Y-R, Que Y-X, Pan Y-Q, et al. (2023) Transcriptomic and proteomic landscape of sugarcane response to biotic and abiotic stressors. *Int. J. Mol. Sci.* 24, 8913. https://doi.org/10.3390/ijms24108913

Li L, Sun W, Hu W, Sun Y. (2021) Impact of natural and social environmental factors on building energy consumption: Based on bibliometrics. *J. Building Eng.* 37, 102136. https://doi.org/10.1016/j.jobe.2020.102136

Linnenluecke MK, Nucifora N, Thompson N. (2018) Implications of climate change for the sugarcane industry. *Wiley Interdiscip. Rev. Clim. Change.* 9(1), e498.

Linnenluecke MK, Zhou C, Smith T, Thompson N, Nucifora N. (2020) The impact of climate change on the Australian sugarcane industry. *J. Cleaner Prod.* 246, 118974. https://doi.org/10.1016/j.jclepro.2019.118974

Liu Z, Saydaliev HB, Lan J, Ali S, Anser MK. (2022) Assessing the effectiveness of biomass energy in mitigating CO2 emissions: Evidence from top-10 biomass energy consumer countries. *Renew. Energy* 191, 842–851. https://doi.org/10.1016/j.renene.2022.03.053

Malanski PD, Dedieu B, Schiavi S. (2021) Mapping the research domains on work in agriculture. A bibliometric review from Scopus database. *J. Rural Stud.* 81, 305–314. https://doi.org/10.1016/j.jrurstud.2020.10.050

Malode SJ, Prabhu KK, Mascarenhas RJ, Shetti NP, Aminabhavi TM. (2021) Recent advances and viability in biofuel production. *Energy Conversion Manag.: X*, 10, 100070. https://doi.org/10.1016/j.ecmx.2020.100070

Malviya MK, Li C-N, Lakshmanan P, Solanki MK, Wang Z, Solanki AC, Nong Q, Verma KK, Singh RK, Singh P, Sharma A, Guo D-J, Dessoky ES, Song X-P, Li Y-R. (2022) High-throughput sequencing-based analysis of rhizosphere and diazotrophic bacterial diversity among wild progenitor and closely related species of sugarcane (*Saccharum* spp. inter-specific hybrids). *Front. Plant Sci.* 13, 829337. https://doi.org/10.3389/fpls.2022.829337.

Malviya MK, Solanki MK, Li CN, Wang Z, Zeng Y, Verma KK, Singh RK, Singh P, Huang HR, Yang LT, Song XP, Li YR. (2021) Sugarcane-legume intercropping can enrich the soil microbiome and plant growth. *Front. Sustain. Food Syst.* 5, 606595. https://doi.org/10.3389/fsufs.2021.606595.

Martíni AF, Valani GP, Boschi RS, Bovi RC, da Silva LFS, Cooper M. (2020) Is soil quality a concern in sugarcane cultivation? A bibliometric review. *Soil Tillage Res.* 204, 104751. https://doi.org/10.1016/j.still.2020.104751

Martíni AF, Valani GP, da Silva LFS, de Paula S, Bolonhezi D, Cooper M. (2024) Soil physical quality response to management systems in a long-term sugarcane trial. *Land Degrad. Develop.* 35(4), 1320–1334. https://doi.org/10.1002/ldr.4988

Moitinho MR, Ferraudo AS, Panosso AR, da Silva Bicalho E, Teixeira DDB, de Andrade Barbosa M, La Scala Jr N. (2021) Effects of burned and unburned sugarcane harvesting systems on soil CO_2 emission and soil physical, chemical, and microbiological attributes. *Catena* 196, 104903. https://doi.org/10.1016/j.catena.2020.104903

Moore CCS, Nogueira AR, Kulay L. (2017) Environmental and energy assessment of the substitution of chemical fertilizers for industrial wastes of ethanol production in sugarcane cultivation in Brazil. *Int. J. Life Cycle Assess.* 22(4), 628–643 https://doi.org/10.1007/s11367-016-1074-0

Morais Junior WG, Pacheco TF, Gao S, Martins PA, Guisán JM, Caetano NS. (2021) Sugarcane bagasse saccharification by enzymatic hydrolysis using endocellulase and β-glucosidase immobilized on different supports. *Catalysts* 11(3), 340. https://doi.org/10.3390/catal11030340

Mugica-Álvarez V, Hernández-Rosas F, Magaña-Reyes M, Herrera-Murillo J, Santiago-De La Rosa N, Gutiérrez-Arzaluz M, González-Cardoso G. (2018) Sugarcane burning emissions: Characterization and emission factors. *Atmospheric Environ.* 193, 262–272.

Nicolletti M, Maschietto F, Moreno T. (2020) Integrating social learning into climate change adaptation public policy cycle: Building upon from experiences in Brazil and the United Kingdom. *Environ. Develop.* 33, 100486. https://doi.org/10.1016/j.envdev.2019.100486

Oliveira MA, Verma KK, Bhatt R, Oliveira TBA. (2022) Impact of green and organic fertilizers on soil fertility and sugarcane productivity. In: Verma KK, et al. (Eds.), *Agro-industrial Perspectives on Sugarcane Production Under Environmental Stress*. https://doi.org/10.1007/978-981-19-3955-6_11.

Patil H, Sudagar IP, Pandiselvam R, Sudha P, Boomiraj K. (2023) Development and characterization of rigid packaging material using cellulose/sugarcane bagasse and natural resins. *Int. J. Biol. Macromol.* 246, 125641. https://doi.org/10.1016/j.ijbiomac.2023.125641

Pérez Medina P, Galindo Mendoza MG, Álvarez Fuentes G, Tenorio Martínez LD, Barrera López VA. (2023) Economic assessment of the impact of the sugarcane industry: An empirical approach with two focuses for San Luis Potosí. *México. J.* 6(2), 342–360. https://doi.org/10.3390/j6020024

Powar RV, Mehetre SA, Powar TR, Patil SB. (2022) End-use applications of sugarcane trash: A comprehensive review. *Sugar Tech.* 24(3), 699–714. https://doi.org/10.1007/s12355-022-01107-5

Rasche L, Sos Del Diego R. (2020) Pros and cons of sugarcane straw recovery in São Paulo. *BioEnergy Res.* 13(1), 147–156. https://doi.org/10.1007/s12155-019-10078-7

Rodionova MV, Bozieva AM, Zharmukhamedov SK, Leong YK, Lan JCW, Veziroglu A, Allakhverdiev SI. (2022) A comprehensive review on lignocellulosic biomass biorefinery for sustainable biofuel production. *Int. J. Hydrogen Energy* 47(3), 1481–1498. https://doi.org/10.1016/j.ijhydene.2021.10.122

Ruths JC, Shikida PFA, Fracarolli IFL. (2023) Rural work in the sugarcane sector and its influences on health: Scoping review. *Rev. Bras. Med. Trab.* 21(1), e2023779. https://doi.org/10.47626/1679-4435-2023-779

Sahni S, Kaurav RPS. (2023) What? Why? When? How? Where? of technology-based bibliometric review. In: *Advancing Methodologies of Conducting Literature Review in Management Domain*. Emerald Publishing Limited, Vol. 2, pp. 79–101. https://doi.org/10.1108/S2754-586520230000002005

Salazar LF, Guerra Noriega A, Bolaños A, Solís JP. (2023) Toward sustainability? Contributions of the Guatemalan Sugarcane Agro-Industry to the Sustainable Development Goals on water and energy. In: *Natural Resources Forum*. Blackwell Publishing Ltd.: Oxford, UK, 47(4), 772–788. https://doi.org/10.1111/1477-8947.12349

Santoyo-Castelazo E, Santoyo E, Zurita-García L, Luengas DC, Solano-Olivares K. (2023) Life cycle assessment of bioethanol production from sugarcane bagasse using a gasification conversion Process: Bibliometric analysis, systematic literature review and a case study. *Appl. Thermal Eng.* 219, 119414. https://doi.org/10.1016/j.applthermaleng.2022.119414

Sharma A, Singh RK, Singh P, Vaishnav A, Guo DJ, Verma KK, Li DP, Song XP, Malviya MK, Khan N, Lakshmanan P, Li YR. (2021) Insights into the bacterial and nitric oxide-induced salt tolerance in sugarcane and their growth-promoting abilities. *Microorganisms* 9, 2203. https://doi.org/10.3390/microorganisms9112203.

Sharma A, Song X-P , Singh RK, Vaishnav A, Gupta S, Singh P, Guo D-J, Verma KK, Li Y-R. (2022) Impact of carbendazim on cellular growth, defence system and plant growth promoting traits of *Priestia megaterium* ANCB-12 isolated from sugarcane rhizosphere. *Front. Microbiol.* 13, 1005942. doi:10.3389/fmicb.2022.1005942

Singh P, Singh RK, Li HB, Guo DJ, Sharma A, Verma KK, Solanki MK, Upadhyay SK, Lakshmanan P, Yang LT, Li YR. (2023) Nitrogen fixation and phytohormone stimulation of sugarcane plant through plant growth promoting diazotrophic *Pseudomonas*. *Biotechnol. Genet. Eng. Rev.* 40(1), 15–35. https://doi.org/10.1080/02648725.2023.2177814

Singh P, Singh RK, Song QQ, Li HB, Guo DJ, Malviya MK, Verma KK, Song XP, Lakshmanan P, Yang LT, Li YR. (2021) Comparative analysis of protein and differential responses of defense-related gene and enzyme activity reveals the long-term molecular responses of sugarcane inoculated with *Sporisorium scitamineum*. *J. Plant Interact.* 16(1), 12–29.

Singh RK, Singh P, Li HB, Song QQ, Guo DJ, Solanki MK, Verma KK, Malviya MK, Song XP, Lakshmanan P, Yang LT, Li YR. (2020) Diversity of nitrogen-fixing rhizobacteria associated with sugarcane: A comprehensive study of plant-microbe interactions for growth enhancement in *Saccharum* spp. *BMC Plant Biol.* 20, 220. https://doi.org/10.1186/s12870-020-02400-9.

Solanki MK, Verma KK, Dastogeer KMG, Mora-Poblete F, Mundra S. (2023) Microbial resilience in plant nutrient management towards sustainable farming. *Front. Microbiol.* 14, 1280811. https://doi.org/10.3389/fmicb.2023.1280811

Solarte-Toro JC, Alzate CAC. (2021) Biorefineries as the base for accomplishing the sustainable development goals (SDGs) and the transition to bioeconomy: Technical aspects, challenges and perspectives. *Bioresour. Technol.* 340, 125626. https://doi.org/10.1016/j.biortech.2021.125626

Solomon S, Quirk RG, Shukla SK. (2019) Green management for sustainable sugar industry. *Sugar Tech.* 21(2), 183–185. https://doi.org/10.1007/s12355-019-00711-2

Song X-P, Mo F, Verma KK, Wei J, Zhang X, Yang L, Li YR. (2019) Effect of sugarcane smut (*Ustilago scitaminea* Syd.) on ultrastructure and biochemical indices of sugarcane. *Biomed J. Sci. Tech. Res.* 17(1), 002950. https://doi.org/10.26717/BJSTR.2019.17.002950.

Song XP, Mo F, Yan M, Zhang X, Zhang B, Huang X, Huang D, Pan Y, Verma KK, Li YR. (2022) Effect of smut infection on the photosynthetic physiological characteristics and related defense enzymes of sugarcane. *Life* 12, 1201. https://doi.org/10.3390/life12081201

Song XP, Verma KK, Tian DD, Zhang XQ, Liang YJ, Huang X, Li CN, Li YR. (2021) Exploration of silicon functions to integrate with biotic stress tolerance and crop improvement. *Biol. Res.* 54, 19. https://doi.org/10.1186/s40659-021-00344-4

Stem AD, Gibb M, Roncal-Jimenez CA, Johnson RJ, Brown JM. (2024) Health burden of sugarcane burning on agricultural workers and nearby communities. *Inhal. Toxicol.* 36(5), 327–342. https://doi.org/10.1080/08958378.2024.2316875

Tam KP, Chan HW, Clayton S. (2023) Climate change anxiety in China, India, Japan, and the United States. *J. Environ. Psychol.* 87, 101991. https://doi.org/10.1016/j.jenvp.2023.101991

Tayade AS, Geetha P, Anusha S, Arunkumar R, Vasantha S. (2023) Agro-technologies to sustain sugarcane productivity under abiotic stresses. In: *Agro-industrial Perspectives on Sugarcane Production under Environmental Stress*, Springer Nature Singapore: Singapore, pp. 231–266. https://doi.org/10.1007/978-981-19-3955-6_13

Tian D, Qin L, Verma KK, Wei L, Li J, Li B, Zhou W, He Z, Wei D, Huang S, Long S, Huang Q, Li C, Wei S. (2023) Transcriptomic and metabolomic differences between banana varieties which are resistant or susceptible to *Fusarium wilt*. *PeerJ* 11, e16549. https://doi.org/10.7717/peerj.16549

Tripathi M, Sahu JN, Ganesan P. (2016) Effect of process parameters on production of biochar from biomass waste through pyrolysis: A review. *Renew. Sust. Energy Rev.* 55, 467–481. https://doi.org/10.1016/j.rser.2015.10.122

Vandenberghe LPS, Valladares-Diestra KK, Bittencourt GA, Torres LZ, Vieira S, Karp SG, Soccol CR. (2022) Beyond sugar and ethanol: The future of sugarcane biorefineries in Brazil. *Renew. Sust. Energy Rev.* 167, 112721. https://doi.org/10.1016/j.rser.2022.112721

Velidandi A, Gandam PK, Chinta ML, Konakanchi S, reddy Bhavanam A, Baadhe RR, Gupta VK. (2023) State-of-the-art and future directions of machine learning for biomass characterization and for sustainable biorefinery. *J. Energy Chem.* 81, 42–63. https://doi.org/10.1016/j.jechem.2023.02.020

Verma KK, Joshi A, Song X-P, Liang Q, Xu L, Huang H-r, Wu K-C, Seth CS, Arora J and Li Y-R (2024a) Regulatory mechanisms of plant rhizobacteria on plants to the adaptation of adverse agroclimatic variables. *Front. Plant Sci.* 15, 1377793. https://doi.org/10.3389/fpls.2024.1377793

Verma KK, Joshi A, Song X-P, Singh S, Kumari A, Arora J, Singh SK, Solanki MK, Seth CS, Li Y-R. (2024b) Synergistic interactions of nanoparticles and plant growth promoting rhizobacteria enhancing soil-plant systems: A multigenerational perspective. *Front. Plant Sci.* 15, 1376214. https://doi.org/10.3389/fpls.2024.1376214

Verma KK, Li DM, Singh M, Rajput VD, Malviya MK, Minkina T, Singh RK, Singh P, Song XP, Li YR. (2020a) Interactive role of silicon and plant-rhizobacteria mitigating abiotic stresses: A new approach for sustainable agriculture and climate change. *Plants* 9, 1055. https://doi.org/10.3390/plants9091055.

Verma KK, Rajput VD, Singh M, Verma CL, Minkina T. (2022a) Influence of waterlogging on plant physiology and molecular biology. In: *Understanding Abiotic Stresses*. Nova Science Publishers, Inc.: New York.

Verma KK, Siddiqui N, Singh M. (2022b) Understanding the plant physiological and molecular adaptation strategies to soil salinity. In: *Understanding Abiotic Stresses*. Nova Science Publishers, Inc.: New York. https://doi.org/10.52305/UBUJ4024

Verma KK, Singh P, Song X-P, Malviya MK, Singh RK, Chen G-L, Solomon S, Li YR. (2020b) Mitigating climate change for sugarcane improvement: Role of silicon in alleviating abiotic stresses. *Sugar Tech.* 22(5), 741–749. https://doi.org/10.1007/s12355-020-00831-0.

Verma KK, Singh RK, Song QQ, Singh P, Zhang B-Q, Song X-P, Chen G-L, Li YR. (2019a) Silicon alleviates drought stress of sugarcane plants by improving antioxidant responses. *Biomed J. Sci. Tech. Res.* 002957. https://doi.org/10.26717/BJSTR.2019.17.002957

Verma KK, Song X, Degu HD, Guo D, Joshi A, Huang H, Xu L, Singh M, Huang D, Rajput VD, Li YR. (2023a) Recent advances in nitrogen and nano-nitrogen fertilizers for sustainable crop production: A mini-review. *Chem. Biol. Technol. Agric.* 10, 1–14. https://doi.org/10.1186/s40538-023-00488-3

Verma KK, Song X, Verma CL, Huang H, Singh M, Xu L, Wu J, Khan ZH, Zhang B, Li Y. (2023b) Mathematical modeling of climate and fluoride effects on sugarcane photosynthesis with silicon nanoparticles. *Plant Physiol. Biochem.* 204, 108089. https://doi.org/10.1016/j.plaphy.2023.108089

Verma KK, Song XP, Budeguer F, Nikpay A, Enrique R, Singh M, Zhang B, Wu, JM, Li YR. (2022c) Genetic engineering: An efficient approach to mitigating biotic and abiotic stresses in sugarcane cultivation. *Plant Signal. Behav.* 17(1), e2108253.
Verma KK, Song XP, Joshi A, Rajput VD, Singh M, Sharma A, Singh RK, Li DM, Arora J, Minkina T, Li YR. (2022d) Nanofertilizer possibilities for healthy soil, water and food in future: An overview. *Front. Plant Sci.* 13, 865048. *https://doi.org/10.3389/fpls.2022.865048*
Verma KK, Song XP, Joshi A, Tian DD, Rajput VD, Singh M, Arora J, Minkina TM, Li YR. (2022e) Recent trends in nano-fertilizer for advancing sustainable agriculture under the era of climate change to ensure food security for future generations. *Nanomaterials* 12, 173. https://doi.org/10.3390/nano12010173
Verma KK, Song XP, Li DM, Singh M, Wu JM, Singh RK, Sharma A, Zhang BQ, Li YR. (2022f) Silicon and soil microorganisms improves rhizospheric soil health with bacterial community, plant growth, performance and yield. *Plant Signal. Behav.* 17(1), e2104004.
Verma KK, Song XP, Lin B, Guo DJ, Singh M, Rajput VD, Singh RK, Singh P, Sharma A, Malviya MK, Chen GL, Li YR. (2022g) Silicon induced drought tolerance in crop plants: Physiological adaptation strategies. *Silicon* 14(6), 2473–2487. https://doi.org/10.1007/s12633-021-01071-x
Verma KK, Song XP, Rajput VD, Boldyreva V, Zhang BQ, Minkina T, Li YR. (2022h) Morpho-physiological, biochemical and ultrastructural modifications on sugarcane to prolonged water deficit. In: Verma KK, et al. (Eds.), *Agro-industrial Perspectives on Sugarcane Production under Environmental Stress*. https://doi.org/10.1007/978-981-19-3955-6_11.
Verma KK, Song XP, Rajput VD, Solomon S, Li YR, Rao GP. (2022i) *Agro-industrial Perspectives on Sugarcane Production under Environmental Stress*. Springer Nature: Singapore, https://doi.org/10.1007/978-981-19-3955-6
Verma KK, Song XP, Singh M, Tian DD, Rajput VD, Minkina T, Li YR. (2023c) Association of silicon and soil microorganisms induces stress mitigation, increasing plant productivity. In: de Mello Prado R, (ed.), *Benefits of Silicon in the Nutrition of Plants*. pp. 299–328. https://doi.org/10.1007/978-3-031-26673-7_17
Verma KK, Song XP, Tian DD, Guo DJ, Chen ZL, Zhong CS, Nikpay A, Singh M, Rajput VD, Singh RK, Minkina T, Li YR. (2021a) Influence of silicon on biocontrol strategies to manage biotic stress for crop protection, performance and improvement. *Plants* 10, 2163. https://doi.org/10.3390/plants10102163
Verma KK, Song XP, Tian DD, Singh M, Verma CL, Rajput VD, Singh RK, Sharma A, Singh P, Malviya MK, Li YR. (2021b) Investigation of defensive role of silicon during drought stress induced by irrigation capacity in sugarcane: Physiological and biochemical characteristics. *ACS Omega* 6, 19811–19821.
Verma KK, Song XP, Verma CL, Chen ZL, Rajput VD, Wu KC, Liao F, Chen GL, Li YR. (2021c) Functional relationship between photosynthetic leaf gas exchange in response to silicon application and water stress mitigation in sugarcane. *Biol. Res.* 54, 15. https://doi.org/10.1186/s40659-021-00338-2
Verma KK, Song XP, Verma CL, Malviya MK, Guo DJ, Rajput VD, Sharma A, Wei KJ, Chen GL, Solomon S, Li YR. (2021d) Predication of photosynthetic leaf gas exchange of sugarcane (*Saccharum* spp.) leaves in response to leaf positions to foliar spray of

potassium salt of active phosphorus under limited water irrigation. *ACS Omega* 6, 2396–2409.

Verma KK, Song X-P, Xu L, Huang H-R, Liang Q, Seth CS, Li Y-R. (2023d) Nano-microplastic and agro-ecosystems: A mini-review. *Front. Plant Sci.* 14, 1283852. https://doi.org/10.3389/fpls.2023.1283852

Verma KK, Song XP, Yadav G, Degu HD, Parvaiz A, Singh M, Huang HR, Mustafa G, Xu L, Li YR. (2022j) Impact of agroclimatic variables on proteo-genomics in sugarcane (*Saccharum* spp.) plant productivity. *ACS Omega* 7, 22997–23008.

Verma KK, Song XP, Zeng Y, Guo DJ, Sing M, Rajput VD, Malviya MK, Wei KJ, Sharma A, Li DP, Chen GL, Li YR. (2021e) Foliar application of silicon boosts growth, photosynthetic leaf gas exchange, antioxidative response and resistance to limited water irrigation in sugarcane (*Saccharum officinarum* L.). *Plant Physiol. Biochem.* 166, 582–592.

Verma KK, Song XP,Zeng Y, Li DM, Guo DJ, Rajput VD, Chen GL, Barakhov A, Minkina TM, Li YR. (2020c) Characteristics and correlation of leaf stomata and its relationship with photosynthesis on *Saccharum* spp. under different irrigation and silicon application. *ACS Omega* 5, 24145–24153.

Verma KK, Vatsal S, Singh M, Rajput VD, Minkina T. (2022k) Coping with water stress: Effects and adaptive mechanisms to plants. In: *Understanding Abiotic Stresses*. Nova Science Publishers, Inc.: New York.

Verma KK, Wu KC, Verma CL, Li DM, Malviya MK, Singh RK, Singh P, Chen GL, Song XP, Li YR. (2020d) Developing mathematical model for diurnal dynamics on photosynthetic responses in sugarcane responsive to different irrigation and silicon application. *PeerJ* 8, e10154. https://doi.org/10.7717/peerj.10154.

Verma KK, Liu X-H, Wu K-C, Singh RK, Song QQ, Malviya MK, Song X-P, Singh P, Verma CL, Li YR. (2020e) The impact of silicon on photosynthetic and biochemical responses of sugarcane under different soil moisture levels. *Silicon* 12, 1355–1367.

Verma KK, Wu K-C, Singh P, Malviya MK, Singh RK, Song X-P, Li YR. (2019b) The protective role of silicon in sugarcane under water stress: Photosynthesis and antioxidant enzymes. *Biomed J. Sci. Tech. Res.* 002685. https://doi.org/0.26717/BJSTR.2019.15.002685

Wang G, Dai Y, Yang H, Xiong Q, Wang K, Zhou J, Wang S. (2020) A review of recent advances in biomass pyrolysis. *Energy Fuels* 34(12), 15557–15578. https://doi.org/10.1021/acs.energyfuels.0c03107

Waqas M, Nizami AS, Aburiazaiza AS, Jabeen F, Arikan OA, Anees A, Rehan M. (2023) Unlocking integrated waste biorefinery approach by predicting calorific value of waste biomass. *Environ. Res.* 237, 116943. https://doi.org/10.1016/j.envres.2023.116943

Wu K-C, Huang C-M, Verma KK, Deng Z-N, Huang H-R, Pang T, Cao H-Q, Luo H-B, Jiang S-L, Xu L. (2022) Transcriptomic responses of *Saccharum spontaneum* roots in response to polyethylene glycol–6000 stimulated drought stress. *Front. Plant Sci.* 13, 992755. https://doi.org/10.3389/fpls.2022.992755

Xu L, Deng Z-N, Wu K-C, Malviya MK, Solanki MK, Verma KK, Pang T, Li Y-J, Liu X-Y, Kashyap BK, Dessoky ES, Wang W-Z and Huang H-R. (2022) Transcriptome analysis reveals a gene expression pattern that contributes to sugarcane bud propagation induced by Indole-3-Butyric acid. *Front. Plant Sci.* 13, 852886. https://doi.org/10.3389/fpls.2022.852886

Zeraibi A, Jahanger A, Adebayo TS, Ramzan M, Yu Y. (2023) Greenfield investments, economic complexity, and financial inclusion-environmental quality nexus in BRICS Countries: Does renewable energy transition matter? *Gondwana Res.* 117, 139–154. https://doi.org/10.1016/j.gr.2022.12.020

Zheng F-J, Lin B, Yang Y-X, Fang X-C, Verma KK, Chen G-L. (2023) Efficacy and functionality of sugarcane original vinegar on mice. *Front. Microbiol.* 14, 1224666. https://doi.org/10.3389/fmicb.2023.1224666

CHAPTER 2

Environmental Diversification as a Measure of Sustainable Sugarcane Growth

TRISHA SINHA,[1], BHASKAR PRATAP SINGH,[2*] and ANSHU GANGWAR[3]

[1]*Department of Plant Physiology, Narayan Institute of Agricultural Sciences, Gopal Narayan Singh University, Rohtas, Bihar, India*

[2]*Department of Agricultural Engineering, ANDUAT-Krishi Vigyan Kendra, Amethi, Uttar Pradesh, India*

[3]*Department of Agricultural Engineering, Krishi Vigyan Kendra, (DRPCAU, Pusa), Bihar, India*

*Corresponding author

ABSTRACT

The intensified agriculture has already destroyed a lot of fertile soil. The surplus in demand and supply of food has not yet ended. The population pressure, along with climate-changing patterns, made it more challenging to achieve the target of zero hunger by 2030. Some conventional cultivation methods have been replaced by sustainable agricultural approaches, and some are yet to be replaced. Complete dependence on exhausting crops like rice is a threat at present, as the groundwater level has been reducing alarmingly. Diversification is of utmost need to save water and other natural resources for a sustainable future. Diversification not only includes a change in crop selection but also combines two or more sustainable approaches. There are multiple sustainable approaches, such as crop

Sugarcane Production: Environmental Impacts and Socioeconomic Issues.
Krishan K. Verma, Xiu-Peng Song, Lin Xu, Rajan Bhatt, and Yang-Rui Li (Eds.)

DOI: 10.1201/9781779641137-2

rotation, integrated nutrient management, and pest management. For crop selection, sugarcane is a great option as the crop is both cash-giving and environmentally sustainable. Though it is also a water-exhausting crop, many by-products are generated during its processing, which can be used for multiple purposes. The sustainable sugarcane initiative was launched to maintain soil fertility and crop production, as well as meet food demands and environmental biodiversity. As a whole, some sustainable sugarcane approaches can minimize water depletion. So, environmental diversification is certainly possible through sugarcane cultivation with the adoption of some smart strategies.

2.1 INTRODUCTION

Climate change is the most discussed topic of the current era. Maintaining environmental diversity along with sustainable production is the most difficult challenge. Agriculture is dependent on many factors, the most important being the climate (Verma et al., 2022a; Kumari et al., 2023a, 2023b; Jagadesh et al., 2024). Environmental components, such as humidity, temperature, wind, etc., have a direct role in deciding the future of a crop for every climatic zone. Tropical and subtropical climatic zones are the most suitable zones for the growth of most of the crops (Song et al., 2019, 2021). It is seen that major agricultural contributions come through the countries belonging to these zones. It is also true that climatic impacts are most prominent in these two zones, as the variations lead to a huge change in the overall crop cycle and yield (Verma et al., 2019a, 2019b; Huang et al., 2024). Tropical and subtropical countries are supremely blessed with maximum crop production as they receive suitable climatic conditions. However, these countries suffer the most if climatic patterns change. Many crops are threatened as global warming is happening at an alarming rate (Verma et al., 2020a, 2020b; Guo et al., 2022). According to Fahad et al. (2022), the most difficult challenge for developing countries to achieve the sustainable development goals (SDGs), especially no poverty (SDG 1), zero hunger (SDG 2), and economic growth (SDG 8). To cope with climate change, a shift in crop as well as cropping methods has been suggested after a lot of discussions in the scientific community (Solanki et al., 2023; Singh et al., 2023). But, here the question comes: what to implement, when, and how to implement? Crop diversification is defined as introducing or adding new crops to the existing farming system. This is possible through both means, that is, at the individual farm level or larger scales (Paroda, 2022; Verma et al., 2024a, 2024b).

Practices like crop rotation and/or intercropping, cultivation of leguminous crops, etc., are some of the most prominent ways of crop diversification. Sugarcane can also be grown sustainably. Sugarcane is grown extensively in the tropical and subtropical belts of the world. It is grown between 35°N and 35°S latitude and altitude of 1600 m above the sea level (Sanghera et al., 2018; Bhatt et al., 2023a, 2023b). The major sugarcane-growing countries include Brazil, India, China, India, Thailand, the USA, Pakistan, Mexico, and others (http://worldpopulationreview.com). This crop has various temperature requirements for different growth stages, that is, 32–38°C during sprouting of stem cuttings and 30°C during ripening. Sugarcane demands an even distribution of 1100 and 1500 mm of rainfall for its optimum growth (Bhatt, 2020). Its best growth in tropical and subtropical climates is prominent in these regions because the required temperature and rainfall ranges are prominent. Sugarcane is a major water-consuming crop, causing a decline in groundwater levels and threatening sustainability, food security, and economic growth (GWP, 2012, FAO, 2012; Verma et al., 2022b; Oliveira et al., 2022). This chapter will be an exhibition of all those possible combinations of sugarcane cultivation that would lead to the achievement of environmental diversification.

2.2 IMPACT OF CLIMATE CHANGE ON AGRICULTURE AND SUGARCANE

The dependence on intensive agriculture over the past few decades has contributed to a huge increase in production. However, at the same time, it costs the environment and cropping system (Tilman et al., 2002; Huang et al., 2022; Joshi et al., 2022). People had easy access to water, fertilizers, and other important inputs. Unfortunately, the unwise application of resources handicapped agriculture. Now, we have technology and everything, but we lack natural resources for smooth production. Overpopulation is a burden on the growth of production. On the other hand, climate change is acting as fuel for it (Verma et al., 2022c, 2022d). Climate change is defined as the slow but constant change in the average weather parameters all over the world. The extreme weather events are the outcomes of climate change (Krishna et al., 2022; Kumari et al., 2022).

Crops of tropical and subtropical climatic zones are under a huge threat as they are not yielding up to expectations. Soil is losing its natural fertility and other important properties. Water scarcity is another big problem (Verma et al., 2022e, 2022f). Different developmental projects have degraded many soils, water bodies, and air quality. The integrated impact of climate change and population pressure on agriculture is alarming (Chaudhary et al., 2023; Chen et al., 2023). The target of ending poverty and hunger needs a strong support system that may promote the simultaneous adoption of two or more approaches (Verma et al., 2022g). Sugarcane is facing climate impacts, too, as it was reported that despite increased demands for sugarcane, its productivity is not satisfactory in countries like India, one of the largest sugarcane-producing countries (Plaisier et al., 2017; Singh et al., 2020; Kumari et al., 2022; Deng et al., 2023). Several unpredictable events, such as the COVID-19 pandemic, economic downfall, the Ukraine war, and climate change, have led to a slower supply of sugarcane in different parts of the world.

2.3 SUSTAINABILITY

The complicated problems of fighting climate change, the upward population curve, and meeting global food demands with a shortage of natural resources are great issues. The unsustainable use of water has caused a huge decline in groundwater levels, raising a question mark in the future of agriculture (Verma et al., 2022h, 2022i). The problem is to be overcome shortly, failing which we might witness another famine in the near future. The concept of sustainability has to be understood deeply (Verma et al., 2020c, 2020d; Guo et al., 2023). It does not merely define good production with good quality; it also emphasizes the uniform distribution of food products to people. Sustainability in every term, such as economic, social, and agricultural aspects, is in demand.

2.3.1 SUSTAINABLE DEVELOPMENT GOALS (SDGS)

SDGs were made to achieve sustainability in all terms, including the environment, which is one of the important ones. The 17 SDGs talk about bringing sustainability, avoiding exploitation, and ensuring balance in all aspects. Some specific goals are based on agriculture, namely, SDG 1 (zero poverty), SDG 2 (zero hunger), and SDG 8 (economic growth), among others. To talk about SDG 1, poverty is not only about lacking property or assets, but it also denotes the wastage of assets randomly. So, it is impossible to end poverty if people remain unaware of their social duties. The second goal is a bigger target. To achieve SDG 2, a lot of efforts have to be implemented. Ending hunger is possible, but ending greed is impossible. Agricultural sustainability does not only rely on more and more production; rather, it focuses on the uniform distribution of all the products among people all over the globe. So, it should be remembered that uniform distribution of foods is assured along with higher quality production.

SDG 8 is aimed at ensuring the economic growth of people. Economic growth is possible with a change in the preference for a profession. Not many people like to choose agriculture or farming as a profession and passion due to the fear of uncertainty. Assuring good economic profit is necessary here. The right place, the right thing, and time should be considered for a sustainable future in agriculture. Sugarcane might be a better crop for both cash and sustainability. It is also true that challenges like biodiversity loss, water shortage, and climate change make it difficult to prefer a crop like sugarcane for a long duration. Therefore, a multifaceted research approach is needed to integrate environmental considerations into development planning, promote sustainable practices, enhance capacity-building efforts, and foster international cooperation. Socioeconomic factors are also to be addressed, with special attention to environmental sustainability and the empowerment of communities to participate in sustainable development efforts.

2.3.2 AGRICULTURAL SUSTAINABILITY CONCERNING THE ENVIRONMENT

Some of the most significant environmental issues of the current era are nitrate pollution of water, eutrophication (phosphate accumulation in water bodies), emissions of greenhouse gases (GHGs) (Therond et al., 2017; Verma et al., 2022j), and loss of habitats and biodiversity (Bommarco et al., 2013; Joshi et al., 2023). Barman et al. (2022) also added that other points like fossil water extraction, contamination of groundwater, and rising atmospheric CO_2 levels are major risks to attaining sustainability. FAO (2022) reported that the post-COVID impacts have increased globally, and a projection has been made that around 670 million people, representing

8% of the world population, will be under hunger by 2030. As a result, there is a need to balance the gap between demand and supply, which will be met through sustainable food production approaches.

2.3.3 CROP DIVERSIFICATION

Raising environmental problems due to farming systems led to concerns about the future functionality of today's cropping systems, climate resilience and adaptability, provisioning of ecosystem services and biodiversity (Wezel et al., 2014; Rusch et al., 2016; Lichtenberg et al., 2017). Crop diversification, in the form of crop rotation, intercropping, and multiple cropping, may prove to be an effective way of improving crop productivity without compromising ecosystem balance (Garbach et al., 2017; Verma et al., 2022k; Li et al., 2023). According to Hufnagel et al. (2020), crop diversification is converting a simplified cropping system into a more diverse one in time and space by adding other crops. Many studies regarding crop diversification have confirmed the positive impacts on crops and the environment. Davis et al. (2012) concluded that diverse cropping systems resulted in similar or even higher yields than simplified systems with the minimum environmental impacts (Malviya et al., 2021, 2022; Tian et al., 2023). Some advantages of crop diversification concerning different aspects are discussed in Table 2.1.

TABLE 2.1 Advantages of Crop Diversification in Sugarcane for Different Aspects

Intercrop	Advantages	References
Peas in two rows	Improved yield	Saini et al. (2003)
Legumes like cowpea, beans, and soybean	Shortened demand for N-fertilizers	Park et al. (2010)
Wheat, sugar beet, onion, potato, toria, berseem, lentil, garlic, coriander, pea, rajmash, kalaunzee, etc.	Increased land utilization efficiency and reduced cost of production	Shukla et al. (2018)
Pulses like rajmash, lentil, mungbean, berseem	Increased benefit: cost ratio	Shukla et al. (2019)
Cowpea+sugarcane, sunhemp+sugarcane and soybean+sugarcane	Weed oppression by 36.3%, 32.3%, and 32% compared to sole sugarcane	Geetha et al. (2019)
French bean and autumn sugarcane	Economic profit for farmers	Singh et al. (2020)
Soybean intercropping	Boosted soil microbial activities along with arbuscular mycorrhizal fungi (AMF)	Zhang et al. (2020)

2.3.4 SUGARCANE AND THE POSSIBILITY OF SUSTAINABILITY

Sugarcane shows the possibility of a rise in its production from around 1.71 billion tons in 2008 to 1.87 billion tons in 2020, which has been reported from 26 million hectares of cultivated area (FAOSTAT, 2022a). It is estimated that by 2030, India will require about 600 million tons of sugarcane (Yadav and Yadav, 2017) to keep up domestic consumption of sugar and ethanol (Duttamajumder et al., 2011; Sharma et al., 2021, 2022). So, there is hope for area expansion for sugarcane, which will entail a productivity requirement of around 100 tons/ha. During the postpandemic era, many nonagricultural-based industries in India suffered due to gaps in production, demand, and supply. However, sugar industries were affected minimally because of several governmental provisions for transporting input and output during lockdown at the country level (Song et al., 2021; Shukla et al., 2022). El Chami et al. (2020) and Khaire et al. (2021) suggested some approaches that need to be adopted on an urgent basis for sustainable sugarcane production. Those include the integrated approach of considering the increase in sugarcane production and productivity after comparing benefits and/or impacts on ecosystem outcomes; experimental methods; studies on benefits/impacts on different natural resources, such as air, water, and land along with biodiversity and wildlife; taking care of environmental, economic, and health issues of workers with social aspects like labor rights and child labor, securing human well-being, and farmers' income (Verma et al., 2020e, 2021a, 2021b; Chen et al., 2020; Zheng et al., 2023).

2.4 CHALLENGES

Sugarcane is a long-duration crop, having a crop cycle of 10–12 months in the subtropics and 12–18 months in the tropics (Singh et al., 2018). So, it is very common for this crop to face various stresses due to seasonal changes, stresses, insect-pest attacks, and the occurrence of diseases. During the long crop cycle, sugarcane passes through water scarcity; temperature fluctuations during summer and winter; waterlogging due to heavy rain; soil salinity and alkalinity; nutritional stress; insect attack by borer, aphids, scales, white flies, etc., and deadly diseases like red rot, rust, grassy shoot disease, smut, and wilt (Vishwanathan and Rao, 2011). Both excess and insufficient rainfall during the reproductive phase could drastically affect the yield potential of sugarcane. Frost is also a natural enemy of sugarcane, and exposure to it results in reduced sugar quantity and poor sugar quality.

Heavy nitrogen application also retards sugar quality (Harender et al., 2021; Anas et al., 2021a, 2021b; Verma et al., 2023a, 2023b).

The increasing cost of cultivation and price fluctuations also act as deciding factors for the future of sugarcane. Chand et al. (2016) also pointed out major challenges like inadequate water availability, nonavailability of easy credit facilities, and high labor costs for sugarcane farmers. Sundara (2011) reported that sugarcane productivity has increased considerably in the past but has been stagnant over the last two decades. Gujja et al. (2009) described some possible reasons for such low and/or stagnant productivity, such as (1) uncertain water availability due to water shortage and improper water management as well, (2) unpredictable climatic patterns, (3) imbalance in nutrient management, (4) lack in plant protection measures, (5) mono-cropping and (6) lack of improved varieties that could thrive well for long term. Some of the major challenges in sugarcane cultivation are summarized in Table 2.2. Linnenluecke et al. (2020) suggested that the practice of different climate-smart cultivation approaches, such as improved weather forecasting, fertilizer management, and water management, must be urgently implemented. Along with these, encouragement from the government through the launching of different climate-friendly strategies, shifting sugarcane growing areas, and emphasizing economic diversification in rural areas will be beneficial.

TABLE 2.2 Major Challenges and Their Impacts on Sugarcane

Challenge	Impacts	Source
Soil acidity or lower soil pH	Impaired uptake and translocation of nutrients like iron, phosphorus, magnesium, and calcium	Mukhopadhyay and Sharma (1991)
Burning of a standing sugarcane field	Air pollution, nutrient loss from leaves, and	Mitchell et al. (2000)
	loss of beneficial soil microorganisms and nutrients leading to soil infertility	Hanboonsong (2018)
Occurrence of pests and diseases	Retarded cane quality	Plaisier et al. (2017), Luo et al. (2022)
Insufficient labor	Higher labor cost	Shanthy (2018)
Plastic mulching	Impaired water uptake by plants and environmental degradation	Luo et al. (2022)
Low genetic improvement	Lack of superior cultivars	Luo et al. (2022)

2.5 SUGARCANE CROP MANAGEMENT

Sugarcane crop management practices focus on implementing new techniques, smart irrigation approaches, and adopting environment-friendly strategies. Sugarcane, a long-duration crop, usually constitutes a plant crop with single or multiple ratoons in its life cycle from planting to ploughing (Loganandhan et al., 2012; Verma et al., 2021c, 2021d; Singh et al., 2021a). Singh et al. (2021b) reported that the sugarcane–potato–ratoon–wheat crop sequence system has improved sugarcane yield. At the IISR, Lucknow, India, sugarcane-based IFS models for marginal and small farmers were developed to improve yield and farmers' income (IISR, 2020).

Shukla et al. (2022) recommended intercropping legumes with sugarcane to enhance soil fertility through the excretion of amino acids into the rhizospheric zone. It has also been observed that treatment with different plant growth regulators resulted in improved yield compared to the conventional method (AICRP Annual Report, 2017–18). Crop management should target one way to obtain a better yield and encourage economic profit for the farmers. Shukla et al. (2011) reported that sugarcane setts, treated with *Trichoderma viride* inoculation in plant crops, recorded enhanced yield for total cane yield owing to its carry-over effect on ratoon crop and also resulted in a higher benefit-cost ratio (Guo et al., 2021; Wu et al., 2022; Xu et al., 2022). Harvesting green and using unburned trash as a blanket over the field can even reduce GHG emissions, as reported by Pryor et al. (2017). Therefore, crop management is an integral part of sugarcane yield improvement and also for sustainability, that is, environmental and economical.

2.6 SUSTAINABLE SUGARCANE INITIATIVE (SSI)

The sustainable sugarcane initiative (SSI) is based on the maximum production from minimum resources. To make it successful, we need to rely on practices, such as improved soil and water management, changes in nutrient choice, effective weed control, integrated disease management, ratoon management, and adoption of improved varieties (Yadav and Yadav, 2017; Harender et al., 2021). Yadav and Shukla (2016) reported that integrated nutrient management (INM) involving organic nutrients, inorganic nutrients, biofertilizers, and green manuring has been very promising in sugarcane-based systems for the sustenance of soil fertility and crop productivity. About 81% of sugarcane produced globally is converted into sugar for consumption purposes. In contrast, the rest is converted into biofuel (Organization for Economic Cooperation and Development [OECD] & Food and Agriculture Organization of the United Nations [FAO] 2021). Accordingly, sugarcane farmers can benefit from ethanol demand due to the strong demand for sugarcane (Voora et al., 2023).

2.6.1 PRINCIPLES OF SUSTAINABLE SUGARCANE INITIATIVE

Gujja et al. (2009) defined SSI as producing sugarcane using less seeds, less water, and optimum utilization of fertilizers and land to achieve more yields. So, in simpler words, SSI is an effort to produce more with less. The noble reasons for launching SSI were to produce more sugarcane from unit areas and meet people's demand with the minimum use of different inputs required. Therefore, some clear differences exist between conventional and SSI cultivation practices. ICRISAT (2009) described the basic principles of SSI, which are:

1. Nursery should be raised using single-budded chips
2. Transplanting should be done with young seedlings of 25–35 days old
3. Wider spacing of 5 × 2 feet in the main field should be practiced
4. Ensuring sufficient moisture and also good drainage for the overflow of water
5. Promoting organic agriculture
6. Intercropping for effective land utilization

There are certain benefits that can be ensured after the implementation of the SSI method. Gujja et al. (2012) mentioned those as below:

1. Reduction in seed inputs by 95%
2. Labor use decreases by 20–30%
3. Weed reduction by even 40–60%
4. Improvement in obtaining yield by 20–50%, depending on the magnitude of effective implementation of SSI
5. The cost of cultivation is also reduced by 20–30%

6. Water saving by 40–70%, based on the irrigation method followed

2.6.2 NUTRIENT MANAGEMENT

Being an essential part of sustainable sugarcane production, nutrient management plays a vital role in determining sugar output, quality, and demand. More specifically, sugarcane demands higher nitrogen use efficiency, failing to which the crop is highly affected due to nitrate accumulation near the root zone/rhizosphere, further disturbing the ecological balance (Bhatt, 2020). It is not wise to apply more nutrients to achieve improved yield. Both nutrient requirements and application methods are important to know for the crop. According to Prasanna and Pattar (2014), sugarcane crop demands different nutrient requirements for different growth stages. It is seen that the requirements of N and P are the least met during the germination and establishment stage. However, the maximum demand for N and P is observed from the tillering stage onwards, with a decline again with advancement toward maturity. Contrary to that, the germination and tillering stages show a lower K requirement, but with time, the requirement reaches its peak during the grand growth stage.

Chaudhari et al. (2010) reported that the use of water-soluble fertilizers combining urea (providing 46% N), poly feed (providing 20% N, 10% P_2O_5, and 10% K_2O), SOP (50% K_2O and 18% S), and mono ammonium phosphate (providing 12% N and 61% P_2O_5) through drip irrigation could enhance plant height, number of internodes, the girth of internodes, leaf area, individual cane weight, and can yield by 2.72, 3.99, 2.14, 5.66, 9.46, and 8.18%, respectively, as compared to normal fertilizer source of urea, single super phosphate and muriate of potash for suru sugarcane. Therefore, it is very important to practice the best nutrient management strategies. Nutrient management in sugarcane can be done through intercropping with legumes, INM, biofertilizers, and others. INM is a well-known practice of combining inorganic and organic fertilizers to ensure sustainability for both the soil and crop. Similarly, the application of biofertilizers improves soil properties and cane quality. Different studies on the positive impacts of effective nutrient management approaches for sugarcane have been reported, which have been discussed in Table 2.3.

TABLE 2.3 Impacts of Different Effective Nutrient Management Approaches on Environmental Sustainability Concerning Sugarcane

Nutrient Management Approach	Results	References
Application of FYM and/or green manure	Improved nitrogen use efficiency (NUE)	Gaur (1992)
Twice or thrice split doses of application of recommended doses	Improved nutrient use efficiency	Gujja et al. (2009)
FYM with biofertilizer *Phosphobacteria* sp.	Improved plant growth	Gujja et al. (2009)
Intercropping with several legumes like green gram, chickpea, soybean, and cowpea	Enhanced fertilizer use efficiency and curtailed the need for recommended doses	Prasanna and Pattar (2014)
Azotobacter or *Azospirillum*, in two equal splits at 30 and 60 days after planting	Saving 25% of N-fertilizers	Shukla et al. (2017)
N application in 3–4 splits	Increased number of millable canes and cane yield	Shukla et al. (2017
Treatment with *Trichoderma*	Enhanced carbon sequestration, improved cane yield, and sugar output	Shukla et al. (2017a)
Biofertilizers	Improved N efficiency and P mineralization	Shanthy and Venkatesaperumal (2018)
Inoculation with biofertilizers like *Azospirillum* sp. and *Azotobacter* sp.	Improved yield	Singh et al. (2018)
INM involving green manure, biofertilizers, crop residues, and organic manure	Improved soil fertility	Shukla et al. (2018a)
Biofertilizers, i.e., *Azotobacter* or *Consortium*	Improved sucrose content and cane yield	PAU (2018–2019)

2.6.3 WATER MANAGEMENT

Water management plays a significant role in SSI. Usually, the water requirement for sugarcane in the case of conventional methods involving flood irrigation is about 80 lakhs liters/acre. Gujja et al. (2012) reported about a 40% saving of water in the SSI method for the required water in the normal or conventional method. Irrigation method and frequency also depend on the crop stage. They have pointed out the overall water management strategies in SSI, mentioned below:

1. The application of irrigation is done at an interval of 10 days during the tillering period (36–100 days after planting). But, during the grand growth period, that is, 101–270 days after planting, an interval of 7 days is maintained. The interval gets broadened up to 15 days when the maturity phase arrives (from 271 days after planting onwards till harvesting).
2. Both furrow irrigation and alternate furrow irrigation should be followed. Alternate furrow irrigation refers to irrigating the furrows initially with odd numbers, followed by irrigating those with even numbers after 7–15 days, depending on the soil moisture content and crop stage. Adopting these irrigation methods can ensure water savings of up to 50%.
3. Drip irrigation also proves to be very effective in SSI. There are certain advantages to adopting drip irrigation for SSI: first, effective irrigation due to wider spacing and the planting of single seedlings followed in SSI; second, water savings of up to 40–70%; and lastly, lesser electricity consumption.

Water management depends on many factors, such as soil quality, water availability, and irrigation methods. As sugarcane is a water-exhausting crop, proper management of water is necessary for sustainable production. It is suggested that special attention must be paid to irrigation water productivity and also water usage efficiency (Tayade et al., 2020; Mawardhi and Debataraja, 2022; Verma et al., 2021e; 2023c, 2023d). Verma et al. (2019) reported that water management tactics in the Upper Gangetic Plains can minimize the effects of climate change on sugarcane production. Improved water management strategies can surpass rainfall shortages. Chouhan et al. (2023) and Naazie et al. (2023) also agreed that the best water management practices must be followed to deal with irregular rainfall patterns. Different strategies have successfully improved water management for sustainable sugarcane production, as described in Table 2.4.

2.7 FUTURE PROSPECTIVE

Sugarcane has much future potential, as proved in many earlier studies. As sugar consumption is in demand with other by-products, sugarcane will never be outdated. As a C_4 tropical grass, sugarcane has excellent biomass

TABLE 2.4 Water Management Strategies for Sustainable Sugarcane Production

Techniques/ Strategies Used	Impacts	References
Wheat sowing in the furrow irrigated raised bed (FIRB) system	Hastens mechanical weeding and saves water.	Bhullar et al. (2006)
Biotechnological strategies such as genetic transformation, micropropagation, and transcriptome studies	Enhanced water stress resilience.	Bermejo et al. (2018)
The least limiting water range (LLWR) under conservation tillage	Improved soil water quality in sugarcane farming and improved sustainable and lucrative sugarcane farming.	Souza et al. (2022)
Improved water use efficiency	It was discovered that irrigation and other water management strategies were ineffective at reducing the negative effects of climate change on South Africa's sugarcane output. Thus, strategies for crop adaptation that emphasize temperature tolerance are recommended.	Deressa et al. (2005)
	Reduced effects of climate change, improved adjustments for fluctuating water quality and availability.	Cai et al. (2015)
	Reduced effects of climate change on sugarcane production and increased resilience to drought.	Zhao and Li (2015); Chohan (2019)
	Enhance water quality through soil mulching, drip irrigation, and creation of locally sensitive water-productivity indicators.	Brauman and Viart (2016)
Climatic adaptation	Maximizing irrigation efficiency and adjusting to shifting rainfall patterns can limit the adverse impact of climate change on sugarcane production.	Uribe et al. (2017)
Improved irrigation infrastructure	Improved resilience to climatic changes.	Srivastava and Rai (2012)
	Developed resistance to drought for sugarcane.	Jones and Singels (2014); Linnenluecke et al. (2018)

TABLE 2.4 *(Continued)*

Techniques/ Strategies Used	Impacts	References
Addressing water stress	By identifying the best irrigation periods for various developmental stages, water stress indices for irrigation control might lessen the negative consequences of climate change on sugarcane output.	Brunini and Turco (2018)
	Reducing water-deficit stress, which causes microbial bio-deterioration in harvested canes and lowers productivity, is a potential way that water management strategies can help lessen the effects of climate change on sugarcane production.	Misra and Mall (2022)
	Water management strategies can improve plant tolerance, growth, and production sustainability while managing the challenges of temperature, drought, and flooding on sugarcane.	Gomathi (2018)
Silicon application	Applying silicon can lessen the impacts of biotic stress generated due to water scarcity.	Costa et al. (2023)
Residue management	Boosted water quality.	Udeigwe (2008)
Enhanced soil organic carbon status	It has been discovered that using micro-irrigation systems and growing irrigated sugarcane crops improves the quality of soil organic carbon, adding to the overall sustainability of sugarcane production.	Soares et al. (2022)

accumulation potential, which is 550 kg ha$_{-1}$d$_{-1}$ (Waclawovsky et al., 2010). It is reported that sugar is the fourth-largest source of human dietary energy and the sixth-largest agricultural export commodity globally (FAO, 2021). To strengthen sugarcane sustainability, the sugarcane sector should not only focus on employment generation and the income potential of the farmers, but it should also contribute to human well-being in terms of food and health security (El Chami et al., 2020). Hassuani et al. (2005) reported that sugarcane residues in the field can also contribute to environmental sustainability by several means, such as protecting the soil from wind and water erosion, maintaining soil temperature, improving biological activity in the soil, acting as mulch, and controlling weeds. The development of stress-tolerant varieties for rainfed conditions could help attain the maximum possible yield of sugarcane and thus boost the sugar industry (Harender et al., 2021). Making farmers aware of new technologies and smart approaches is also necessary. Plaisier et al. (2017) reported that farmers should be well trained with different techniques, such as soil testing, trash shredding and mulches, good agricultural practices, and efficient irrigation methods.

2.8 CONCLUSION

The concern for environmental degradation due to sugarcane is strong enough to urge for. But, a lot of revolutionary steps have been adopted in current agriculture. Modern agriculture keeps sustainability and secured production in mind. Environmental sustainability is also possible through sugarcane if the right choice is made at the proper time and place. SSIs have to gain farmers' trust. Sugarcane can be produced globally with curtailed use of natural resources and inputs. No crop is of complete demerits; people need to minimize the loss due to environmental and manual factors. Besides these, other approaches like the release of improved varieties, genetic diversification, improved resistance to biotic and abiotic stresses, better nutrient use efficiency, and economic sustainability should be ensured. Thus, a sustainable future can be dreamt of in terms of food production and food distribution. The demand for sugar never declined; it increased constantly among people of different nations, especially those of tropical and subtropical countries. Apart from that, sugarcane provides gur, khandsari, ethanol, and other byproducts, such as molasses, bagasse, and press mud. This will also generate a good number of employees. So, economic sustainability is also possible. Therefore, sustainable sugarcane growth is possible with the expansion of the industrial usage of it.

KEYWORDS

- **climate change**
- **zero hunger**
- **diversification**
- **by-products**
- **sustainable sugarcane initiative**
- **smart approaches**

REFERENCES

AICRP Annual Report. (2017–18) ICAR-Indian Institute of Sugarcane Research: Lucknow.

Anas M, Verma KK, Haq IU, Naeem M, Li Q, Liao F, Xiang YX, Li YR. (2021a) Characterization of exotic, native and wild-type genotypes of sugarcane (*Saccharum* spp. Hybrids) for internal nitrogen use efficiency under different nitrogen levels. *Sugar Tech.* 23, 1258–1267. http://dx.doi.org/10.1007/s12355-021-01014-1

Anas M, Verma KK, Riaz M, Qiang L, Liao F, Liu Y, Li YR. (2021b) Physio-morphological and biochemical mechanism of nitrogen use efficiency in sugarcane (*Saccharum* spp.) genotypes under different growth stages and nitrogen levels. *J. Plant Interact.* 16, 332–343.

Barman A, Saha P, Patel S, Bera A. (2022) *Crop Diversification An Effective Strategy for Sustainable Agriculture Development.* IntechOpen. http://dx.doi.org/10.5772/intechopen.102635

Bermejo EAC, Alamillo MAH, Tuz SDG, et al. (2018) Transcriptome, genetic transformation and micropropagation: Some biotechnology strategies to diminish water stress caused by climate change in sugarcane. *Plant, Abiotic Stress and Responses to Climate Change* 89. http://dx.doi.org/10.5772/intechopen.72438

Bhatt R, Verma KK, Jamwal NS. (2023a) Can paddy-compost eliminate N fertilization for climate-smart sugarcane? *Sugar Tech.* 25, 1036–1046. http://dx.doi.org/10.1007/s12355-023-01270-3

Bhatt R, Verma KK, Kumar R, Sanghera GS. (2023b) Foliar application of potassium salt of active phosphorus (PSAP) mitigates insect pests and improves yield along with sugarcane quality in response to agroclimatic conditions of Punjab. *Sugar Tech.* 25(3), 660–669.

Bhatt R. (2020) Resources management for sustainable sugarcane production. In: Kumar S., et al. (Eds.), *Resources Use Efficiency in Agriculture*, Springer Nature Singapore Pte Ltd. https://doi.org/10.1007/978-981-15-6953-1_18

Bhullar MS, Kamboj A, Singh GP. (2006) Weed management in spring planted sugarcane-based intercropping systems. *Ind. J. Agron.* 51(3), 183–185.

Bommarco R, Kleijn D, Potts SG. (2013) Ecological intensification: Harnessing ecosystem services for food security. *Trends Ecol. Evol.* 28(4), 230–238. https://doi.org/10.1016/j.tree.2012.10.012

Brauman KA, Viart N. (2016) Development of a regionally sensitive water-productivity indicator to identify sustainable practices for sugarcane growers. *Integr. Environ. Assess. Manage.* 12(4), 811–820.

Brunini RG, Turco JEP. (2018) Water stress index on sugarcane in different developmental phases. *Ciência e Agrotecnologia* 42, 204–215.

Cai X, Zhang X, Noël PH et al. (2015) Impacts of climate change on agricultural water management: A review. *Wiley Interdiscip. Rev.: Water* 2(5), 439–455.

Chand RA, Pawar BR, Krishna M. (2016) Constraints faced in production and marketing of sugarcane in Parbhani district of Maharashtra. *Agric. Update* 11(3), 209–213. https://doi.org/10.15740/HAS/AU/11.3/209-213

Chaudhari BC, Patil VN, Patel PL. (2010) Influence of liquid fertilizers through drip irrigation on growth and yield of Suru sugarcane. *An Asian J. Soil Sci.* 5(1), 209–211.

Chaudhary P, Xu M, Ahamad L, Chaudhary A, Kumar G, Adeleke BS, Verma KK, Hu D-M, Siric I, Kumar P, Popescu SM, Fayssal SA. (2023) Application of synthetic consortia for improvement of soil fertility, pollution remediation, and agricultural productivity: A review. *Agronomy* 13, 643. https://doi.org/10.3390/agronomy13030643.

Chen GL, Zheng FJ, Lin B, Lao SB, He J, Huang Z, Zeng Y, Sun J, Verma KK. (2020) Phenolic and volatile compounds in the production of sugarcane vinegar. *ACS Omega* 5, 30587–30595.

Chen G-L, Zheng F-J, Lin B, Yang Y-X, Fang X-C, Verma KK, Yang L-F. (2023) Vinegar: A potential source of healthy and functional food with special reference to sugarcane vinegar. *Front. Nutr.* 10, 1145862. http://dx.doi.org/10.3389/fnut.2023.1145862

Chohan M. (2019) Impact of climate change on sugarcane crop and remedial measures—A review. *Pak. Sugar J.* 34(1), 15–22.

Chouhan S, Kumari S, Kumar R, et al. (2023) Climate resilient water management for sustainable agriculture. *Int. J. Env. Clim. Change* 13(7), 411–426.

Costa MG, de M Prado R, Sarah MMS, et al. (2023) New approaches to the effects of Si on sugarcane ratoon under irrigation in Quartzipsamments, Eutrophic Red Oxisol, and Dystrophic Red Oxisol. *BMC Plant Biol.* 23(1), 51.

Davis AS, Hill JD, Chase CA, et al. (2012) Increasing cropping system diversity balances productivity, profitability and environmental health. *PLoS One* 7(10), e47149. https://doi.org/10.1371/journal.pone.0047149

Deng ZN, Huang HR, Huang CM, Verma KK, Luo HB, Jiang SL, Li YJ, Cao HQ, Wu KC, Xu L. (2023) Short-term limited water irrigation influences the photosynthetic pigments, enzymatic and non-enzymatic activities in *Saccharum spontaneum* L. at vegetative stage. *Sugar Tech.* 25, 110–118. http://dx.doi.org/10.1007/s12355-022-01175-7

Deressa T, Hassan R, Poonyth D. (2005) Measuring the impact of climate change on South African agriculture: The case of sugarcane growing regions. *Agrekon* 44(4), 524–542. 10.1080/03031853.2005.9523726

Duttamajumder SK, Sharma AK, Brahm P. (2011) *Vision 2030*. Indian Institute of Sugarcane Research: Lucknow, India. http://www.iisr.nic.in/iisrvision2030.pdf

El Chami D, Daccache A, Moujabber M-E. (2020) What are the impacts of sugarcane production on ecosystem services and human well-being? A review. *Ann. Agric. Sci.* 65(2), 188–199.

Fahad S, Hossain MS, Huong NTL, et al. (2022) An assessment of rural household vulnerability and resilience in natural hazards: Evidence from flood prone areas. *Environ. Dev. Sustain.* 25, 5561–5577. https://doi.org/10.1007/s10668-022-02280-z

FAO. (2012) The State of Food Insecurity in the World. Economic growth is necessary but not sufficient to accelerate reduction of hunger and malnutrition. http://www.fao.org/docrep/016/i3027e/i3027e.pdf

FAO. (2021) *Statistical Database*. FAO: Rome, 2021. Available at FAO website on January 20, 2021.

FAO. (2022) The state of food security and nutrition in the world 2022. In: *Repurposing Food and Agricultural Policies to Make Healthy Diets More Affordable*. FAO: Rome. https://doi.org/10.4060/cc0639en

FAOSTAT (2022a) Crops and livestock products. https://www.fao.org/faostat/en/#data/QCL

Garbach K, Milder JC, DeClerck FAJ, et al. (2017) Examining multi-functionality for crop yield and ecosystem services in five systems of agroecological intensification. *Int. J. Agric. Sustain.* 15(1), 11–28. https://doi.org/10. 1080/14735903.2016.1174810

Gaur AC. (1992) In: Tondon HLS. (Ed.), *Fertilizer, Organic Manures, Recyclable Wastes and Biofertilizers*. EFCO: New Delhi, India, pp. 36–51.

Geetha P, Tayade AS, Chandrasekar CA, et al. (2019) Agronomic response, weed smothering efficiency, and economic feasibility of sugarcane and legume intercropping system in tropical India. *Sugar Tech.* 21, 838–842.

Gomathi R. (2018) Water induced stresses in sugarcane: Responses and management. In: *Sustainable Sugarcane Production*. Apple Academic Press, pp. 211–238.

Gujja B, Goud V, Natarajan US, et al. (2012) *Sustainable Sugarcane Initiative-Producing 'MORE WITH LESS'*. In: Agarwal M, Echanbadi AS. (Eds.), AgSri team.

Gujja B, Loganandhan N, Goud VV, et al. (2009) Sustainable sugarcane initiative—Improving sugarcane cultivation in India. Training Manual by ICRISAT-WWF Project.

Guo D-J, Li D-P, Chen Z-D, Yang B, Verma KK, Singh RK, Singh P, Khan Q, Sharma A, Qin Y, Zhang B-Q, Song X-P, Li Y-R. (2023) Effect of endophytic diazotroph *Enterobacter roggenkampii* ED5 on nitrogen-metabolism-related microecology in the sugarcane rhizosphere at different nitrogen levels. *Front. Microbiol.* 14, 1132016. http://dx.doi.org/10.3389/fmicb.2023.1132016

Guo D-J, Li D-P, Singh RK, Singh P, Sharma A, Verma KK, Qin Y, Khan Q, Lu Z, Malviya MK, Song X-P, Xing Y-X and Li Y-R. (2021) Differential protein expression analysis of two sugarcane varieties in response to diazotrophic plant growth-promoting endophyte *Enterobacter roggenkampii* ED5. *Front. Plant Sci.* 12, 727741. http://dx.doi.org/10.3389/fpls.2021.727741

Guo D-J, Li DP, Singh RK, Singh P, Verma KK, Sharma A, Qin Y, Khan Q, Song XP, Malviya MK, Xing YX, Li YR. (2022) Comparative transcriptome analysis of two sugarcane varieties in response to diazitrophic plant growth promoting endophyte *Enterobacter roggenkampii* ED5. *J. Plant Interact.* 17(1), 75–84.

GWP. (2012) Groundwater resources and irrigated agriculture—Making a beneficial relation more sustainable. https://www.gwp.org/global/the%20challenge/resource%20 material/perspectives%20 paper_groundwater_web.pdf

Hanboonsong Y. (2018) Insect vector and sugarcane disease management. In: Singh P, Tiwari AK. (Eds.), *Sustainable Sugarcane Production*. Apple Academic Press Inc., pp. 287–298.

Harender, Nimbrayan PK, Sunil, et al. (2021) Sustainability of sugarcane cultivation and production in India—A review. *AMA* 51(1), 1051–1071.

Hassuani SJ, Leal MRLV, de Carvalho M-I. (2005) *Biomass Power Generation: Sugar Cane Bagasse and Trash*. CTC, p. 217.

Huang CM, Wu KC, Verma KK, Luo HB, Deng ZN, Shi GY, Cao HQ, Wu XJ, Ye LP, Wei YW, Xu L. (2024) Isolation and expression analysis of cellulose synthase 3 (*Ces3*) genes from sugarcane (*Saccharum officinarum* L.). *Chem. Biol. Technol. Agric.* 11, 71. http://dx.doi.org/10.1186/s40538-024-00597-7

Huang HR, Gao YJ, Malviya MK, Verma KK, Solanki MK, Huang YX, Li X, Deng Y, Yan J, Tang SY, Wang LW, Xu L. (2022) Genetic diversity analysis of sugarcane (*Saccharum* spp. Hybrids) among high-sucrose clones of GT series and commonly used parents by using microsatellite markers in Guangxi, China. *Sugar Tech.* 24, 397–407. http://dx.doi.org/10.1007/s12355-021-01033-y

Hufnagel J, Reckling M, Ewert F. (2020) Diverse approaches to crop diversification in agricultural research. A review. *Agron. Sustain. Dev.* 40, 14. https://doi.org/10.1007/s13593-020-00617-4

ICRISAT. (2009) Training Manual on Sustainable Sugarcane Initiative: Improving Sugarcane Cultivation in India, an Initiative of ICRISAT-WWF Project, ICRISAT.

IISR. (2020) *Annual Report*. Indian Institute of Sugarcane Research: Lucknow, pp. 7–13.

Jagadesh M, Dash M, Kumari A, Singh SK, Verma KK, Kumar P, Bhatt R, Sharma SK. (2024) Revealing the hidden world of soil microbes: Metagenomic insights into plant, bacteria, and fungi interactions for sustainable agriculture and ecosystem restoration. *Microbiol. Res.* 285, 127764. http://dx.doi.org/10.1016/j.micres.2024.127764

Jones MR, Singels A. (2014) A preliminary assessment of mid-century climate change impacts on sugarcane production in South Africa. *Int. Sugar J.* 117, 1394.

Joshi A, Verma KK, Rajput VD, Arora J. (2023) Metabolic engineering of microorganisms in advancing biofuel production. In: *Advances in Lignocellulosic Biofuel Production Systems*. pp. 181–202. http://dx.doi.org/10.1016/B978-0-323-91192-4.00016-X

Joshi A, Verma KK, Rajput VD, Minkina T, Arora J. (2022) Recent advances in metabolic engineering of microorganisms for advancing lignocellulose-derived biofuels. *Bioengineered* 13(4), 8135–8163. http://dx.doi.org/10.1080/21655979.2022.2051856.

Khaire KC, Moholkar VS, Goyal A. (2021) Bioconversion of sugarcane tops to bioethanol and other value-added products: An overview. *Mater. Sci. Energy Technol.* 4, 54–68.

Krishna D, Sachan HK, Verma KK. (2022) Abiotic stresses in plants: An overview. In: *Understanding Abiotic Stresses*. Nova Science Publishers, Inc.: New York.

Kumari A, Dash M, Singh SK, Jagadesh M, Mathpal B, Mishra PK, Pandey SK, Verma KK. (2023a) Soil microbes: A natural solution for mitigating the impact of climate change. *Environ. Monit. Assess.* 195, 1436. https://doi.org/10.1007/s10661-023-11988-y

Kumari A, Lakshmi GA, Krishna GK, Patni B, Prakash S, Bhattacharyya M, Singh SK, Verma KK. (2022) Climate change and its impact on crops: A comprehensive investigation for sustainable agriculture. *Agronomy* 12, 3008. https://doi.org/10.3390/agronomy12123008

Kumari A, Singh SK, Mathpal B, Verma KK, Garg VK, Bhattacharyya M, Bhatt R. (2023b) The biosynthesis, mechanism of action, and physiological functions of melatonin in horticultural plants: A review. *Horticulturae* 9, 913. http://dx.doi.org/10.3390/horticulturae9080913

Li A-M, Liao F, Wang M, Chen Z-L, Qin C-X, Huang R-Q, Verma KK, Li Y-R, Que Y-X, Pan Y-Q, et al. (2023) Transcriptomic and proteomic landscape of sugarcane response to biotic and abiotic stressors. *Int. J. Mol. Sci.* 24, 8913. http://dx.doi.org/10.3390/ijms24108913

Lichtenberg EM, Kennedy CM, Kremen C, et al. (2017) A global synthesis of the effects of diversified farming systems on arthropod diversity within fields and across agricultural landscapes. *Glob. Change Biol.* 23(11), 4946–4957. https://doi.org/10.1111/gcb.13714

Linnenluecke MK, Nucifora N, Thompson N. (2018) Implications of climate change for the sugarcane industry. *Wiley Interdiscip. Rev.: Clim. Change* 9(1), e498.

Linnenluecke MK, Zhou C, Smith T, et al. (2020) The impact of climate change on the Australian sugarcane industry. *J. Cleaner Prod.* 246, 118974. https://doi.org/10.1016/j.jclepro.2019.118974

Loganandhan N, Gujja B, Goud V, et al. (2012) Sustainable Sugarcane Initiative (SSI): A methodology of 'More with Less'. *Sugar Tech.* 15(1), 98–102.

Luo T, Lakshmanan P, Zhou Z, et al. (2022) Sustainable sugarcane cropping in China. *Front. Agr. Sci. Eng.* 9(2), 272–283

Malviya MK, Li C-N, Lakshmanan P, Solanki MK, Wang Z, Solanki AC, Nong Q, Verma KK, Singh RK, Singh P, Sharma A, Guo D-J, Dessoky ES, Song X-P, Li Y-R. (2022) High-throughput sequencing-based analysis of rhizosphere and diazotrophic bacterial diversity among wild progenitor and closely related species of sugarcane (*Saccharum* spp. inter-specific hybrids). *Front. Plant Sci.* 13, 829337. http://dx.doi.org/10.3389/fpls.2022.829337.

Malviya MK, Solanki MK, Li CN, Wang Z, Zeng Y, Verma KK, Singh RK, Singh P, Huang HR, Yang LT, Song XP, Li YR. (2021) Sugarcane-legume intercropping can enrich the soil microbiome and plant growth. *Front. Sustain. Food Syst.* 5, 606595. http://dx.doi.org/10.3389/fsufs.2021.606595.

Mawardhi AD, Debataraja CO. (2022) Simulating improved field management for sugarcane using AquaCrop model to respond water scarcity. In: *IOP Conference Series: Earth and Environmental Science*, IOP Publishing. 1105, 012020. https://doi.org/10.1088/1755-1315/1105/1/012020

Misra V, Mall AK. (2022) Productivity losses through bio-deterioration in water-deficit harvested sugarcane. In: *Microbial Based Land Restoration Handbook*. CRC Press, 2, pp. 261–272.

Mitchell RDJ, Thorburn PJ, Larsen P. (2000) Quantifying the loss of nutrients from the immediate area when sugarcane residues are burnt. In: *Proceedings of the Australian Society of Sugar Cane Technology,* Bundaberg, Queensland, Australia, 2–5 May, pp. 206–211.

Mukhopadhyay MJ, Sharma A. (1991) Manganese in cell metabolism of higher plants. *Bot. Rev.* 57(2), 117–149.

Naazie GK, Dakyaga F, Derbile EK. (2023) Agro-ecological intensification for climate change adaptation: Tales on soil and water management practices of smallholder farmers in rural Ghana. *Discov. Sustain.* 4(1), 27.

Oliveira MA, Verma KK, Bhatt R, Oliveira TBA. (2022) Impact of green and organic fertilizers on soil fertility and sugarcane productivity. In: Verma KK. et al. (Eds.), *Agro-industrial Perspectives on Sugarcane Production Under Environmental Stress.* http://dx.doi.org/10.1007/978-981-19-3955-6_11.

Organisation for Economic Co-operation and Development & Food and Agriculture Organization of the United Nations. (2021) OECD–FAO agricultural outlook 2021–2030: Sugar. https://www.oecd-ilibrary.org/sites/969526b0-en/index.html?itemId=/content/component/969526b0-en

Park SE, Webster TJ, Horan HL, et al. (2010) A legume rotation crop lessens the need for nitrogen fertiliser throughout the sugarcane cropping cycle. *Field Crops Res.* 119(2–3), 331–341.

Paroda R. (2022) Crop diversification for sustainable agriculture. *Ecol. Econ. Soc.–INSEE J.* 5(1), 15–21. https://doi.org/10.37773/ees.v5i1.611

PAU (2018–2019) Package of practices for crops of Punjab. *Rabi* 2018–19, released by Punjab Agricultural University, Ludhiana.

Plaisier C, van Rijn F, Van der Ende H, et al. (2017) Towards a sustainable sugarcane industry in India. Baseline results on Solidaridad's programme: Increasing water use efficiency in sugarcane growing in India. *Wageningen, Wageningen Economic Research.* Report 2017-051. http://dx.doi.org/10.18174/413767

Prasanna SM, Pattar PS. (2014) Successful ratoon management in sugarcane. *Res. Rev.: J. Agric. Allied Sci.* 3(4), 39–47.

Pryor SW, Smithers J, Lyne P, et al. (2017) Impact of agricultural practices on energy use and greenhouse gas emissions for South African sugarcane production. *J. Cleaner Prod.* 141, 137–145. https://doi.org/10.1016/j.jclepro.2016.09.069

Rusch A, Chaplin-Kramer R, Gardiner MM, et al. (2016) Agricultural landscape simplification reduces natural pest control: A quantitative synthesis. *Agric. Ecosyst. Environ.* 221, 198–204. https://doi.org/10.1016/j.agee.2016.01.039

Saini LK, Singh M, Kapur ML. (2003) Relative profitability of intercropping vegetable crops in autumn planted sugarcane. *Sugar Tech.* 5, 95–97.

Sanghera G S, Singh R P, Singh O, et al. (2018) Initiatives for sustainable sugarcane farming in Punjab. *J. Plant Sci. Res.* 34(2), 137–152.

Shanthy TR, Venkatesaperumal M. (2018) Bio-fertilizers for sustainable sugarcane production: A socio-economic analysis. *J. Sugarcane Res.* 8(2), 127–137

Shanthy TR. (2018) Transfer of technology approaches for sustained sugarcane productivity. In: Singh P, Tiwari AK. (Eds.), *Sustainable Sugarcane Production.* Apple Academic Press Inc., pp. 385–404.

Sharma A, Singh RK, Singh P, Vaishnav A, Guo DJ, Verma KK, Li DP, Song XP, Malviya MK, Khan N, Lakshmanan P, Li YR. (2021) Insights into the bacterial and nitric oxide-induced salt tolerance in sugarcane and their growth-promoting abilities. *Microorganisms* 9, 2203. http://dx.doi.org/10.3390/microorganisms9112203.

Sharma A, Song X-P , Singh RK, Vaishnav A, Gupta S, Singh P, Guo D-J, Verma KK, Li Y-R. (2022) Impact of carbendazim on cellular growth, defence system and plant growth promoting traits of *Priestia megaterium* ANCB-12 isolated from sugarcane rhizosphere. *Front. Microbiol.* 13, 1005942. http://dx.doi.org/10.3389/fmicb.2022.1005942

Shukla SK, Lalan S, Awasthi SK, et al. (2017) *Sugarcane in India: Package of Practices for Different Agro-climatic Zones*. pp. 1–64.

Shukla SK, Sharma L, Jaiswal VP, et al. (2022) Diversification options in sugarcane-based cropping systems for doubling farmers' income in subtropical India. *Sugar Tech.* 24, 1212–1229. https://doi.org/10.1007/s12355-022-01127-1

Shukla SK, Sharma L, Jaiswal VP. (2018a) Sugarcane ratoon management. In: Singh P, Tiwari AK. (Eds.), *Sustainable Sugarcane Production*, Apple Academic Press Inc., pp. 299–331.

Shukla SK, Shee S, Maity SK, et al. (2017a) Soil carbon sequestration and crop yields in rice–wheat and sugarcane–ratoon–wheat cropping systems through crop residue management and inoculation of Trichoderma viride in subtropical India. *Sugar Tech.* 19, 347–358.

Shukla SK, Singh PN, Chauhan RS. (2011) Effect of organic wastage amended with Trichoderma and Gluconacetobacter on physiological properties of soil and sugarcane ratoon yield and in Udic Ustochrept. *Ind. J. Agron.* 56(3), 254–259.

Shukla SK, Yadav SK, Lal M, et al. (2018) Low cost technologies in sugarcane agriculture. In: *ICAR-All India Coordinated Research Project on Sugarcane*. IISR: Lucknow, pp. 1–55.

Singh AK, Lal M, Singh E. (2018) Headways agro-techniques for heightened yield of sugarcane: Indian perspective. In: Singh P, Tiwari AK. (Eds.), *Sustainable Sugarcane Production*. Apple Academic Press Inc., pp. 17–75.

Singh AK, Pathak AD, Srivastava TK, et al. (2020) Sugarcane based integrated farming system models for enhancing farmers' income and livelihoods. *Ind. Farming* 70(05), 02–06.

Singh P, Singh RK, Li HB, Guo DJ, Sharma A, Verma KK, Solanki MK, Upadhyay SK, Lakshmanan P, Yang LT, Li YR. (2023) Nitrogen fixation and phytohormone stimulation of sugarcane plant through plant growth promoting diazotrophic *Pseudomonas*. *Biotechnol. Genet. Eng. Rev*. 40(1), 15–35. http://dx.doi.org/10.1080/02648725.2023.2177814

Singh P, Singh RK, Song QQ, Li HB, Guo DJ, Malviya MK, Verma KK, Song XP, Lakshmanan P, Yang LT, Li YR. (2021a) Comparative analysis of protein and differential responses of defense-related gene and enzyme activity reveals the long-term molecular responses of sugarcane inoculated with *Sporisorium scitamineum*. *J. Plant Interact*. 16(1), 12–29.

Singh RK, Singh P, Li HB, Song QQ, Guo DJ, Solanki MK, Verma KK, Malviya MK, Song XP, Lakshmanan P, Yang LT, Li YR. (2020) Diversity of nitrogen-fixing rhizobacteria associated with sugarcane: A comprehensive study of plant-microbe interactions for growth enhancement in *Saccharum* spp. *BMC Plant Biol*. 20, 220. http://dx.doi.org/10.1186/s12870-020-02400-9.

Singh SR, Yadav P, Singh D, et al. (2021b) Intercropping in sugarcane improves functional diversity, soil quality, and crop productivity. *Sugar Tech*. 23, 794–810. https://doi.org/10.1007/s12355-021-00955-x

Soares VASF, Frazão LA, Pegoraro RF, et al. (2022) Semina: Ciênc. *Agrár. Londrina*. 43(1), 283–296. https://doi.org/10.5433/1679-0359.2022v43n1p283

Solanki MK, Verma KK, Dastogeer KMG, Mora-Poblete F, Mundra S. (2023) Microbial resilience in plant nutrient management towards sustainable farming. *Front. Microbiol*. 14, 1280811. http://dx.doi.org/10.3389/fmicb.2023.1280811

Song X-P, Mo F, Verma KK, Wei J, Zhang X, Yang L, Li YR. (2019) Effect of sugarcane smut (*Ustilago scitaminea* Syd.) on ultrastructure and biochemical indices of sugarcane. *Biomed J. Sci. Tech. Res*. 002950. http://dx.doi.org/10.26717/BJSTR.2019.17.002950.

Song XP, Verma KK, Tian DD, Zhang XQ, Liang YJ, Huang X, Li CN, Li YR. (2021) Exploration of silicon functions to integrate with biotic stress tolerance and crop improvement. *Biol. Res*. 54, 19. http://dx.doi.org/10.1186/s40659-021-00344-4

Souza CMA, Suárez AHT, Arcoverde SNS, et al. (2022) Least limiting water range in Oxisol under two conservation tillage systems in sugarcane farming. *Rev. Ceres* 69, 586–593.

Srivastava AK, Rai MK. (2012) Sugarcane production: Impact of climate change and its mitigation. *Biodiversitas J. Biol. Div*. 13(4), 214–227.

Sundara B. (2011) Agrotechnologies to enhance sugarcane productivity in India. *Sugar Tech*. 13(4), 281–298.

Tayade AS, Vasantha S, Anusha S, et al. (2020) Irrigation water use efficiency and water productivity of commercial sugarcane hybrids under water-limited conditions. *Trans. ASABE* 63(1), 125–132.

Therond O, Duru M, Roger-Estrade J, et al. (2017) A new analytical framework of farming system and agriculture model diversities. A review. *Agron. Sustain. Dev.* 37(3), 21. https://doi.org/10.1007/s13593- 017-0429-7

Tian D, Qin L, Verma KK, Wei L, Li J, Li B, Zhou W, He Z, Wei D, Huang S, Long S, Huang Q, Li C, Wei S. (2023) Transcriptomic and metabolomic differences between banana varieties which are resistant or susceptible to *Fusarium wilt. PeerJ* 11, e16549. http://dx.doi.org/10.7717/peerj.16549

Tilman D, Cassman KG, Matson PA, et al. (2002) Agricultural sustainability and intensive production practices. *Nature* 418(6898), 671–677. https://doi.org/10.1038/nature01014

Udeigwe TK. (2008) *Evaluation of Management Practices and Remediation Techniques for Improving Water Quality in Agricultural Systems*. Louisiana State University and Agricultural & Mechanical College.

Uribe RAM, Lupino LM, Silvério PC, et al. (2017) Effects of Climate Change on the Sugarcane Productivity. *Int. J. Innov. Edu. Res.* 5(4), 150–156.

Verma KK, Joshi A, Song X-P, Liang Q, Xu L, Huang H-R, Wu K-C, Seth CS, Arora J and Li Y-R (2024a) Regulatory mechanisms of plant rhizobacteria on plants to the adaptation of adverse agroclimatic variables. *Front. Plant Sci.* 15, 1377793. http://dx.doi.org/10.3389/fpls.2024.1377793

Verma KK, Joshi A, Song X-P, Singh S, Kumari A, Arora J, Singh SK, Solanki MK, Seth CS, Li Y-R. (2024b) Synergistic interactions of nanoparticles and plant growth promoting rhizobacteria enhancing soil-plant systems: A multigenerational perspective. *Front. Plant Sci.* 15, 1376214. http://dx.doi.org/10.3389/fpls.2024.1376214

Verma KK, Li DM, Singh M, Rajput VD, Malviya MK, Minkina T, Singh RK, Singh P, Song XP, Li YR. (2020a) Interactive role of silicon and plant-rhizobacteria mitigating abiotic stresses: A new approach for sustainable agriculture and climate change. *Plants* 9, 1055. http://dx.doi.org/10.3390/plants9091055.

Verma KK, Rajput VD, Singh M, Verma CL, Minkina T. (2022a) Influence of waterlogging on plant physiology and molecular biology. In: *Understanding Abiotic Stresses*. Nova Science Publishers, Inc.: New York.

Verma KK, Siddiqui N, Singh M. (2022b) Understanding the plant physiological and molecular adaptation strategies to soil salinity. In: *Understanding Abiotic Stresses*. Nova Science Publishers, Inc.: New York. http://dx.doi.org/10.52305/UBUJ4024

Verma KK, Singh P, Song X-P, Malviya MK, Singh RK, Chen G-L, Solomon S, Li YR. (2020b) Mitigating climate change for sugarcane improvement: Role of silicon in alleviating abiotic stresses. *Sugar Tech.* 22(5), 741–749. http://dx.doi.org/10.1007/s12355-020-00831-0.

Verma KK, Singh RK, Song QQ, Singh P, Zhang B-Q, Song X-P, Chen G-L, Li YR. (2019a) Silicon alleviates drought stress of sugarcane plants by improving antioxidant responses. *Biomed. J. Sci. Tech. Res.* 002957. http://dx.doi.org/10.26717/BJSTR.2019.17.002957

Verma KK, Song X, Degu HD, Guo D, Joshi A, Huang H, Xu L, Singh M, Huang D, Rajput VD, Li YR. (2023a) Recent advances in nitrogen and nano-nitrogen fertilizers for sustainable crop production: A mini-review. *Chem. Biol. Technol. Agric.* 10, 1–14. https://doi.org/10.1186/s40538-023-00488-3

Verma KK, Song X, Verma CL, Huang H, Singh M, Xu L, Wu J, Khan ZH, Zhang B, Li Y. (2023b) Mathematical modeling of climate and fluoride effects on sugarcane photosynthesis with silicon nanoparticles. *Plant Physiol. Biochem.* 204, 108089. https://doi.org/10.1016/j.plaphy.2023.108089

Verma KK, Song XP, Budeguer F, Nikpay A, Enrique R, Singh M, Zhang B, Wu, JM, Li YR. (2022c) Genetic engineering: An efficient approach to mitigating biotic and abiotic stresses in sugarcane cultivation. *Plant Signal. Behav.* 17(1), e2108253.

Verma KK, Song XP, Joshi A, Rajput VD, Singh M, Sharma A, Singh RK, Li DM, Arora J, Minkina T, Li YR. (2022d) Nanofertilizer possibilities for healthy soil, water and food in future: An overview. *Front. Plant Sci.* 13, 865048. http://dx.doi.org/10.3389/fpls.2022.865048

Verma KK, Song XP, Joshi A, Tian DD, Rajput VD, Singh M, Arora J, Minkina TM, Li YR. (2022e) Recent trends in nano-fertilizer for advancing sustainable agriculture under the era of climate change to ensure food security for future generations. *Nanomaterials* 12, 173. *https://doi.org/10.3390/nano12010173.*

Verma KK, Song XP, Li DM, Singh M, Wu JM, Singh RK, Sharma A, Zhang BQ, LI YR. (2022f) Silicon and soil microorganisms improves rhizospheric soil health with bacterial community, plant growth, performance and yield. *Plant Signal. Behav.* 17(1), e2104004.

Verma KK, Song XP, Lin B, Guo DJ, Singh M, Rajput VD, Singh RK, Singh P, Sharma A, Malviya MK, Chen GL, Li YR. (2022g) Silicon induced drought tolerance in crop plants: Physiological adaptation strategies. *Silicon* 14(6), 2473–2487. http://dx.doi.org/10.1007/s12633-021-01071-x

Verma KK, Song XP, Rajput VD, Boldyreva V, Zhang BQ, Minkina T, Li YR. (2022h) Morpho-physiological, biochemical and ultrastructural modifications on sugarcane to prolonged water deficit. In: Verma KK. et al. (Eds.), *Agro-industrial Perspectives on Sugarcane Production Under Environmental Stress.* http://dx.doi.org/10.1007/978-981-19-3955-6_11.

Verma KK, Song XP, Rajput VD, Solomon S, Li YR, Rao GP. (2022i) *Agro-industrial Perspectives on Sugarcane Production Under Environmental Stress.* Springer Nature: Singapore, http://dx.doi.org/10.1007/978-981-19-3955-6

Verma KK, Song XP, Singh M, Tian DD, Rajput VD, Minkina T, Li YR. (2023c) Association of silicon and soil microorganisms induces stress mitigation, increasing plant productivity. In: de Mello Prado R. (Ed.), *Benefits of Silicon in the Nutrition of Plants.* pp. 299–328. http://dx.doi.org/10.1007/978-3-031-26673-7_17

Verma KK, Song XP, Tian DD, Guo DJ, Chen ZL, Zhong CS, Nikpay A, Singh M, Rajput VD, Singh RK, Minkina T, Li YR. (2021a) Influence of silicon on biocontrol strategies to manage biotic stress for crop protection, performance and improvement. *Plants* 10, 2163. http://dx.doi.org/10.3390/plants10102163

Verma KK, Song XP, Tian DD, Singh M, Verma CL, Rajput VD, Singh RK, Sharma A, Singh P, Malviya MK, Li YR. (2021b) Investigation of defensive role of silicon during drought stress induced by irrigation capacity in sugarcane: Physiological and biochemical characteristics. *ACS Omega* 6, 19811–19821.

Verma KK, Song XP, Verma CL, Chen ZL, Rajput VD, Wu KC, Liao F, Chen GL, Li YR. (2021c) Functional relationship between photosynthetic leaf gas exchange in response to silicon application and water stress mitigation in sugarcane. *Biol. Res.* 54, 15. http://dx.doi.org/10.1186/s40659-021-00338-2

Verma KK, Song XP, Verma CL, Malviya MK, Guo DJ, Rajput VD, Sharma A, Wei KJ, Chen GL, Solomon S, Li YR. (2021d) Predication of photosynthetic leaf gas exchange of sugarcane (*Saccharum* spp.) leaves in response to leaf positions to foliar spray of potassium salt of active phosphorus under limited water irrigation. *ACS Omega* 6, 2396–2409.

Verma KK, Song X-P, Xu L, Huang H-R, Liang Q, Seth CS, Li Y-R. (2023d) Nano-microplastic and agro-ecosystems: A mini-review. *Front. Plant Sci.* 14, 1283852. http://dx.doi.org/10.3389/fpls.2023.1283852

Verma KK, Song XP, Yadav G, Degu HD, Parvaiz A, Singh M, Huang HR, Mustafa G, Xu L, Li YR. (2022j) Impact of agroclimatic variables on proteo-genomics in sugarcane (*Saccharum* spp.) plant productivity. *ACS Omega* 7, 22997–23008.

Verma KK, Song XP, Zeng Y, Guo DJ, Sing M, Rajput VD, Malviya MK, Wei KJ, Sharma A, Li DP, Chen GL, Li YR. (2021e) Foliar application of silicon boosts growth, photosynthetic leaf gas exchange, antioxidative response and resistance to limited water irrigation in sugarcane (*Saccharum officinarum* L.). *Plant Physiol. Biochem.* 166, 582–592.

Verma KK, Song XP,Zeng Y, Li DM, Guo DJ, Rajput VD, Chen GL, Barakhov A, Minkina TM, Li YR. (2020c) Characteristics and correlation of leaf stomata and its relationship with photosynthesis on *Saccharum* spp. under different irrigation and silicon application. *ACS Omega* 5, 24145–24153.

Verma KK, Liu X-H, Wu K-C, Singh RK, Song QQ, Malviya MK, Song X-P, Singh P, Verma CL, Li YR. (2020e) The impact of silicon on photosynthetic and biochemical responses of sugarcane under different soil moisture levels. *Silicon* 12, 1355–1367.

Verma RR, Srivastava TK, Singh P. (2019) Climate change impacts on rainfall and temperature in sugarcane growing Upper Gangetic Plains of India. *Theoretical Appl. Climatol.* 135, 279–292.

Verma KK, Vatsal S, Singh M, Rajput VD, Minkina T. (2022k) Coping with water stress: Effects and adaptive mechanisms to plants. In: *Understanding Abiotic Stresses*. Nova Science Publishers, Inc.: New York.

Verma KK, Wu K-C, Singh P, Malviya MK, Singh RK, Song X-P, Li YR. (2019b) The protective role of silicon in sugarcane under water stress: Photosynthesis and antioxidant enzymes. *Biomed. J. Sci. Tech. Res.* 002685. http://dx.doi.org/0.26717/BJSTR.2019.15.002685

Verma KK, Wu KC, Verma CL, Li DM, Malviya MK, Singh RK, Singh P, Chen GL, Song XP, Li YR. (2020d) Developing mathematical model for diurnal dynamics on photosynthetic responses in sugarcane responsive to different irrigation and silicon application. *PeerJ* 8, e10154. http://dx.doi.org/10.7717/peerj.10154.

Vishwanathan R, Rao GP. (2011) Disease scenario and management of major sugarcane diseases in India. *Sugar Tech.* 13(4), 336–353.

Voora V, Bermúdez S, Le H, et al. (2023) Global Market Report: Sugar cane prices and sustainability.

Waclawovsky AJ, Sato PM, Lembke CG, et al. (2010) Sugarcane for bioenergy production: An assessment of yield and regulation of sucrose content. *Plant Biotech. J.* 8(3), 263–276.

Wezel A, Casagrande M, Celette F. (2014) Agroecological practices for sustainable agriculture. A review. *Agron. Sust. Dev.* 34(1), 1–20.

Wu K-C, Huang C-M, Verma KK, Deng Z-N, Huang H-R, Pang T, Cao H-Q, Luo H-B, Jiang S-L, Xu L. (2022) Transcriptomic responses of *Saccharum spontaneum* roots in response to polyethylene glycol-6000 stimulated drought stress. *Front. Plant Sci.* 13, 992755. http://dx.doi.org/10.3389/fpls.2022.992755

Xu L, Deng Z-N, Wu K-C, Malviya MK, Solanki MK, Verma KK, Pang T, Li Y-J, Liu X-Y, Kashyap BK, Dessoky ES, Wang W-Z and Huang H-R. (2022) Transcriptome analysis reveals a gene expression pattern that contributes to sugarcane bud propagation induced by Indole-3-Butyric acid. *Front. Plant Sci.* 13, 852886. http://dx.doi.org/10.3389/fpls.2022.852886

Yadav RL, Shukla SK. (2016) Good agronomic practice (GAP) for efficient use of soil, water and nutrients in sugarcane-based system. *Ind. J. Ferti.* 12(11), 98–109.

Yadav S, Yadav S. (2017) Enhancement in sustainable sugarcane and sugar production in U.P. towards secure food and yawning production gaps with future strategies. *Rashtrita Krishi* 12(2), 65–67.

Zhang R, Mu Y, Li X, et al. (2020) Response of the arbuscular mycorrhizal fungi diversity and community in maize and soybean rhizosphere soil and roots to intercropping systems with different nitrogen application rates. *Sci. Tot. Environ.* 740, 139810. https://doi.org/10.1016/j.scitotenv.2020.139810

Zhao D, Li YR. (2015) Climate change and sugarcane production: Potential impact and mitigation strategies. *Int. J. Agron.* 2015, 547386.

Zheng F-J, Lin B, Yang Y-X, Fang X-C, Verma KK, Chen G-L. (2023) Efficacy and functionality of sugarcane original vinegar on mice. *Front. Microbiol.* 14, 1224666. http://dx.doi.org/10.3389/fmicb.2023.1224666

CHAPTER 3

Understanding the Biochemical and Physiological Traits for Selecting Drought-Tolerant Sugarcane Genotypes

VARUCHA MISRA* and A. K. MALL

ICAR-Indian Institute of Sugarcane Research, Lucknow, Uttar Pradesh, India

*Corresponding author

ABSTRACT

Drought stress poses a significant threat to global sugarcane production, particularly impacting the vital sugarcane industry, which serves as a primary source of sucrose and bioenergy worldwide. The adverse effects of drought occur both early and mid-season, including compromised plant growth, reduced tillering, and the development of sparse and underdeveloped millable stalks, consequently diminishing cane and sugar yields. Developing drought-tolerant sugarcane varieties is paramount to counteract these challenges and ensure the sustainability of rainfed sugarcane farming. Central to this endeavor is identifying effective screening tools and measurable traits that facilitate the selection of resilient cultivars. Physiological traits, such as expansion of leaf area development, stomatal conductance, photosynthetic pigments, relative water content, chlorophyll fluorescence efficiency (Fv/Fm), and photosynthetic rate, have emerged as key targets in breeding programs aimed at enhancing drought resistance in sugarcane. Similarly, sugarcane has identified biochemical traits like proline, glycine betaine, trehalose, etc., as drought-tolerant traits. Understanding and utilizing these screening tools and biochemical traits

Sugarcane Production: Environmental Impacts and Socioeconomic Issues.
Krishan K. Verma, Xiu-Peng Song, Lin Xu, Rajan Bhatt, and Yang-Rui Li (Eds.)

DOI: 10.1201/9781779641137-3

are pivotal in developing robust sugarcane cultivars that maintain high yields even under drought-stress conditions. By utilizing advancements in physiological and biochemical screening methodologies, researchers and breeders can accelerate the breeding process, ultimately enhancing the productivity and sustainability of sugarcane farming systems worldwide.

3.1 INTRODUCTION

Drought stress represents an escalating crisis, significantly curtailing crop production globally and inflicting substantial agricultural losses, particularly in arid and semiarid regions (Boyer, 1982; Verma et al., 2021a, 2021b; Vadez et al., 2024). Sugarcane holds significant economic importance worldwide, being a key source of sucrose and bioenergy (Lakshman et al., 2013; Guo et al., 2022; Huang et al., 2024). The prevalence of drought in sugarcane cultivation stands out as a primary obstacle, significantly deteriorating both production levels and overall productivity worldwide (Misra et al., 2020b; Verma et al., 2022a, 2022b, 2022c). Drought poses a significant threat across India, impacting approximately 2.97 lakh hectares of sugarcane fields (Misra et al., 2020b; Bhatt et al., 2023a, 2023b). Sugarcane yield is frequently compromised by adverse growth conditions, particularly drought stress during the critical formative growth stage, which is highly vulnerable to moisture deficits (Song et al., 2019, 2021; Kumar et al., 2021; Mahadevaiah et al., 2021; Reyes et al., 2020, 2021). This challenge is prevalent in tropical regions (Khonghintaisong et al., 2021), resulting in considerable yield losses of up to 60% in rainfed areas (Ferreira et al., 2017; Verma et al., 2019a, 2019b; Misra et al., 2020a). Furthermore, it even causes heavy losses in sugarcane during postharvest situations (Misra et al., 2016). Early-season drought and mid-season drought can hinder plant growth, leading to stunted growth and reduced tillering, resulting in sparse and poorly developed millable stalks and losses of crop production (Dinh et al., 2017; Huang et al., 2022; Joshi et al., 2022). Previous studies have documented the above-ground response to early-season drought during the tillering phase (Robertson et al., 1999; Inman-Bamber, 2004; Zhao et al., 2010; Jangpromma et al., 2012; Singh et al., 2021). Drought significantly diminished sugarcane stalk diameter and biomass in pot conditions, exerting notable effects on leaf area, tillering, biomass uptake, and total mass (Jangpromma et al., 2012; Verma et al., 2022d, 2022e; 2023a, 2023b; Zheng et al., 2023). However, these responses varied across different locations and sugarcane cultivars. Furthermore, the response during the re-watering phase also influenced growth and yield (Robertson et al., 1999). Water stress reduced stomatal conductance, photosystem II (PSII) photochemical efficiency (Fv/Fm), leaf photosynthesis rate, tiller count, and green leaf area, decreasing shoot biomass (Zhao et al., 2010; Verma et al., 2020a, 2020b, 2020c; 2021c; 2023c).

Developing drought-tolerant sugarcane varieties capable of sustaining high yields with minimal reduction under drought stress is crucial for bolstering the productivity and sustainability of rainfed sugarcane farming (Verma et al., 2023d). Discovering effective screening tools and measurable characteristics would streamline the process of enhancing crop resilience to drought (Anas et al., 2021a, 2021b; Krishna et al., 2022). Key physiological traits, such as leaf area, stomatal conductance, chlorophyll content, relative water content (RWC), Fv/Fm, and photosynthetic rate, have been targeted to enhance drought resistance in sugarcane breeding programs (Silva et al., 2008; Verma et al., 2021d, 2021e). Furthermore, strategies involving the reduction of stomatal conductance to minimize water loss and enhancing root density and deep root traits to improve water uptake capacity have been identified as adaptation mechanisms in sugarcane during limited water supply (Wasson et al., 2012; Verma et al., 2020d, 2020e; Malviya et al., 2021, 2022). It leads to downregulate of stomatal conductance, reducing water loss from the leaves. This reduction in water loss also diminishes the influx of carbon dioxide through the stomata, consequently lowering photosynthesis rates, which can further contribute to decreased sugarcane yields (Wu et al., 2022; Xu et al., 2022). The selection of physiological traits associated with drought tolerance (Grover et al., 2014; Zang et al., 2014; Guo et al., 2021; Kumari et al., 2022) may enhance the success of breeding efforts for drought-tolerant sugarcane varieties. The ability of plants to adjust their roots to access water for transpiration could serve as a crucial mechanism for drought avoidance (Fukai and Cooper, 1995; Songsri et al., 2009; Singh et al., 2020; Sharma et al., 2021). Cultivars with high levels of drought avoidance are expected to maintain yield stability under drought conditions (Pinto and Reynolds, 2015; Serraj et al., 2004; Chen et al., 2020, 2023).

This chapter explores the intricate mechanisms of sugarcane's response to drought stress, elucidating the biochemical and physiological traits crucial for identifying and selecting drought-tolerant sugarcane genotypes.

3.2 STRESS TOLERANCE MECHANISM IN SUGARCANE

Plants subjected to drought stress exhibit diverse responses, contingent upon the species or cultivars (Bhargava and Sawant, 2013). Plants, including sugarcane, employ three primary strategies to cope with drought: escape, avoidance, and tolerance (Fang and Xiong, 2015). Sugarcane cultivars equipped with drought-tolerant mechanisms effectively manage water-limited environments by maintaining a high-water status. Overall, drought stress disrupts numerous physio-biochemical pathways through intricate mechanisms (Verma et al., 2022f, 2022g, 2022h; Song et al., 2022; Jagadesh et al., 2024). Adapting to drought presents one of the most complicated biological challenges for plants. This process encompasses a multitude of alterations, such as diminished growth, the activation or deactivation of specific genes at the transcriptional level, temporary spikes in ABA levels, the accumulation of compatible solutes and protective enzymes, heightened levels of antioxidants, and the suppression of energy-intensive pathways (Wassem et al., 2011; Verma et al., 2022i, 2022j, 2022k; Oliveira et al., 2022).

3.2.1 ESCAPE STRATEGY

Drought escape entails plants adjusting their life cycle, often by shortening it, to avoid prolonged exposure to drought stress. The escape strategy involves hastening reproduction through a shortened life cycle well before the onset of stress, coupled with enhanced growth rates and efficient storage (Campos et al., 2004; Sharma et al., 2022; Solanki et al., 2023). The time of flowering occurrence plays a critical role in a plant's ability to withstand drought, as a shorter life cycle can facilitate escape from drought conditions (Araus et al., 2002; Tian et al., 2023). The crop growth duration is influenced by genetic factors and environmental conditions, ultimately affecting the crop's capacity to evade climatic pressures such as drought. Sugarcane is often planted in drought-prone areas during rainy climatic conditions, ensuring it receives sufficient water during its initial growth stages.

By the time the dry season arrives, the plants are well-established and better able to withstand drought stress (Kumari et al., 2023a, 2023b). Water stress poses a significant risk to sugarcane, particularly during its initial developmental phases, such as tillering and grand growth, where approximately 70–80% of its yield is generated (Li et al., 2023; Singh et al., 2023; Joshi et al., 2023). Drought occurrences during these crucial stages frequently lead to decreased biomass and sucrose yields, with potential consequences extending to subsequent ratoon crops (Silva et al., 2012, 2013; Ramesh and Mahadevaswamy, 2000; Castro et al., 2002; Deng et al., 2023).

Sugarcane possesses a remarkable compensatory mechanism as an escape strategy, which helps in facing drought stress (Shrivastava et al., 2017; Guo et al., 2023). This capability stems from several factors: tillering, which involves the performance and timing of tillers, including late shoots; the presence of two distinct root types, sett roots and shoot roots, along with their emergence from root primordia at different intervals; leaf development; and extensive ratooning potential. This adaptive capacity allows sugarcane to survive in adverse conditions and mitigate losses caused by insect pests, mechanical damage, delayed planting, or various stressors (Chaudhary et al., 2023). By enhancing specific physiological and morphological characteristics, sugarcane can offset incurred losses or deficiencies. Moreover, sugarcane exhibits a higher tolerance to elevated temperatures than C_3 plants, except for sugar storage processes. This resilience to higher temperatures further contributes to its ability to thrive in diverse environments and withstand temperature fluctuations (Shrivastava et al., 2017; Verma et al., 2024a, 2024b).

3.2.2 AVOIDANCE STRATEGY

Drought avoidance involves the continuation of physiological processes, such as stomatal regulation and root system development, despite water deficits. The avoidance strategy focuses on maintaining water content in the plant without significant loss during stress, thereby favoring plant growth and development. This strategy may involve minimizing water loss through mechanisms, such as senescence of older leaves, stomatal closure, or reduction in leaf area through leaf rolling, or maximizing water uptake by extending root length to reach available water (Araus et al., 2008; Blum, 2009). In sugarcane, leaf rolling, the presence of thick cuticles, stomatal closure, water uptake regulation, and leaf growth regulation were observed during drought stress to avoid the effects of the stress (Mall et al., 2022; Misra et al., 2020, 2022a).

Leaf rolling is a crucial indicator for evaluating the effect of stress on plants. It is the major concept influenced by the variation in the potential of water in the bulliform cells shown in the plant leaves. It contributes a major response in slowing down the process of transpiration and enhancing the yield-contributing traits in crops that suffer from limited water supply (Youzhong et al., 2004). The upper surface of the plant leaf loses water, which reduces the potential of pressure, leading to turgidity and diminished osmotic adjustment (Hsiao et al., 1984). It is significant for minimizing the ambient temperature of leaves (O'Toole and Chang, 1978; O'Toole et al., 1979; Begg, 1980), excess absorption of light, and upregulating transpiration efficiency (Turner et al., 1986; Kadioglu and Terzi, 2007). The initial response of crop plants is rolling of leaves during stressed conditions associated with the thickness of lamina and epidermal cells swelling (Cal et al., 2019). This process involves bulliform and hypodermis cells, with bulliform cells, particularly those near the midrib or vascular bundles in the upper epidermis, primarily responsible for leaf rolling in sugarcane.

A higher count of bulliform cells increases leaf curling, reducing the evaporative surface. Instead of senescence, leaf rolling initiation enables continued photosynthesis during high precipitation periods (Knapp, 1985). Studies have shown that this trait enhances photosynthetic efficiency under water stress conditions by minimizing water loss and respiration rates. Also, leaf rolling may vary among sugarcane genotypes (Richards et al., 2002; Rutherford, 1989). The association of bulliform cells with the leaf-rolling response was normally observed in grass leaves during limited water irrigation (Matschi et al., 2020). Plants require a sufficient water supply for better growth during the entire life cycle.

Water requirements fluctuate throughout the different stages of the sugarcane life cycle. During the germination phase (0–45 days), approximately 300 nm of water is necessary. The tillering phase (45–120 days) demands an increased water supply of around 550 nm. As the sugarcane enters the grand growth stage (120–270 days), it requires a substantial amount of water, totaling approximately 1000 nm. Transitioning into the ripening phase (270–360 days), the water requirement decreases slightly to about 650 nm. The variation in water needs during growth is the main reason leaf rolling is a vital adaptation under drought stress (Zou et al., 2011; Hussain et al., 2019; Cal et al., 2019; Matschi et al., 2020). Adjusting the leaf rolling manages the photosynthetic CO_2 assimilation response per unit leaf area (Gunnula et al., 2022). It decreases the loss affected by light by downregulating the exposed area by sunlight (Xu et al., 2013; Yuan et al., 2015). The maximum leaf rolling expression assists in increasing the water usage efficiency as compared to a totally rolled leaf (Juarez et al., 2004).

Stomatal closure, aimed at minimizing water loss, is closely linked to leaf rolling. Evidence suggests that this closure, induced by a combination of adjacent cell water status and photon flux intensity, regulates transpiration. In sugarcane under water stress, stomatal closure reduces carbon dioxide and water vapor efflux. In the case of IACSP96-2042, stomatal closure and a decrease in PN/Ci happen earlier than IACSP94-2094 (Sales et al., 2015). In an experiment comparing the Badila clone of *Saccharum officinarum* (drought-susceptible) and Co 312 (drought-resistant variety), the closure of 50% of stomata served as a criterion for screening drought-tolerant varieties (Dwivedi, 2000). Stomatal limitation emerges as a crucial factor in the decline of photosynthesis in sugarcane, disrupting stomatal conductance (gs), transpiration rate (E), and internal CO_2 concentration (Ci), and resulting in lower photosynthetic rates (Inman-Bamber and Smith, 2005; Silva et al., 2007; Da Graca et al., 2010; Endres et al., 2010; da Silva et al., 2012; Medeiros et al., 2013; Ribeiro et al., 2013; Basnayake et al., 2015).

Sugarcane plant leaves under water deficiency employ two strategies: increasing and decreasing thickness. The thickening of leaves in drought conditions is primarily attributed to an increase in upper epidermis thickness, which serves as a barrier to reduce water loss (Verma et al., 2022a, 2022c). Mild to severe stress-tolerance cultivars exhibited increased lamina thickness in arid conditions. Additionally, the thickness of the cuticle and cell wall of epidermal cells in moderately drought-resistant sugarcane varieties increased. Interestingly, stomata size increased with drought in moderately drought-tolerant sugarcane, whereas the reduction was monitored in stress-tolerant cultivars (Taratima et al., 2020; Verma et al., 2020a, 2020b; 2021a, 2021e). These interactions associated with stress and the thickness of leaves highlight the mechanisms plants employ to survive in harsh environments. Furthermore, the deposition of cuticles on epidermal cells is among the modifications plants adopt to enhance tolerance to water-deficient conditions (Kosma et al., 2009). These adaptations not only enable plants to endure immediate water stress but also regulate long-duration resilience in the agroecosystems.

Roots are likely the first organs to perceive drought stress, playing a crucial role in the plant's response. Longer roots are typically more advantageous for drought tolerance compared to shorter ones. Drought stress often leads to a significant reduction in the number of roots. Plant roots with desirable drought-tolerant traits, such as longer length, higher density, and a vigorous root system, contribute to the plant's adaptability. Plants require long roots with a dense network to sustain productivity in water-limited conditions, especially when water is deeper in the soil (Comas et al., 2013). A denser root system can absorb more water than thinner roots, as more roots can come into contact with more soil water vapor (Abdelraheem et al., 2019). In sugarcane, cultivars with a propensity for developing deep root systems exhibit greater drought tolerance. Root system properties, such as root length, root volume, and root-shoot ratio, may serve as selection criteria for drought tolerance and water use efficiency. The typical sugarcane root system consists of superficial roots, downward-oriented buttress roots, and deep vertically penetrating root clusters referred to as rope roots (Evans, 1935). Sugarcane boasts a deep-rooted nature, attributed to its extended growth cycle and the longevity of its root system across multiple rotations, distinguishing it from other crops. Root systems can extend to depths of 1.5 m and even up to 6 m (Ball-Codho et al., 1992; Evans, 1936; Misra et al., 2022b). Sett roots emerge from nodes after planting, characterized by their thinness and extensive branching. These roots darken, decompose rapidly, and degrade within approximately eight weeks of planting.

Primary shoot roots become visible about one-week postplanting, initially exhibiting slow growth but accelerating in growth rate thereafter. Studies by Baran et al. (1974) and Kingston (1977) have demonstrated that short irrigation intervals prevent surface soil from drying out, promoting a higher percentage of roots to develop near the soil surface. In rainfed conditions, roots tend to shift to deeper soil layers to access more water, whereas under irrigation, root length density is primarily distributed in the upper soil layer (Laclau and Laclau, 2009). A 10 days dehydration period has been observed to significantly reduce root length, surface area, root volume, and root dry weight. However, it does not notably affect the root/shoot ratio (Jangpromma et al., 2012). However, it is important to note that the drought stress period experienced by sugarcane in natural conditions is typically longer than the 10 days reported by Jangpromma et al. (2012). In regions like Thailand, the drought period can extend for approximately 2–4 months during the early crop cycle of sugarcane production systems.

Enhanced root activity at deeper soil layers is promoted following decreased soil surface moisture content (Inman-Bamber et al., 2003). A larger root system, including deeper roots and increased root surface area, enhances nutrient extraction and water uptake capacity from soil layers (Jongrungklang et al., 2011; Inman-Bamber and McGlinchey, 2012). The efficient development of root systems in response to drought conditions has been addressed in drought-tolerant sugarcane cultivars (Smith et al., 2005). Drought stress decreases soil water potential, reducing hydraulic conductance in the entire root system of sugarcane (Saliendra and Meinzer, 1992). Consequently, sugarcane must develop adaptive root traits to enhance water uptake and meet the plant's demands under drought conditions.

Nevertheless, drought conditions have been shown to stimulate root system expansion in search of water in moisture-deficient soil environments. Root traits associated with drought avoidance aim to explore deeper soil layers for water. Under well-watered conditions, sugarcane root systems predominantly occupy the upper soil surface (Smith et al., 2005). However, soil surface moisture declines and prompts root growth in deeper soil surface. This adapted root system facilitates water absorption, maintaining water balance. Additionally, the adapted root system increases transpiration, aiding sugarcane in avoiding dehydration risks.

The maintenance of water uptake may be facilitated by improving the root/shoot ratio, enabling drought-resistant cultivars to maintain a high proportion of root and shoot during drought, similar to control (Jangpromma et al., 2012). Although roots are affected and respond to early drought from the onset, the impact of water deficit on roots is relatively lower than on above-ground parts (Inman-Bamber and Smith, 2005; Smit and Singels, 2006). Early season drought, characterized by water deficit during tillering and elongation stages, can significantly impair stem and leaf growth of susceptible sugarcane cultivars (Ramesh, 2000; Inman-Bamber, 2004; Inman-Bamber et al., 2008; Machado et al., 2009; Lakshmanan and Robinson, 2013), as division and elongation of cells are disrupted (Machado et al., 2009).

3.2.3 TOLERANCE STRATEGY

Drought tolerance refers to a plant's capability to withstand dehydration through various physiological activities, such as osmotic adjustment facilitated by osmoprotectants (Luo, 2010). The tolerance strategy enables the plant to sustain its metabolism under water deficit conditions, accompanied by decreased tissue water potential (Mitra, 2001). This strategy involves the interaction of various physiological, biochemical, and regulatory mechanisms within plants, as well as microbial interactions (Khan et al., 2011). Drought tolerance encompasses a set of mechanisms enabling plants to sustain their metabolic processes despite decreased tissue water potential (Mitra, 2001). This involves intricate physiological, biochemical, and gene regulatory interactions within the plant (Valliyodan and Nguyen, 2006; Bernstein et al., 2010; Rogiers et al., 2011; del Pozo et al., 2012), as well as interactions with microorganisms (Vallino et al., 2009; Andrews et al., 2010; Dodd et al., 2011; Khan et al., 2011). Plant adaptations include the accumulation of compatible solutes and enhanced antioxidant capacity, enabling them to withstand soil water potential reduction (Sairam et al., 2002; Moussa and Abdel-Aziz, 2008). In sugarcane, drought tolerance traits observed were leaf water potential, solute/osmoprotectant accumulation, rise in stress proteins, osmotic adjustment, and regulation of reactive oxygen species (ROS).

Leaf water potential serves as a crucial indicator of drought avoidance. Ludlow (1980) observed that drought in C_4 plants leads to increased water loss and a subsequent decrease in leaf water potential. Sugarcane plants, being amphistomatous, possess twice as many stomata on the underside of leaves compared to the upper epidermis (Van Dillewijn, 1952). Thind (1996) demonstrated that stomata frequency was relatively higher on the lower epidermis of leaves in drought canes compared to unstressed cane plants. Laffray and Louguet (1990) revealed that under conditions of null stomatal conductance, water potential values in drought plants decrease significantly. Drought-tolerant sugarcane varieties like Co 285 and Co 312 exhibited decreased relative turgidity (approximately 60%) due to increased evaporative surface area and stomatal closure after five days of drought exposure (Shrivastava and Srivastava, 2006). Conversely, although relative turgidity decreased in drought-susceptible *S. officinarum* clones, stomatal closure occurred after seven days of exposure to water stress (Naidu and Bhagyalakshmi, 1973).

Osmoprotectants also play a crucial role in osmotic adjustment during drought stress by safeguarding the enzymes in sugarcane (Hsiao et al., 1984; Mall et al., 2020). Sugarcane naturally contains osmoprotectants, such as glycine betaine (GB), trehalose, proline, and myo-inositol, whose levels rise during drought conditions (Silva et al., 2011). These compounds shield the metabolic system and partially mitigate morphological alterations caused by drought stress. Incorporating these osmoprotectants into the breeding process is vital for developing drought-tolerant sugarcane cultivars (Cha-um and Kirdmanee, 2008). Proline, GB, trehalose, and similar osmolytes can maintain the structure and function of macromolecules even at increased concentrations (Hoekstra et al., 2001; Rogiers et al., 2011).

Similarly, osmotic adjustment is also considered as a drought-tolerant trait in sugarcane. The osmotic adjustment refers to the reduction in osmotic potential caused by increased intracellular solutes in response to drought conditions. It is seen as an adaptive mechanism to enhance drought tolerance. According to Huff and Ross (1975), sugars play a crucial role in osmotic adjustment by contributing to the accumulation of solutes, thereby enhancing protoplasmic tolerance under drought. In sugarcane, the accumulation of metabolites like proline, soluble sugars, phenolic compounds, GB, and carotenoids is essential for developing drought tolerance. Additionally, the buildup of low molecular weight carbohydrates, such as inositol, glucose, sucrose, and fructose, in drought-sugarcane leaves aids in water stress tolerance by preventing stomatal closure, thus maintaining increased levels of photosynthesis.

Furthermore, the production of ROS is also triggered by the exposure of sugarcane to drought. This is even regarded as a screening tolerant trait for sugarcane. ROS, such as the superoxide radical (O^{2-}), H_2O_2, and hydroxyl radical (OH^-), are generated through the reduction of molecular O_2, while singlet oxygen ($^1O_2$) forms via the reaction of O_2 with excited chlorophyll (Gratão et al., 2005). Arora et al. (2002) noted that stomatal closure during water stress reduces internal CO_2 concentration, leading to ROS accumulation, which activates mechanisms to mitigate oxidative stress, a secondary stress often induced under such conditions. Gratão et al. (2008) further proposed that increased ROS levels disrupt cellular redox balance, interfere with the electron transport chain, and induce oxidative stress through their toxic effects.

3.3 PHYSIOLOGICAL PARAMETERS FOR SCREENING DROUGHT TOLERANCE IN SUGARCANE

The aim of genetic enhancement initiatives for sugarcane has included the creation of cultivars resilient to drought conditions (Inman-Bamber and Smith, 2005). This endeavor necessitates pinpointing crucial physiological mechanisms that can serve as benchmarks for selection (Smit and Singels, 2006). Sugarcane varieties exhibit varying reactions to drought stress, enabling the identification of those more adept at tolerating such conditions. This discernment holds significant importance, especially in regions characterized by a prolonged duration of limited water irrigation. Physiological indicators are crucial for assessing the drought resistance of different genotypes (Buckley, 2005; Vinocur and Altman, 2005; Shao et al., 2008; Tezara et al., 2008). Hotta et al. (2010) demonstrated that these indicators, combined with identified genes, serve as a fundamental basis for developing new drought-tolerant varieties. Azevedo et al. (2011) and Silva et al. (2007) found that specific physiological and biochemical markers effectively distinguish between sensitive and tolerant cane cultivars under drought conditions. Prioritizing physiological characteristics associated with drought tolerance could enhance the success of sugarcane breeding efforts.

Utilizing physiological markers provides a relatively more precise means of evaluating drought tolerance, particularly in field trials. Four physiological markers, namely leaf temperature (LT), SPAD unit, PSII photochemical efficiency (Fv/Fm), and RWC, were developed as rapid tools for screening drought tolerance in eight field-grown sugarcane genotypes (Silva et al., 2007). These markers are significant for swiftly identifying drought-tolerant genotypes in plant breeding programs and constructing models to ensure optimal crop growth for sustained high-yield productivity. The selection of drought-tolerant plants is typically challenging due to the intricate genetic nature of the drought trait (Silva et al., 2011).

LT serves as a reliable physiological indicator that demonstrates a robust correlation with drought conditions. LT has been shown to strongly correlate with susceptibility to drought (Silva et al., 2007). LT is particularly sensitive to fluctuations in wind and the reflection of sunlight. O'Neill et al. (2006) presented evidence supporting LT as a reliable indicator of plant stress; the closure of stomata represents one of the initial adaptive responses to escalating water stress, as plants endeavor to minimize transpiration losses through their leaves and prevent the onset of potentially lethal water deficits in their tissues.

RWC is also considered a drought-tolerant identifying trait. As per Colom and Vazzana (2003), plants exhibiting greater drought tolerance tend to possess a heightened ability to conserve water during dry spells, resulting in elevated RWC values. This phenomenon occurs due to a reduction in their uptake of CO_2 and the associated biochemical processes of photosynthesis, leading to a decrease in photoinhibition and damage to PSII. Silva et al. (2011) affirm that RWC serves as a marker associated with drought, enabling the categorization of drought-tolerant and susceptible sugarcane varieties. Silva et al. (2007) showed the correlation between RWC and drought tolerance and productivity. Selecting genotypes with higher RWC values under water stress corresponds to choosing plants with greater productivity.

Stomatal closure represents one of the initial responses to drought stress, mitigating water loss by reducing transpiration. However, the reversibility of this closure depends on the severity of the stress (Jangpromma et al., 2012; Guan et al., 2015). In cases of severe stress, stomata remain closed, leading to a decline in photosynthesis due to limited CO_2 diffusion, ultimately compromising biomass accumulation (Medeiros et al., 2013; Zivcak et al., 2013; Barbosa et al., 2015). Drought stress also triggers photoinhibition and photodestruction of PSII pigment-protein complexes, alongside destabilization of photosynthetic membranes. This can result from the excess energy not utilized in photosynthesis, leading to damage to the photochemical apparatus through the photoinactivation of PSII reaction centers (Kato et al., 2003; Sales et al., 2013). Stomatal conductance decreases with plant size and age, and maintaining a balance between transpiration water loss and hydraulic conductance is crucial for preserving leaf water status amid changing environmental variables (Meinzer and Grant, 1990).

The SPAD unit serves as a reliable physiological indicator, demonstrating a positive correlation with drought tolerance. This reinforces the findings of Silva et al. (2007), who established a consistent association between SPAD unit values and the classification of sugarcane genotypes as tolerant or susceptible to drought. They suggested that this technique holds promise for swiftly and nondestructively selecting genotypes with enhanced drought tolerance. Chlorophyll fluorescence, SPAD index, and thermal imaging emerge as the most favorable options for swift and noninvasive screening to assess drought tolerance. Furthermore, the correlation between drought resilience and chlorophyll fluorescence utilizing portable fluorometers has been firmly established in sugarcane (Luo et al., 2004; Molinari et al., 2007; Silva et al., 2007).

3.4 BIOCHEMICAL PARAMETERS FOR SCREENING DROUGHT TOLERANCE IN SUGARCANE

Biochemical parameters for assessing drought tolerance in sugarcane varieties often involve monitoring changes in proline content, trehalose content, GB, and malondialdehyde (MDA). GB and proline are prominent organic osmolytes that accumulate in various plant species under drought stress. Numerous studies have established a positive correlation between the accumulation of GB and proline and plant stress tolerance (Hamdia and Shaddad, 2010; Ashraf and Foolad, 2005).

3.4.1 GLYCINE BETAINE

Glycine betaine (GB), an amphoteric quaternary amine, is a crucial compatible solute synthesized under water stress conditions, acting as a beneficial agent (Rhodes and Hanson, 1993; Sakamoto and Murata, 2002). Its primary role includes adjusting the osmotic potential within cells to regulate water levels and shielding proteins from damage caused by water stress-induced pressure (Sakamoto and Murata, 2002). Additionally, GB helps stabilize the structure of macromolecules and ensures proper membrane function (Sugiharto, 2018). Estimating GB levels serves as a physiological marker for stress and protects against oxidative stress, thereby preventing damage to cell membranes, proteins, and DNA structures and inhibiting photosynthesis under water stress conditions (Sairam and Saxena, 2000). The levels of this osmolyte typically increase with higher concentrations of PEG, indicating a response to drought stress. Sugarcane accumulation increases with the severity of drought stress (Abdul Jaleel et al., 2007). Under control conditions, GB production ranges from 15.9 to 42.2 μmol/g in sugarcane (Abbas et al., 2014). According to Chen and Murata (2011), GB serves to stabilize membranes, safeguard chloroplast membranes and proteins, preserve PSII structures, and mitigate oxidative damage under abiotic stress conditions. Notably, GB has been implicated not only in stress tolerance but also in plant recovery during rehydration (Wang et al., 2010; Cheng et al., 2013). Similar to proteins, GB expression is contingent upon genes involved in biosynthesis (Cheng et al., 2013).

3.4.2 PROLINE

The accumulation of proline stands as a crucial physio-biochemical marker for plants undergoing water stress. Proline accumulation serves as a crucial indicator, aiding in osmotic adjustment and acting as an osmo-protectant and osmotic regulator during stress while also being associated with scavenging ROS (Molinari et al., 2007; Vendruscolo et al., 2007). These benefits may be due to subcellular compartmentalization between cytosol, chloroplasts, and mitochondria, enabling specific functions in these different cell compartments (Kaur and Asthir, 2015). Elevated proline levels in stressed plants may serve as an adaptive response to combat the stress conditions, providing a continuous energy supply for growth and survival, thus aiding the plants' stress tolerance (Abdul Jaleel et al., 2008). Drought stress manifests in two ways: through the augmentation of proline synthesis, involving increased levels of proline enzymes, and the reduction of detrimental proline enzyme activities (Rontein et al., 2002). In plant stress, the proline content can surge dramatically, up to a hundred-fold, compared to plants grown under normal conditions (Verbruggen and Hermans, 2008). This buildup constitutes a primary response to water stress in plants and is closely linked to their tolerance levels (Carlin and Santos, 2009). The reduction of molecular oxygen leads to the production of ROS, that is, superoxide radical (O^{2-}), hydrogen peroxide (H_2O_2), and hydroxyl radical (OH^-), while singlet oxygen ($^1O_2$) is produced through the interaction of oxygen with excited chlorophyll (Gratão et al., 2005). Water stress amplifies ROS production, inducing a redox imbalance within cells. This disrupts the electron transport chain and initiates oxidative stress due to the detrimental effects of ROS (Gratao et al., 2008). This impacts cell survival and death across various species (Liang et al., 2013).

Furthermore, proline serves as a nonenzymatic antioxidant, aiding in the removal of ROSs and thwarting damage caused by lipid peroxidation (Verbruggen and Hermans, 2008; Das and Roychoudhury, 2014).

Kishor et al. (2005) demonstrated that proline serves as the most effective scavenger of hydroxyl radicals. Additionally, proline appears to play a role in regulating pH levels, acting as a mechanism to alleviate cytosolic acidity, often associated with stress conditions. Hare and Cress (1997) indicated that the decrease in intracellular pH has been identified as a factor triggering the accumulation of proline in plants. The synthesis of proline and subsequent removal of H^+ ions help prevent a decline in plant respiration during drought stress. Proline is a vital component in the molecular signaling pathway responding to stress, and it is a significant constituent of plant cell walls (Nepomuceno et al., 2001; Deuschle et al., 2004).

Several studies have associated this accumulation with drought tolerance in sugarcane (Paquet et al., 1994; Wahid, 2004; Guimarães et al., 2008). Additionally, proline plays a role in osmotic adjustment (Handa et al., 1986; Gzik, 1996), storage of carbon and nitrogen for poststress recovery, protein and membrane stabilization, and scavenging of ROS (Kishor et al., 2005). The elevation of free proline levels indicates damage resulting from disruptions in other metabolic pathways (Bhaskaran et al., 1985; Perez-Alfocea and Larher, 1995). Moreover, the beneficial effects of proline in conferring osmotolerance have been well-documented. High proline concentrations under water stress reflect the drought resistance potential of sugarcane genotypes (Molinari et al., 2007; Vendruscolo et al., 2007). Drought-tolerant sugarcane varieties often exhibit a rapid rate of proline accumulation (Sivaramakrishnan et al., 1988; Gzik, 1996), aiding leaf rolling (Kadioğlu and Turgut, 1999). Comparative studies between genotypes have shown that those with higher proline accumulation and elasticity rates tend to be more water-stress resistant (Shrivastava and Srivastava, 2006). In Taiwan, certain sugarcane varieties exhibited minimal cane yield reduction and the highest free proline accumulation under drought stress. Proline accumulation in young plants may serve as an indicative trait for water stress tolerance (Hong et al., 1987), and a positive correlation between increased proline content and biomass yield has been observed in transgenic cane plants (Molinari et al., 2007). Endres et al. (2018) showed that the drought-tolerant variety (RB92579) maintained high proline content after stress and recovery; it is important for restoration and preserving the chloroplasts' functional integrity. This form of these plants displayed a high PSII photochemical efficiency, with an increase in ETR, F_v/F_m, and ΦPSII. Proline accumulated after drought stress can be used as a carbon and nitrogen skeleton source that helps in the energy redistribution to restore plant physiological activities (Hemaprabha et al., 2013). Studies with different crops report that proline accumulation is significant during rewatering to increase their recovery capacity after drought (Vanková et al., 2012; An et al., 2013). Plants rewatered rapidly, expressing proline-degrading enzyme-encoding genes until they remain proline-biosynthesis-encoding genes. These genes remain active until the plant fully recovers, thus making proline important for restoring chloroplast activity after drought stress (Vanková et al., 2012). The high proline concentration recorded in the RB92579 variety under drought stress may have contributed to maintaining turgescence and protecting cell membranes, as observed in electrolyte leakage, leading to better drought tolerance (Santos and Silva, 2015; Anjum et al., 2017; Endres et al., 2018). This may have helped to maintain the integrity of chloroplast structures, minimizing damage to the photochemical mechanisms of PSII caused by water shortages. It suggested that the positive correlation between proline, ETR, and F_v/F_m, as well as the negative association with EL, exists under water stress.

Moreover, studies with sugarcane report that high proline concentrations act as antioxidants, lowering oxidative stress and contributing to the photochemical equilibrium of chloroplasts' (Hemaprabha et al., 2013; Santos and Silva, 2015). In sugarcane, the concentration of free proline can range from 0.4 to 1.6 μg/gm of fresh weight under water-restriction conditions (Guimarães et al., 2008). Similar to other compatible solutes, proline has a low molecular weight, carries no charge at neutral pH, is highly soluble, and in high concentrations, it minimally interferes with macromolecular interactions in the solvent (Chen and Murata, 2002). Additionally, proline serves as a reservoir for carbon, nitrogen, and energy during stress recovery, as it is rapidly converted to glutamate (Raymond and Smirnoff, 2002). Contrarily, mature canes exhibit increased amino acid concentrations under drought conditions but the lowest proline content (Iskander et al., 2011), suggesting that amino acids could be markers for assessing drought tolerance in sugarcane (Ferreira et al., 2017).

3.4.3 TREHALOSE

Trehalose, a nonreducing disaccharide (1,1 α-D-glucopyranosyl α-D-glucopyranoside), known for its role in lipid stabilization and enzyme protection during dehydration, has been implicated in drought tolerance in sugarcane (Pilon-Smits et al., 1998; Goddijn and van Dun, 1999; Wingler et al., 2000; Garg et al., 2002). Trehalose phosphate synthase, an enzyme involved in trehalose formation, has been identified in drought-tolerant sugarcane varieties (Vantini et al., 2015), with its production notably higher in plants experiencing stress, particularly drought (Paul et al., 2008). Trehalose is produced through a two-step pathway. Initially, trehalose-6-phosphate synthase (TPS) catalyzes the first step, utilizing uridine diphosphoglucose and glucose 6-phosphate as substrates to generate trehalose 6-phosphate (Tre6P). Subsequently, trehalose-6-phosphate phosphatase (TPP) dephosphorylates Tre6P to yield trehalose. Trehalase can then further hydrolyze trehalose to produce two molecules of glucose.

Trehalose in sugarcane plays a crucial role in stabilizing lipids and dehydrated enzymes (Pilon-Smits et al., 1998; Goddijn and van Dun, 1999; Wingler et al., 2000; Garg et al., 2002). Additionally, drought-tolerant canes exhibit elevated expression of trehalose genes (Vantini et al., 2015). Transgenic sugarcane plants engineered to have increased trehalose content via transformation from Grifola frondosa trehalose synthase (TSase) exhibited enhanced drought tolerance due to elevated trehalose accumulation (Zhang et al., 2006). Moreover, Tre6P serves as both a signal of sugar availability and an indicator of sucrose status (Lunn et al., 2006; Yadav et al., 2014). Tre6P functions as both a signal and a negative feedback regulator of sucrose levels (Figueroa and Lunn, 2016; Yadav et al., 2014). Consequently, changes in sucrose levels affect Tre6P, which in turn, alters the plant's metabolism to maintain sucrose homeostasis. Conversely, fluctuations in Tre6P levels reciprocally influence sucrose levels as part of a negative feedback loop. In sugarcane leaf and internodal tissue, trehalose has been reported to be either undetectable or present at low concentrations (0.31–3.91 nmol/g FW), contrasting with the considerably higher accumulation of sucrose, reaching up to 650 mM (approximately 650 μmol/g FW) in the stem (Bosch, 2005; Welbaum and Meinzer, 1990; Zhang et al., 2006). Moreover, transcripts of putative TPS and TPP genes have been identified in sugarcane across various tissues and genotypes (Bosch, 2005; Hu et al., 2020; McCormick et al., 2008; Nicolau et al., 2013; Thirugnanasambandam et al., 2017). However, a fungal TSase gene, which encodes an enzyme synthesizes trehalose directly from glucose and glucose 1-phosphate in sugarcane. This enhanced drought tolerance by increasing trehalose concentrations from undetectable levels to 12.86 mg/g FW in the transgenic plants (Zhang et al., 2006). Tre6P influences the allocation of photoassimilates among sucrose, organic acids, and amino acids in source leaves, and it regulates the breakdown of transitory starch in leaves during the night, potentially connecting starch turnover to sucrose demand in sink organs (Debast et al., 2011; Figueroa et al., 2016; Figueroa and Lunn, 2016; O'Leary et al., 2011).

Due to its ability to preserve membranes and inhibit protein denaturation during drought stress, trehalose is known to accumulate in large quantities and facilitate survival of total dehydration (Drennan et al., 1993; Crowe et al., 1984). Experiments conducted *in vitro* and *in vivo* revealed strong evidence that trehalose protects membranes from lipid peroxidation. Trehalose directly interacts with the oxidation-sensitive regions of unsaturated fatty acids, inhibiting auto-oxidation, based on experimental evidence (Oku et al., 2003). It has recently been shown for the first time that external trehalose provides *in vivo* membrane protection (da Costa Morato Nery et al., 2008). Trehalose directly beneficial impact on membranes and macromolecules may be attributed to a variety of mechanisms, including glass formation, chemical stability, and water replacement (Luo et al., 2008; Lun, 2007). The water replacement theory states that water generally stabilizes all biological macromolecules through the formation of hydrogen bonds surrounding them. According to specific concepts, trehalose takes the role of water molecules during dehydration, inhibiting membrane fusion and protein denaturation (Clegg, 1985; Jain and Roy, 2009).

The glassy state hypothesis is an alternative explanation that postulates that trehalose develops glasses in the dry state, which may be necessary to maintain dry macromolecules. The third explanation is the chemical stabilization hypothesis, which states that trehalose does not undergo Maillard browning with proteins and is stable at high temperatures and low pH (Wingler, 2002). These pathways are not mutually exclusive and could all play a part in trehalose's protective effect. While the glassy state hypothesis is more frequently accepted, the chemical stability and the water replacement concept have been the topic of discussion (Sussich et al., 2001). An alternate mechanism of trehalose anhydrobiosis is proposed by Sussisch et al. (2001). This approach comprises the nucleation and generation of layered epitaxial crystals of dihydrate trehalose on the surface of cellular membranes. According to this theory, the dihydrate trehalose captures the water molecules in the same hydrogen bonding network as the solvated trehalose, rather than substituting them. More dihydrate gradually accumulates at the membrane surface as dehydration increases without causing structural changes to the cells. Ultimately, anhydrous trehalose is produced by the gradual dehydration of the stacked crystals. Through the preservation of membrane structure and active molecular conformations, this strategy would safeguard vital processes during dehydration. Zhu (2001) suggests that proline, mannitol, GB, trehalose, and other substances are effective at scavenging ROS. This has been confirmed by Luo et al. (2008), who discovered that trehalose possesses a direct role in removing $O_2^{-\bullet}$ (superoxide anion) and H_2O_2 (hydrogen peroxide) in a concentration-dependent manner, with the highest rate of $O_2^{-\bullet}$ and H_2O_2 scavenging occurring at 50 mM trehalose (95% and 78%, respectively). Trehalose not only has antioxidant properties but also indirectly protects against oxidative stress by moderating superoxide dismutase activity. Trehalose has been proposed to protect plants against the detrimental effects of several forms of abiotic stress, either directly or indirectly, by scavenging ROS. Since ROS accumulates as a result of all stresses, the trehalose and other protective chemicals' protective effect may mainly stem from their ability to scavenge these oxygen radicals.

3.4.4 MALONDIALDEHYDE

Oxidative stress in plants manifests visibly through lipid peroxidation, where lipids undergo oxidative degradation. Lipid peroxidation is a common oxidative stress marker characterized by ROS attacking the polyunsaturated fatty acids in cellular membranes. This process can severely compromise cell integrity and functionality, leading to irreversible damage (Gratão et al., 2005). This phenomenon is triggered by ROS (Huang et al., 2012), and it primarily involves oxygen molecules originating from PSII. These oxygen molecules interact with lipoxygenases within plastid membranes, catalyzing their conversion into lipid hydroperoxides. This transformation renders the membrane susceptible to fragmentation, initiating a detrimental cascade of responses (Barclay and McKersie, 1994; Skorzynska-Polit, 2007). The fragmentation process not only compromises membrane integrity but also propagates the generation of new radicals. For example, peroxyl radicals (LOO*) generated during lipid peroxidation give rise to numerous intermediates, ultimately altering membrane structure and inducing severe cell damage (Scandalios, 1993; Spiteller, 2003).

Kaur et al. (2013) had shown the association of SOD, CAT, MDA, and proline activities in drought tolerance in chickpea, and these traits could be related to drought tolerance.

Among the various byproducts, MDA significantly alters the properties of cell membranes, that is, fluidity, transport of ions, and enzymatic activities (Sharma et al., 2012). Monitoring changes in aldehydic products, including MDA and 4-HNE, is crucial for assessing oxidative stress in plants. Thiobarbituric acid reactive substances are reliable markers (Hodges et al., 1999; Shulaev and Oliver, 2006; Pagariya et al., 2012). During the initial growth phase of young sugarcane plants subjected to severe water stress, Boaretto et al. (2014) noted elevated levels of H_2O_2 alongside an increase in lipid peroxidation. This observation underscores the impact of water stress on oxidative processes.

Furthermore, studies by Cia et al. (2012) and Sales et al. (2015) have indicated that the interaction between stress avoidance and decreased MDA content. The lipid peroxidation (MDA) could be a reliable parameter for identifying water stress-tolerant sugarcane varieties. This implies that the extent of lipid peroxidation could be utilized to assess sugarcane cultivars' resilience to water stress conditions (Abbas et al., 2014). Generally, the range of MDA levels varied from nearly 27 to 46 nmol/g of fresh leaves across different cultivars under irrigated situations. Sugarcane genotypes, namely, HSF-242, NSG-45, and Lho-83, exhibited the smallest increase in MDA levels, suggesting their potential drought tolerance (Abbas et al., 2014).

3.5 CONCLUSION

Drought poses a formidable obstacle to global sugarcane productivity, prompting the quest for efficient screening techniques and measurable characteristics to bolster crop resilience against water scarcity. Within this pursuit, key physiological markers, such as expansion of leaf area, stomatal conductance, photosynthetic pigments, RWC, PSII efficiency, and photosynthetic rate, have emerged as focal points in efforts to enhance drought resistance within sugarcane breeding endeavors. These critical attributes indicate the sugarcane's ability to withstand water stress and provide valuable insights for selecting drought-tolerant sugarcane genotypes. Additionally, strategies aimed at reducing stomatal conductance to curtail water loss and augmenting root density alongside deep root traits to improve water uptake capacity have been identified as pivotal adaptations in sugarcane under water scarcity conditions. By targeting these physiological traits and employing adaptive measures, researchers aim to cultivate sugarcane varieties that are better equipped to thrive in drought-prone environments.

Moreover, it is crucial to acknowledge that while above-ground traits are pivotal in responding to early-season drought, rooting characteristics also play a significant role in the plant's ability to adapt to such adverse conditions. Root traits, including root density and depth, are essential to water uptake efficiency and drought tolerance in sugarcane and other crops. Enhancing rooting traits can facilitate improved access to soil moisture reserves, enabling plants to withstand water scarcity better. Thus, a comprehensive approach to breeding for drought resilience in sugarcane must encompass above-ground and below-ground traits, recognizing the intertwined relationship between these physiological aspects in the plant's response to limited water supply. By integrating these insights into breeding programs, researchers can develop sugarcane varieties that are better equipped to thrive in the face of increasingly unpredictable climatic conditions, ultimately contributing to the resilience and sustainability of agricultural systems worldwide.

KEYWORDS

- **limited water supply**
- **sugarcane**
- **stress tolerance**
- **physio-biochemical activities**

REFERENCES

Abbas S, Ahmad S, Sabir S, Shah A. (2014) Detection of drought-tolerant sugarcane genotypes (*Saccharum officinarum*) using lipid peroxidation, antioxidant activity, glycine-betaine and proline contents. *J. Soil Sci. Plant Nutr.* 14, 233–243. https://doi.org/10.4067/s0718-95162014005000019

Abdelraheem A, Esmaeili N, O'Connell M, Zhang J. (2019) Progress and perspective on drought and salt stress tolerance in cotton. *Ind. Crop Prod.* 130, 118–129.

Abdul Jaleel C, Manivannan P, Kishorekumar A, Sridharan R, Anneersevam RP. (2008) Salt stress mitigation by calcium chloride in *Phyllanthus amarus*. *Acta Botanica Croatica* 67, 53–62.

Abdul Jaleel C, Sankar B, Manivannan P, Kishorekumar A, Sridharan R, Panneersevam R. (2007) Calcium chloride effect on salinity induced oxidative stress, proline metabolism and indole alkaloid accumulation in *Catharanthus roseus*. *C.R. Biol*. 330, 674–683.

An Y, Zhang M, Liu G, Han R, Liang Z. (2013) Proline accumulation in leaves of periplocasepium via both biosynthesis up-regulation and transport during recovery from severe drought. *PLoS One* 8, 1–10.

Anas M, Verma KK, Haq IU, Naeem M, Li Q, Liao F, Xiang YX, Li YR. (2021a) Characterization of exotic, native and wild-type genotypes of sugarcane (*Saccharum* spp. Hybrids) for internal nitrogen use efficiency under different nitrogen levels. *Sugar Tech*. 23, 1258–1267. https://doi.org/10.1007/s12355-021-01014-1

Anas M, Verma KK, Riaz M, Qiang L, Liao F, Liu Y, Li YR. (2021b) Physio-morphological and biochemical mechanism of nitrogen use efficiency in sugarcane (*Saccharum* spp.) genotypes under different growth stages and nitrogen levels. *J. Plant Interact*. 16, 332–343.

Andrews M, Hodge S, Raven JA. (2010) Positive plant microbial interactions. *Ann. Appl. Biol*. 157, 317–320.

Anjum SA, Ashraf U, Tanveer M, Khan I, Hussain S, Shahzad B, Zohaib A, Abba F, Saleem MF, Ali I, Wang LC. (2017) Drought induced changes in three maize hybrids' growth, osmolyte accumulation, and antioxidant metabolism. *Front. Plant Sci*. 8, 69.

Araus JL, Slafer GA, Royo C, Serret MD. (2008) Breeding for yield potential and stress adaptation in cereals. *Crit. Rev. Plant Sci*. 27, 1–36.

Ashraf M, Foolad MR. (2005) Pre sowing seed treatment: A shotgun approach to improve germination, plant growth and crop yield under saline and non-saline conditions. *Adv. Agron*. 88, 825–831.

Azevedo MCB, Chopart JL, Medina CC. (2011) Sugarcane root length density and distribution from root intersection counting on trench profile. *Sci. Agric*. 68, 94–101.

Ball-Codho, Sampaio, Tiessen H, Stewart JWB. (1992) Root dynamics in plant and ratoon crops of sugarcane. *Plant Soil* 142, 297–305.

Baran R, Bassereau D, Gillet N. (1974) Measurement of available water and root development on an irrigated sugar cane crop in the Ivory Coast. *Proc. Int. Soc. Sugar Cane Technol*., 15, 726–735.

Barbosa AM, Guidorizi KA, Catuchi TA, Marques TA, Ribeiro RV, Souza GM. (2015) Biomass and bioenergy partitioning of sugarcane plants under water deficit. *Acta Physiologiae Plant*. 37, 142.

Barclay KD, McKersie BD. (1994) Peroxidation reactions in plant membranes: Effects of free fatty acids. *Lipids* 29, 877–882. https://doi.org/10.1007/BF02536256

Basnayake J, Jackson PAN, Man-Bamber G, Lakshmanan P. (2015) Sugarcane for water-limited environments. Variation in stomatal conductance and its genetic correlation with crop productivity. *J. Exp. Bot*. 66, 3945–3958.

Begg JE. (1980) Morphological adaptation of leaves to water stress. In: Turner NC, Kramer PJ. (Eds.), *Adaptation of Plants to Water and High Temperature Stress*. John Wiley & Sons: New York, pp. 33–42

Bernstein N, Kravchik M, Dudai N. (2010) Salinity-inducedchanges in essential oil, pigments and salts accumulationin sweet basil (*Ocimum basilicum*) in relation to alterations of morphological development. *Ann. Appl. Biol.* 156, 167–177

Bhargava S, Sawant K. (2013) Drought stress adaptation: Metabolic adjustment and regulation of gene expression. *Plant Breeding* 132(1), 21–32

Bhaskaran S, Smith RH, Newton RJ. (1985) Physiological changes in cultured sorghum cells in response to induced water stress. 1. Free proline. *Plant Physiol.* 79, 266–269.

Bhatt R, Verma KK, Jamwal NS. (2023a) Can paddy-compost eliminate N fertilization for climate-smart sugarcane? *Sugar Tech.* 25, 1036–1046. https://doi.org/10.1007/s12355-023-01270-3

Bhatt R, Verma KK, Kumar R, Sanghera GS. (2023b) Foliar application of potassium salt of active phosphorus (PSAP) mitigates insect pests and improves yield along with sugarcane quality in response to agroclimatic conditions of Punjab. *Sugar Tech.* 25(3), 660–669.

Blum A. (2009) Effective use of water (EUW) and not water-use efficiency (WUE) is the target of crop yield improvement under drought stress. *Field Crops Res.* 112, 119–123.

Boaretto LF, Carvalho G, Borgo L, Creste S, Landell MG, Mazzafera P, Azevedo RA. (2014) Water stress reveals differential antioxidant responses of tolerant and non-tolerant sugarcane genotypes. *Plant Physiol. Biochem.* 74, 165–175. https://doi.org/10.1016/j.plaphy.2013.11.016

Bosch S. (2005) Trehalose and carbon partitioning in sugarcane. PhD Thesis, University of Stellenbosch, Stellenbosch.

Boyer JS. (1982) Plant productivity and environment. *Science* 218, 443–448.

Buckley TN. (2005) The control of stomata by water balance. *New Phytologist* 168(2), 275–292. https://doi.org/10.1111/j.1469-8137.2005.01543.x

Cal AJ, Sanciangco M, Rebolledo MC, Luquet D, Torres RO, Mcnally KL, Henry A. (2019) Leaf morphology, rather than plant water status, underlies genetic variation of rice leaf rolling under drought. *Plant, Cell Environ.* 42(5), 1532.

Campos H, Cooper A, Habben JE, Edmeades GO, Schussler JR. (2004) Improving drought tolerance in maize: A view from industry. *Field Crops Res.* 90, 19–34.

Carlin SM, Santos DMM. (2009) Indicadoresfisiológicos da interação entre déficithídrico e acidez do solo emcanade-açúcar. *PesqAgropec Bras.* 44, 1106–1113.

Castro PRC, Zambon S, Sansigolo MA, Beltrame JA, Nogueira MCS. (2002) Açãocomparada de ethrel, fuzilade e roundup, em duas épocas de aplicação, namaturação e produtividade da cana-de-açúcar SP 70-1143. *Rev. Agric.* 77, 23–38.

Chaudhary P, Xu M, Ahamad L, Chaudhary A, Kumar G, Adeleke BS, Verma KK, Hu D-M, Siric I, Kumar P, Popescu SM, Fayssal SA. (2023) Application of synthetic consortia for improvement of soil fertility, pollution remediation, and agricultural productivity: A review. *Agronomy* 13, 643. https://doi.org/10.3390/agronomy13030643.

Cha-um S, Kirdmanee C. (2008) Effect of osmotic stress on proline accumulation, photosynthetic abilities and growth of sugarcane plantlets (*Saccharum officinarum* L.) *Pak. J. Bot.* 40(6), 2541–2552.

Chen GL, Zheng FJ, Lin B, Lao SB, He J, Huang Z, Zeng Y, Sun J, Verma KK. (2020) Phenolic and volatile compounds in the production of sugarcane vinegar. *ACS Omega* 5, 30587–30595.

Chen G-L, Zheng F-J, Lin B, Yang Y-X, Fang X-C, Verma KK, Yang L-F. (2023) Vinegar: A potential source of healthy and functional food with special reference to sugarcane vinegar. *Front. Nutr.* 10, 1145862. https://doi.org/10.3389/fnut.2023.1145862

Chen THH, Murata N. (2002) Enhancement of tolerance to abiotic stress by metabolic engineering of betaines and other compatible solutes. *Curr. Opin. Plant Biol.* 5, 250–257.

Chen THH, Murata N. (2011) Glycine betaine protects plants against abiotic stress: Mechanism and biotechnological applications. *Plant, Cell Environ.* 34, 1–20.

Cheng YJ, Deng XP, Kwak SS, Chen W, Emeji AE. (2013) Enhanced tolerance of transgenic potato plants expressing choline oxidase in chloroplasts against water stress. *Bot. Stud.* 54, 1–9.

Cia MC, Guimaraes ACR, Medici LO, Chabregas SM, Azevedo RA. (2012) Antioxidant responses to water deficit by drought tolerant and sensitive sugarcane varieties. *Ann. Appl. Biol.* 161, 313–324.

Clegg JS. (1985) The physical properties and metallic status of Artemia cysts at low water contents: The "water replacement hypothesis," In: Leopold, AC. (Ed.), *Membranes, Metabolism and Dry Organisms*. Cornell University Press: Ithaca. NY, USA, pp. 169–187.

Colom MR, Vazzana C. (2003) Photosynthesis and PSII functionality of drought-resistant and drought sensitive weeping lovegrass plants. *Environ. Exp. Bot.* 49, 135–144.

Comas LH, Becker SR, Cruz VMV, Byrne PF, Dierig DA. (2013) Root traits contributing to plant productivity under drought. *Front. Plant Sci.* 4, https://doi.org/10.3389/fpls.2013.00442

Crowe JH, Crowe LM, Chapman D. (1984) Preservation of membranes in anhydrobiotic organisms: The role of trehalose. *Science* 223(4637), 701–703

da Costa Morato Nery D, da Silva CG, Mariani D, Fernandes PN, Pereira MD, Panek AD, Eleutherio EC. (2008) The role of trehalose and its transporter in protection against reactive oxygen species. Eleutherio. *Biochim. Biophys. Acta* 1780, 1408.

Da Graca J, Rodrigues F, Farias J, Oliveira Md, Hoffmann Campo C, Zingaretti S. (2010) Physiological parameters in sugarcane cultivars submitted to water deficit. *Braz. J. Plant Physiol.* 22, 189–197.

da Silva PP, Soares L, Costa JG, da Viana LDS, Andrade JC, Farias de, Goncalves ER, Santos JM dos, de Verıssimo G. (2012) Path analysis for selection of drought tolerant sugarcane genotypes through physiological components. *Ind. Crops Prod.* 37, 11–19.

Das K, Roychoudhury A. (2014) Reactive oxygen species (ROS) and response of antioxidants as ROS scavengers during environmental stress in plants. *Front. Environ. Sci.* 2. https://doi.org/10.3389/fenvs.2014.00053

Debast S, Nunes-Nesi A, Hajirezaei M-R, Hofmann J, Sonnewald U, Fernie AR, Börnke F. (2011) Altering trehalose-6-phosphate content in transgenic potato tubers affects tuber growth and alters responsiveness to hormones during sprouting. *Plant Physiol.* 156, 1754–1771.

del Pozo A, Castillo D, Inostroza L, Matus I, Mendez AM, Morcuende R. (2012) Physiological and yield responses of recombinant chromosome substitution lines of barley to terminal drought in a Mediterranean-type environment. *Ann. Appl. Biol.* 160, 157–167. https://doi.org/10.1111/j.1744-7348.2011.00528.x.

Deng ZN, Huang HR, Huang CM, Verma KK, Luo HB, Jiang SL, Li YJ, Cao HQ, Wu KC, Xu L. (2023) Short-term limited water irrigation influences the photosynthetic pigments, enzymatic and non-enzymatic activities in *Saccharum spontaneum* L. at vegetative stage. *Sugar Tech.* 25, 110–118. https://doi.org/10.1007/s12355-022-01175-7

Deuschle K, Funck D, Forlani G, Stransky H, Biehl A, Leister D, Graaff EVD, Kunze R, Frommer WB. (2004) The role of Δ1-pyrroline-5-carboxylate dehydrogenase in proline degradation. *Plant Cell* 16, 3413–3425.

Dinh TH, Takargawa H, Watanbe K, Nakabaru M, Kawamitsu Y. (2017) Leaf photosynthesis response to the change of soil moisture content in sugarcane. *Sugar Tech.* 21, 949–958.

Dodd IC, Whalley WR, Ober ES, Parry MAJ. (2011) Genetic and management approaches to boost UK wheat yields by ameliorating water deficits. *J. Exp. Bot.* 62(15), 5241–5248.

Drennan PM, Smith MT, Goldsworthy D, Staden JV. (1993) The occurrence of trhalose in the leaves of the dessication tolerant angiosperm *Myrothamnus flabellifolius* welw. *J. Plant Physiol.* 142(4), 493–496.

Dwivedi RS. (2000) Physiology of sugarcane. In: Shahi HN, Shrivastava AK, Sinha OK., (Eds.), *Fifty Years of Sugarcane Research in India*, Indian Institute of Sugarcane Research: Lucknow, pp. 73–110.

Endres L, Silva JV, Ferreira VM, Barbosa GVDS. (2010) Photosynthesis and water relations in Brazilian sugarcane. *Open Agric. J.* 4, 31–37.

Endres L, Santos CMD, Souza GVD, Menossi M, Santos JCMD. (2018) Morphological changes recorded in different phenophases of sugarcane plants subjected to water stress in tropical field conditions. *Aust. J. Crop Sci.* 12(07), 1041–1050.

Evans H. (1935) Investigation on the root system of sugarcane varieties. Sugarcane Res. Stational Mauritius, Bull. No. 6, pp. 44.

Evans H. (1936) The root system of sugarcane. II some typical root systems. *Empire. J. Exp. Agric.* 4, 208–221.

Fang Y, Xiong L. (2015) General mechanisms of drought response and their application in drought resistance improvement in plants. *Cell Mol. Life Sci.* 72(4), 673–689.

Ferreira THS, Tsunada MS, Bassi D, Araújo P, Mattiello L, Guidelli GV, Righetto GL, Gonçalves VR, Lakshmanan P, Menossi M. (2017) Sugarcane water stress tolerance mechanisms and its implications on developing biotechnology solutions. *Front. Plant Sci.* 8, 1077

Figueroa CM, Feil R, Ishihara H, Watanbe M, Kolling K, Krause U, Hohne M, Encke B, Plaxton WC, Zeeman SC, Li Z, Schulze WX, Hoefgen R, Stitt M, Lunn JE. (2016) Trehalose 6-phosphate coordinates organic and amino acid metabolism with carbon availability. *The Plant J.* 85(3), 410–423.

Figueroa CM, Lunn JE. (2016) A tale of two sugars: Trehalose 6-phosphate and sucrose. *Plant Physiol.* 172(1), 7–27

Fukai S, Cooper M. (1995) Development of drought-resistant cultivars using physiomorphological traits in rice. *Field Crop Res.* 40, 67–86

Garg AK, Kim JK, Owens TG, Ranwala AP, Choi YD, Kochiam LV, Wu RJ. (2002) Trehalose accumulation in rice plants confers high tolerance levels to different abiotic stresses. *PNAS* 99(25), 15903.

Goddijn OJM, van Dun K. (1999) Trehalose metabolism in plants. *Trends Plant Sci.* 4(8), P315–319.

Gratao PL, Monteiro CC, Antunes AM, Peres LEP, Azevedo RA. (2008) Acquired tolerance of tomato (Lycopersicon esculentum cv. Micro-Tom) plants to cadmium induced stress. *Ann. Appl. Biol.* 153, 321–333.

Gratão PL, Polle A, Lea PJ, Azevedo RA. (2005) Making the life of heavy metal-stressed plants a little easier. *Funct. Plant Biol.* 32, 481–494.

Grover M, Madhubala R, Ali SZ, Yadav SK, Venkateswarlu B. (2014) Influence of Bacillus spp. strains on seedling growth and physiological parameters of sorghum under moisture stress conditions. *J. Basic Microbiol.* 54(9), 951–961.

Guan XK, Song L, Wang TC, Turner NC, Li FM. (2015) Effect of drought on the gas exchange, chlorophyll fluorescence and yield of six different era spring cultivars. *J. Agronomy Crop Sci.* 201(4), 253–266.

Guimarães ER, Mutton MA, Mutton MJR, Ferro MIT, Ravaneli GC, da Silva JA. (2008) Free proline accumulation in sugarcane under water restriction and spittlebug infestation. *Sci. Agric.* 65, 628–633. https://doi.org/10.1590/S0103-90162008000600009

Gunnula W, Kanawapee N, Somta P, Phansak P. (2022) Evaluating anatomical characteristics associated with leaf rolling in northeastern Thai rice cultivars during drought by decision tree. *Acta Agrobotanica* 75(1).

Guo D-J, Li D-P, Chen Z-D, Yang B, Verma KK, Singh RK, Singh P, Khan Q, Sharma A, Qin Y, Zhang B-Q, Song X-P, Li Y-R. (2023) Effect of endophytic diazotroph *Enterobacter roggenkampii* ED5 on nitrogen-metabolism-related microecology in the sugarcane rhizosphere at different nitrogen levels. *Front. Microbiol.* 14, 1132016. https://doi.org/10.3389/fmicb.2023.1132016

Guo D-J, Li D-P, Singh RK, Singh P, Sharma A, Verma KK, Qin Y, Khan Q, Lu Z, Malviya MK, Song X-P, Xing Y-X and Li Y-R. (2021) Differential protein expression analysis of two sugarcane varieties in response to diazotrophic plant growth-promoting endophyte *Enterobacter roggenkampii* ED5. *Front. Plant Sci.* 12, 727741. https://doi.org/10.3389/fpls.2021.727741

Guo DJ, Li DP, Singh RK, Singh P, Verma KK, Sharma A, Qin Y, Khan Q, Song XP, Malviya MK, Xing YX, Li YR. (2022) Comparative transcriptome analysis of two sugarcane varieties in response to diazitrophic plant growth promoting endophyte *Enterobacter roggenkampii* ED5. *J. Plant Interact.* 17(1), 75–84.

Gzik A. (1996) Accumulation of proline and pattern of alpha-amino acids in sugar beet plants in response to osmotic, water, and salt stress. *Environ. Exp. Bot.* 36(1), 29–38.

Hamdia MA, Shaddad MAK. (2010) Salt tolerance of crop plants. *J. Stress Physiol. Biochem.* 6, 64–90, 2010.

Handa S, Handa AK, Hasegawa PM, Bressan RA. (1986) Proline accumulation and the adaptation of cultures plant cells to water stress. *Plant Physiol.* 80, 938–945.

Hare PD, *Cress* WA. (*1997*) Metabolic implications of stress-induced *proline* accumulation in plants. *Plant Growth Regul.* 21, 79–102.

Hemaprabha G, Swapna S, Lavanya DL, Sajitha B, Venkataramana S. (2013) Evaluation of drought tolerance potential of elite genotypes and progenies of sugarcane (*Saccharum* sp. hybrids). *Sugar Tech.* 15, 9–16.

Hodges DM, DeLong JM, Forney CF, Prange RK. (1999) Improving the thiobarbituric acid-reactive-substances assay for estimating lipid peroxidation in plant tissues containing anthocyanin and other interfering compounds. *Planta* 207, 604–611. https://doi.org/10.1007/s004250050524

Hoekstra FA, Golovina EA, Buitink J. (2001) Mechanisms of plant desiccation tolerance. *Trends Plant Sci.* 6, 431–438.

Hong JC, Nagao RT, Key JL. (1987) Characterization and sequence analysis of a developmentally regulated putative cell wall protein gene isolated from soybean. *J. Biol. Chem.* 262, 8367–8376.

Hotta CT, Lembke CG, Domingues DS, Ochoa EA, Cruz GMQ, Melotto-Passarin DM, Marconi TG, Margarido GRA, Crivellari AC, dos Santos WD, de Souza AP, Hoshino AA, Carrer H, Souza AP, Garcia AAF, Buckeridge MS, Menossi M, Sluys MAV, Souza GM. (2010) The biotechnology roadmap for sugarcane. *Tropical Plant Biol.* 3, 75–87.

Hsiao TC, O'Toole JC, Yambao EB, Turner NC. (1984) Influence of osmotic adjustment on leaf rolling and tissue death in rice. *Plant Physiol.* **75**, 328.

Hu X, Wu Z-D, Luo Z-Y, Burner DM, Pan Y-B, Wu C-W. (2020) Genome-wide analysis of the trehalose 6-phosphate synthase (TPS) gene family and expression profiling of ScTPS genes in sugarcane. *Agronomy* 10, 969. https://doi.org/10.3390/agronomy10070969

Huang CM, Wu KC, Verma KK, Luo HB, Deng ZN, Shi GY, Cao HQ, Wu XJ, Ye LP, Wei YW, Xu L. (2024) Isolation and expression analysis of cellulose synthase 3 (*Ces3*) genes from sugarcane (*Saccharum officinarum* L.). *Chem. Biol. Technol. Agric.* 11, 71. https://doi.org/10.1186/s40538-024-00597-7

Huang G-T, Ma S-L, Bai L-P, Zhang L, Ma H, Jia P, Liu J, Zhong M, Guo ZF. (2012) Signal transduction during cold, salt, and drought stresses in plants. *Mol. Biol. Rep.* 39, 969–987. https://doi.org/10.1007/s11033-011-0823-1

Huang HR, Gao YJ, Malviya MK, Verma KK, Solanki MK, Huang YX, Li X, Deng Y, Yan J, Tang SY, Wang LW, Xu L. (2022) Genetic diversity analysis of sugarcane (*Saccharum* spp. Hybrids) among high-sucrose clones of GT series and commonly used parents by using microsatellite markers in Guangxi, China. *Sugar Tech.* 24, 397–407. https://doi.org/10.1007/s12355-021-01033-y

Huff AK, Ross CW. (1975) Promotion of radish cotyledon enlargement and reducing sugar content by zeatin and red light. *Plant Physiol.* 56, 429–433.

Hussain HA, Men S, Hussain S, Chen Y, Ali S, Zhang S, Zhang K, Li Y, Xu Q, Liao C, Wang L. (2019) Interactive effects of drought and heat stresses on morphophysiological attributes, yield, nutrient uptake and oxidative status in maize hybrids. *Sci. Rep.* 9(1), 3890.

Inman-Bamber N, Bonnett G, Spillman M, Hewitt M, Jackson J. (2008) Increasing sucrose accumulation in sugarcane by manipulating leaf extension and photosynthesis with irrigation. *Aust. J. Agric. Res.* 59, 13–26.

Inman-Bamber N, Smith D. (2005) Water relations in sugarcane and response to water deficits. *Field Crops Res.* 92, 185–202.

Inman-Bamber N. (2004) Sugarcane water stress criteria for irrigation and drying off. *Field Crops Res.* 89, 107–122.

Inman-Bamber NG, Lakshmanan P, Park S. (2003) Sugarcane for water-limited environments: Theoretical assessment of suitable traits. *Field Crops Res.* 134, 95–104.

Inman-Bamber NG, McGlinchey MG. (2012) Crop coefficients and water-use estimates for sugarcane based on long-term Bowen ratio energy balance measurements. *Field Crop Res.* 83(2), 125–138.

Iskander HM, Casu RE, Fletcher AT, Schmidt S, Xu J, Maclean DJ, Manners JM, Bonnett GD. (2011) Identification of drought response genes and a study of their expression during sucrose accumulation and water deficit in sugarcane culms. *BMC Plant Biol.* 11, 12.

Jagadesh M, Dash M, Kumari A, Singh SK, Verma KK, Kumar P, Bhatt R, Sharma SK. (2024) Revealing the hidden world of soil microbes: Metagenomic insights into plant, bacteria, and fungi interactions for sustainable agriculture and ecosystem restoration. *Microbiol. Res.* 127764. https://doi.org/10.1016/j.micres.2024.127764

Jain NK, Roy I. (2009) Effect of trehalose on protein structure. *Protein Sci.* 18(1), 24–36.

Jangpromma N, Thammasirirak S, Jaisil P, Songsri P. (2012) Effects of drought and recovery from drought stress on above ground and root growth, and water use efficiency in sugarcane (*Saccharum officinarum* L.). *Aust. J. Crop Sci.* 6(8), 1298–1304.

Jongrungklang N, Toomsan B, Vorasoot N, Jogloy S, Boote KJ, Hoogenboom G, Patanothai A. (2011) Rooting traits of peanut genotypes with different yield responses to pre-flowering drought stress. *Field Crop Res.* 120, 262–270.

Joshi A, Verma KK, Rajput VD, Arora J. (2023) Metabolic engineering of microorganisms in advancing biofuel production. In: *Advances in Lignocellulosic Biofuel Production Systems*. pp. 181–202. https://doi.org/10.1016/B978-0-323-91192-4.00016-X

Joshi A, Verma KK, Rajput VD, Minkina T, Arora J. (2022) Recent advances in metabolic engineering of microorganisms for advancing lignocellulose-derived biofuels. *Bioengineered* 13(4), 8135–8163. https://doi.org/10.1080/21655979.2022.2051856.

Juarez MT, Twigg RW, Timmermans MC. (2004) Specification of adaxial cell fate during maize leaf development. *Nature* 131, 4544.

Kadioglu A, Terzi R. (2007) A dehydration avoidance mechanism: Leaf rolling. *Bot. Rev.* 73(4), 290–302.

Kadioğlu A, Turgut R. (1999) Some biochemical changes during leaf rolling in *Ctenanthe setosa* (Marantaceae). *Acta Physiol. Plant.* 21, 209–214. https://doi.org/10.1007/s11738-999-0034-8

Kato MC, Hikosaka K, Hirotsu N, Makin A, Hirose T. (2003) The excess light energy that is neither utilized in photosynthesis nor dissipated by photoprotective mechanisms determines the rate of photoinactivation in photosystem II. *Plant Cell Physiol.* 44, 318–325.

Kaur G, Asthir B. (2015) Proline: A key player in plant abiotic stress tolerance. *Biologia Plantarum* 59, 609–619.

Kaur K, Kaur N, Gupta AK, Singh I. (2013) exploration of the antioxidative defense system to characterize chick pea genotypes showing differential response in water deficit conditions. *Plant Growth Regul.* 70, 49–60.

Khan MA, Iqbal M, Jameel M, Nazeer W, Shakir S, Aslma MT, Iqbal B. (2011) Potentials of molecular based breeding to enhance drought tolerance in wheat (*Triticum aestivum* L.). *African J. Biotechnol.* 10(55), 11340–11344.

Khonghintaisong J, Songsri P, Jongrungklang N. (2021) Understanding growth rate patterns among different drought resistant sugarcane cultivars during plant and ratoon crops encountered water deficit at early growth stage under natural field conditions. *Agronomy* 11, 2083.

Kingston G. (1977) The influence of accessibility on moisture extraction by sugarcane. In: *Proceedings of the XVI Congress of the International Society of Sugar Cane Technologists,* Sao Paulo, Brazil, pp. 1239–1250.

Kishor KPB, Sangam S, Amrutha RN, Sri Laxmi PS, Naidu KR, Rao KRSS, Rao S, Reddy KJ, Theriappan KJ, Sreenivasulu N. (2005) Regulation of proline biosynthesis, degradation, uptake and transport in higher plants: Its implications in plant growth and abiotic stress tolerance. *Curr. Sci.* 88(3), 424–438.

Knapp AK. (1985) Effect of fire and drought on the ecophysiology of Andropogon gerardii and Panicum virgatum in a tallgrass prairie. *Ecology* 66, 1309–1320.

Kosma DK, Bourdenx B, Bernard A, Parsons EP, Lu S, Joubes J, Jenks MA. (2009) The impact of water deficiency on leaf cuticle lipids of Arabidopsis. *Plant Physiol.* 151(4), 1918.

Krishna D, Sachan HK, Verma KK. (2022) Abiotic stresses in plants: An overview. In: *Understanding Abiotic Stresses*. Nova Science Publishers, Inc.: New York.

Kumar D, Malik N, Sengar RS, Yadav B, Singh AK, Yadav CL, Yadav MK. (2021) Transcriptional regulation in sugarcane under water deficit during formative growth stage. *Cytol. Genet.* 55, 598–605.

Kumari A, Dash M, Singh SK, Jagadesh M, Mathpal B, Mishra PK, Pandey SK, Verma KK. (2023a) Soil microbes: A natural solution for mitigating the impact of climate change. *Environ. Monit. Assess.* 195, 1436. https://doi.org/10.1007/s10661-023-11988-y

Kumari A, Lakshmi GA, Krishna GK, Patni B, Prakash S, Bhattacharyya M, Singh SK, Verma KK. (2022) Climate change and its impact on crops: A comprehensive investigation for sustainable agriculture. *Agronomy* 12, 3008. https://doi.org/10.3390/agronomy12123008

Kumari A, Singh SK, Mathpal B, Verma KK, Garg VK, Bhattacharyya M, Bhatt R. (2023b) The biosynthesis, mechanism of action, and physiological functions of melatonin in horticultural plants: A review. *Horticulturae* 9, 913. https://doi.org/10.3390/horticulturae9080913

Laclau BP, Laclau JP. (2009) Growth of the whole root system for a plant crop of sugarcane under rainfed and irrigated environments in Brazil. *Field Crops Res.* 114, 351–360. http://dx.doi.org/10.1016/j.fcr.2009.09.004

Laffray D, Louguet P. (1990) Stomatal responses and drought resistance. *Bull. Bot. Soc. France, Bot. News* 137(1), 47–60.

Lakshmanan P, Robinson N. (2013) Stress physiology: Abiotic stresses. 1st edn., In: Paul HM, Frederik CB. (Eds.), *Sugarcane: Physiology, Biochemistry, and Functional Biology*, Wiley-Blackwell: Hoboken, NJ, USA, pp. 411–434.

Lakshmanan P, Robinson N. (2014) Stress physiology: Abiotic stresses. In: Moore PH, Botha FC., (Eds.), *Sugarcane: Physiology, Biochemistry, and Functional Biology*, Wiley: Chichester, pp. 411–434.

Li A-M, Liao F, Wang M, Chen Z-L, Qin C-X, Huang R-Q, Verma KK, Li Y-R, Que Y-X, Pan Y-Q, et al. (2023) Transcriptomic and proteomic landscape of sugarcane response to biotic and abiotic stressors. *Int. J. Mol. Sci.* 24, 8913. https://doi.org/10.3390/ijms24108913

Liang X, Zhang L, Natarjan SK, Becker DF. (2013) Proline mechanisms of stress survival. *Antioxid Redox Signal* 19(9), 998–1011.

Ludlow MM. (1980) Stress physiology of tropical pasture plants. *Tropical Grasslands*, 14, 136–145.

Lun JE. (2007) Gene families and evolution of trehalose metabolism in plants. *Funct. Plant Biol.* 34(6), 550–563.

Lunn JE, Feil R, Hendricks JHM, Gibon Y, Morcuende R, Osuna D, Scheible WR, Carillo P, Hajiresaei MR, Stitt M. (2006) Sugar-induced increases in trehalose 6-phophate are correlated with redox activation of ADP glucose pyrophosphorylase and higher rates of starch synthesis in *Arabidopsis thaliana*. *Biochem. J.* 397, 139–148.

Luo J, Zhang MQ, Lin YQ. (2004) Studies on the relationship of chlorophyll fluorescence characters and drought tolerance in seedling of sugarcane under water stress. *Sci. Agric. Sin.* 37, 1718–1721.

Luo LJ. (2010) Breeding for water-saving and drought-resistance rice (WDR) in China. *J. Exp. Bot.* 61, 3509–3517.

Luo Y, Li WM, Wang W. (2008) Trehalose: Protector of antioxidant enzymes or reactive oxygen species scavenger under heat stress? *Environ. Exp. Bot.* 63(1–3), 378–384.

Machado R, Ribeiro R, Marchiori P, Machado D, Machado E, Landell M. (2009) Biometric and physiological responses to water deficit in sugarcane at different phenological stages. *Pesquisa Agropecua'ria Brasileira* 44, 1575–1582.

Mahadevaiah C, Hapase P, Sreenivasa V, Hapase R, Swamy HKM, Anilkumar C, Mohanraj K, Hemaprabha G, Ram B. (2021) Delineation of genotype × environment interaction for identification of stable genotypes for tillering phase drought stress tolerance in sugarcane. *Sci. Rep.* 11, 18649

Mall AK, Misra V, Pathak AD, Srivastava S. (2022) Breeding for drought tolerance in sugarcane: Indian perspective. *Sugar Tech.* 6, 1625–1635.

Mall AK, Misra V, Singh BD, Kumar M, Pathak AD. (2020) Drought tolerance: Breeding efforts in sugarcane. In: Hasaanzuman M. (Ed.), *Agronomic Crops*. Springer Nature Singapore Pte Ltd. Publications, pp. 157–172. doi10.1007/978-981-15-0025-1_10

Malviya MK, Li C-N, Lakshmanan P, Solanki MK, Wang Z, Solanki AC, Nong Q, Verma KK, Singh RK, Singh P, Sharma A, Guo D-J, Dessoky ES, Song X-P, Li Y-R. (2022) High-throughput sequencing-based analysis of rhizosphere and diazotrophic bacterial diversity among wild progenitor and closely related species of sugarcane (*Saccharum* spp. inter-specific hybrids). *Front. Plant Sci.* 13, 829337. https://doi.org/10.3389/fpls.2022.829337.

Malviya MK, Solanki MK, Li CN, Wang Z, Zeng Y, Verma KK, Singh RK, Singh P, Huang HR, Yang LT, Song XP, Li YR. (2021) Sugarcane-legume intercropping can enrich the soil microbiome and plant growth. *Front. Sustain. Food Syst.* 5, 606595. https://doi.org/10.3389/fsufs.2021.606595.

Matschi S, Vasquez MF, Bourgault R, Steinbach P, Chamness J, Kaczmar N, Gore MA, Molina I, Smith LG. (2020) Structure-function analysis of the maize bulliform cell cuticle and its potential role in dehydration and leaf rolling. *Plant Direct* 4(10), 00282.

McCormick AJ, Cramer MD, Watt DA. (2008) Regulation of photosynthesis by sugars in sugarcane leaves. *J. Plant Physiol.* 165, 1817–1829 https://doi.org/10.1016/j.jplph.2008.01.008

Medeiros DB, da Silva EC, Mansur Custodio Nogueira RJ, Teixeira MM, Buckeridge MS. (2013) Physiological limitations in two sugarcane varieties under water suppression and after recovering. *Theoretical Exp. Plant Physiol.* 25, 213–222.

Meinzer FC, Grant DA. (1990) Stomatal and hydraulic conductance in growing sugarcane adjustment to water transport capacity. *Plant Cell Environ.* 13, 383–388.

Misra V, Mall AK, Pathak AD. (2022a) Avoidance and tolerance strategy of sugarcane in drought stress: Response. In: Gaur RK. (Ed.), *Omics Approaches for Sugarcane Crop Improvement*. CRC Press. doi10.1201/9781003292425-11

Misra V, Solomon S, Ansari MI. (2016) Impact of drought on post-harvest quality of sugarcane crop. *Adv. Life Sci.* 20(5), 9496–9505.

Misra V, Solomon S, Hassan A, Allah EFE, Al-Arjani AF, Mall AK, Prajapati CP, Ansari MI. (2020a) Minimization of post harvest sucrose losses in drought affected sugarcanes using chemical formulation. *Saudi J. Biol. Sci.* 27(1), 309–317. https://doi.org/10.1016/j.sjbs.2019.09.028

Misra V, Solomon S, Mall AK, Abid M, Khan MMAA, Ansari MI. (2022b) Drought stress and sustainable sugarcane production. In: Arora, NK, Bouizgarne B., (Eds.), *Microbial Biotechnology for Sustainable Agriculture, Vol. 1.* Springer: Singapore, pp. 353–368 https://doi.org/10.1007/978-981-16-4843-4_11

Misra V, Solomon S, Mall AK, Prajapati CP, Hassan A, Allah EFE, Ansari MI. (2020b) Morphological assessment of water stressed sugarcane: A comparison of waterlogged and drought affected crop. *Saudi J. Biol. Sci.* 27(5), 1228–1236. https://doi.org/10.1016/j.sjbs.2020.02.007

Mitra J. (2001) Genetics and genetic improvement of drought tolerance in crop plants. *Curr. Sci.* 80, 758–762.

Molinari HB, Marur CJ, Daros E, Carvalho J, Bespalhok Filho JC, Pereira L, Vieira LGE. (2007) Evaluation on the stress-inducible production of proline in transgenic sugarcane (*Saccharum* spp.): Osmotic adjustment, chlorophyll fluorescence and oxidative stress. *Physiol. Plant.* 130, 218–229.

Moussa HR, Abdel-Aziz SM. (2008) Comparative response of drought tolerant and drought sensitive maize genotypes to water stress. *Aust. J. Crop Sci.* 1(1), 31–36.

Naidu KM, Bhagyalakshmi KV. (1973) Relative turgidity and stomatal movement of sugarcane varieties in relation to drought resistance. *Ind. J. Agric. Sci.* 43(11), 1016–1018.

Nepomuceno AL, Neumaier N, Farias JRB, Oya T. (2001) Tolerancia a secaem plants: Mecanismosfidiologicos e moleculares. *Biotechnologia Ciencia Desenvolvimento* 23, 12–18.

Nicolau NJ, Nicolau MSAP, Mantovanini LJ, Zingaretti SM. (2013) Expression analysis of two genes coding for trehalose 6-phosphate synthase (TPS), in sugarcane (*Saccharum* spp.) under water stress. *American J. Plant Sci.* 4, 91–99. https://doi.org/10.4236/ajps.2013.412A3011

O'Leary B, Park J, Plaxton WC. (2011) The remarkable diversity of plant PEPC (phosphoenolpyruvate carboxylase): Recent insights into the physiological functions and post-translational controls of non-photosynthetic PEPCs. *Biochem. J.* 436, 15–34.

O'Neill PM, Shanahan JF, Schepers JS. (2006) Use of chlorophyll fluorescence assessments to differentiate corn hybrid response to variable water conditions. *Crop Sci.* 46, 681–687.

O'Toole JC, Chang TT. (1978) Drought and rice improvement in perspective. IRRI Research Paper Series, No. 14. International Rice Research Institute.

Oku H, Watanabe H, Kubota M, Fukuda S, Kurimoto M, Tsujisaka Y, Komori M, Inoue Y, Sakurai M. (2003) NMR and quantum chemical study on the OH–π and CH–O interactions between trehalose and unsaturated fatty acids: Implication for the mechanism of antioxidant function of trehalose. *J. Am. Chem. Soc.* 125, 12739–12748.

Oliveira MA, Verma KK, Bhatt R, Oliveira TBA. (2022) Impact of green and organic fertilizers on soil fertility and sugarcane productivity. In: Verma KK. et al. (Eds.), *Agro-industrial Perspectives on Sugarcane Production Under Environmental Stress*. https://doi.org/10.1007/978-981-19-3955-6_11.

O'Toole JC, Cruz RT, Singh TN. (1979) Leaf rolling and transpiration. *Plant Sci. Lett.* 16(1), 111–114.

Pagariya MC, Devarumath RM, Kawar PG. (2012) Biochemical characterization and identification of differentially expressed candidate genes in salt stressed sugarcane. *Plant Sci.* 184, 1–13. https://doi.org/10.1016/j.plantsci.2011.12.002

Paquet L, Rathinasabapathi B, Saini H, Zamir L, Gage DA, Huang ZH, Hanson AD. (1994) Accumulation of the compatible solute 3-dimethylsulfoniopropionate in sugarcane and its relatives but not other gramineous crops. *Aust. J. Plant Physiol.* 21(1), 37–48.

Perez-Alfocea F, Larher F. (1995) Sucrose and proline accumulation and sugar efflux in tomato leaf discs affected by NaCl and polyethylene glycol 6000 iso-osmotic stresses. *Plant Sci.* 107, 9–15.

Pilon-Smits EAH, Terry N, Sears T, Kim H, Zayed A, Hwang S, van Dun K, Voogd E, Verwoerd TC, Krutwagen RW, Goddijn OJM. (1998) Trehalose-producing transgenic tobacco plants show improved growth performance under drought stress. *J. Plant Physiol.* 152**,** 525–532.

Pinto RS, Reynolds MP. (2015) Common genetic basis for canopy temperature depression under heat and drought stress associated with optimized root distribution in bread wheat. *Theoretical Appl. Genet.* 128(4), 575–585.

Ramesh P, Mahadevaswamy M. (2000) Effect of formative phase drought on different classes of shoots, shoot mortality, cane attributes, yield and quality of four sugarcane cultivars. *J. Agron. Crop Sci.* 185, 249–258.

Ramesh P. (2000) Effect of different levels of drought during the formative phase on growth parameters and its relationship with dry matter accumulation in sugarcane. *J. Agronomy Crop Sci.* 185, 83–89.

Raymond MJ, Smirnoff N. (2002) Proline metabolism and transport in maize seedlings at low water potential. *Ann. Bot.* 89(7), 813–823.

Reyes JA, Casas D, Gandia J, Delfin E. (2021) Drought impact on sugarcane production. In: Gorawala P, Mandhatri S. (Eds.), *Agricultural Research Updates*. Nova Science Publishers, Inc.: Hauppauge, NY, USA, 35, pp. 53–93.

Reyes JAO, Carpentero AS, Santos PJA, Delfin EF. (2020) Effects of water regime, genotype, and formative stages on the agro-physiological response of sugarcane (*Saccharum officinarum* L.) to drought. *Plants* 9, 661.

Rhodes D, Hanson AD. (1993) Quaternary ammonium and tertiary sulfonium compounds in higher plants. *Annu. Rev. Plant Physiol. Plant Mol. Biol.* 44, 357–384.

Ribeiro RV, Machado RS, Machado EC, Machado DFSP, Magalhaes Filho JR, Landell MGA. (2013) Revealing drought resistance and productive patterns in sugarcane genotypes by evaluating both physiological responses and stalk yield. *Exp. Agricul.* 49, 212–224.

Richards RA, Rebetzke GJ, Condon AG, van Herwaarden AF. (2002) Breeding opportunities for increasing the efficiency of water use and crop yield in temperate cereals. *Crop Sci.* 42, 111–121.

Robertson MJ, Inman-Bamber NG, Mochow RC, Wood AW. (1999) Physiology and productivity of sugarcane with early and mid-season water deficit. *Field Crops Res.* 64, 211–227.

Rogiers SY, Greer DH, Hatfield JM, Hutton RJ, Clarke SJ, Hutchinson PA, Somers A. (2011) Stomatal response of anisohydric grapevine cultivar to evaporative demand available soil moisture and abscisic acid. *Tree Physiol.* 32(3), 249–261.

Rontein D, Basset G, Hason AD. (2002) Metabolic engineering of osmo-protectant accumulation in plants. *Metab. Eng.* 4, 49–56.

Rutherford RS. (1989) The assessment of proline accumulation as a mechanism of drought resistance in sugarcane. *Proc. South African Sugar Technol.* pp. 136–141.

Sairam RK, Rao KV, Srivastava GC. (2002) Differential response of wheat genotypes to long term salinity stress in relation to oxidative stress, antioxidant activity and osmolyte concentration. *Plant Sci.* 163, 1037–1046. https://doi.org/10.1016/S0168-9452(02)00278-9

Sairam RK, Saxena DC. (2000) Oxidative stress and Antioxidants in wheat genotypes: Possible mechanism of water stress tolerance. *J. Agronomy Crop Sci.* 184, 55–61.

Sakamoto A, Murata N. (2002) The role of glycine betaine in the protection of plants from stress: Clues from transgenic plants. *Plant Cell Environ.* 25, 163–171. https://doi.org/10.1046/j.0016-8025.2001.00790.x

Sales CRG, Marchiori PER, Machado RS, Fontenele AV, Machado EC, Silveira JAG, Riveiro RV. (2015) Photosynthetic and antioxidant responses to drought during sugarcane ripening. *Photosynthetica* 53(4), 547–554.

Sales CRG, Ribeiro RV, Silveira JAG, Machado EC, Martins MO, Lagoa AMMA. (2013) Superoxide dismutase and ascorbate peroxidase improve the recovery of photosynthesis in sugarcane plants subjected to water deficit and low substrate temperature. *Plant Physiol. Biochem.* 73, 326–336.

Saliendra NZ, Meinzer FC. (1992) Genotypic, developmental and drought-induced differences in root hydraulic conductance of contrasting sugarcane cultivars. *J. Exp. Bot.* 43, 1209– 1217.

Santos CM, Silva MA. (2015) Physiological and biochemical responses of sugarcane to oxidative stress induced by water deficit and paraquat. *Acta Physiol. Plant.* 37, 1–14.

Scandalios JG. (1993) Oxygen stress and superoxide dismutases. *Plant Physiol.* 101, 7–12. https://doi.org/10.1104/pp.101.1.7

Serraj R, Krishnamurthy L, Kashiwagi J, Kumar J, Chandra S, Crouch JH. (2004) Variation in root traits of chickpea (*Cicer arietinum* L.) grown under terminal drought. *Field Crop Res.* 88, 115–127.

Shao HB, Chu LY, Jaleel CA, Zhao CX. (2008) Water deficit stress induced anatomical changes in higher plants. *C. R. Biol.* 331, 215–225.

Sharma A, Singh RK, Singh P, Vaishnav A, Guo DJ, Verma KK, Li DP, Song XP, Malviya MK, Khan N, Lakshmanan P, Li YR. (2021) Insights into the bacterial and nitric oxide-induced salt tolerance in sugarcane and their growth-promoting abilities. *Microorganisms* 9, 2203. https://doi.org/10.3390/microorganisms9112203.

Sharma A, Song X-P , Singh RK, Vaishnav A, Gupta S, Singh P, Guo D-J, Verma KK, Li Y-R. (2022) Impact of carbendazim on cellular growth, defence system and plant growth promoting traits of *Priestia megaterium* ANCB-12 isolated from sugarcane rhizosphere. *Front. Microbiol.* 13, 1005942. https://doi.org/10.3389/fmicb.2022.1005942

Sharma P, Jha AB, Dubey RS, Pessarakli M. (2012) Reactive oxygen species, oxidative damage, and antioxidative defense mechanism in plants under stressful conditions. *J. Bot.* 26. https://doi.org/10.1155/2012/217037

Shrivastava AK, Pathak AD, Misra V, Srivastava S, Swapna M, Shukla SP. (2017) Sugarcane Crop: It's tolerance towards abiotic stresses. In: Minas PS, Rane J, Pasala RK. (Eds.), *Abiotic Stress Management for Resilient Agriculture*. Springer Nature Singapore Pte Ltd. Publications, pp. 375–397. https://doi.org/10.1007/978-981-10-5744-1_17

Shrivastava AK, Srivastava MK. (2006) *Abiotic Stresses Affecting Sugarcane: Sustaining Productivity*. International Book distributing Co., Lucknow: India, pp. 1–332.

Shulaev V, Oliver DJ. (2006) Metabolic and proteomic markers for oxidative stress. New tools for reactive oxygen species research. *Plant Physiol.* 141, 367–372. https://doi.org/10.1104/pp.106.077925

Silva MA, Jifon JL, Santos CM, Jadoski CJ, Silva JAG. (2013) Photosynthetic capacity and water use efficiency in sugarcane genotypes subject to water deficit during early growth phase. *Braz. Arch. Biol. Technol.* 56, 735–748.

Silva MDA, Jifon JL, de Silva JAG, Sharma V. (2007) Use of physiological parameters as fast tools to screen for drought tolerance in sugarcane. *Braz. J. Plant Physiol.* 19, 193–201.

Silva MDA, Jifon JL, Sharma V, Silva JAGD, Caputo MM, Domaj MB, Guimaraes ER, Ferro MIT. (2011) Use of physiological parameters in screening drought tolerance in sugarcane genotypes. *Sugar Tech.* 13(3), 191–197.

Silva MDA, Silva JA, Enciso J, Sharma V, Jifon J. (2008) Yield components as indicators of drought tolerance of sugarcane. *Sci. Agricola* 65, 620–627.

Silva PP, Soares L, Costa JG, Viana LS, Andrade JCF, Goncalves ER, Santos JM, Barbosa GVS, Nascimento VX, Todaro AR, et al. (2012) Path analysis for selection of drought tolerant sugarcane genotypes through physiological components. *Ind. Crops Prod.* 2012, 37, 11–19.

Singh P, Singh RK, Li HB, Guo DJ, Sharma A, Verma KK, Solanki MK, Upadhyay SK, Lakshmanan P, Yang LT, Li YR. (2023) Nitrogen fixation and phytohormone stimulation of sugarcane plant through plant growth promoting diazotrophic *Pseudomonas*. *Biotechnol. Genetic Eng. Rev.* 40(1), 15–35. https://doi.org/10.1080/02648725.2023.2177814

Singh P, Singh RK, Song QQ, Li HB, Guo DJ, Malviya MK, Verma KK, Song XP, Lakshmanan P, Yang LT, Li YR. (2021) Comparative analysis of protein and differential responses of defense-related gene and enzyme activity reveals the long-term molecular responses of sugarcane inoculated with *Sporisorium scitamineum*. *J. Plant Interact.* 16(1), 12–29.

Singh RK, Singh P, Li HB, Song QQ, Guo DJ, Solanki MK, Verma KK, Malviya MK, Song XP, Lakshmanan P, Yang LT, Li YR. (2020) Diversity of nitrogen-fixing rhizobacteria associated with sugarcane: A comprehensive study of plant-microbe interactions for growth enhancement in *Saccharum* spp. *BMC Plant Biol.* 20, 220. https://doi.org/10.1186/s12870-020-02400-9.

Sivaramakrishnan S, Patell VZ, Flower DJ, Peacock JM. (1988) Proline accumulation and nitrate reductase activity in contrasting sorghum lines during mid-season drought stress. *Physiol. Plant.* 74, 418–426

Skorzynska-Polit E. (2007) Lipid peroxidation in plant cells, its physiological role and changes under heavy metal stress. *Acta Soc. Bot. Pol.* 76, 49–54. https://doi.org/10.5586/asbp.2007.006

Smit MA, Singels A. (2006) The response of sugarcane canopy development to water stress. *Field Crops Res*. 98, 91–97.

Smith D, Inman-Bamber N, Thorburn P. (2005) Growth and function of the sugarcane root system. *Field Crop Res*. 92, 169–183.

Solanki MK, Verma KK, Dastogeer KMG, Mora-Poblete F, Mundra S. (2023) Microbial resilience in plant nutrient management towards sustainable farming. *Front. Microbiol*. 14, 1280811. https://doi.org/10.3389/fmicb.2023.1280811

Song X-P, Mo F, Verma KK, Wei J, Zhang X, Yang L, Li YR. (2019) Effect of sugarcane smut (*Ustilago scitaminea* Syd.) on ultrastructure and biochemical indices of sugarcane. *Biomed J. Sci. Tech. Res*. 17(1), 002950. https://doi.org/10.26717/BJSTR.2019.17.002950.

Song XP, Mo F, Yan M, Zhang X, Zhang B, Huang X, Huang D, Pan Y, Verma KK, Li YR. (2022) Effect of smut infection on the photosynthetic physiological characteristics and related defense enzymes of sugarcane. *Life* 12, 1201. https://doi.org/10.3390/life12081201

Song XP, Verma KK, Tian DD, Zhang XQ, Liang YJ, Huang X, Li CN, Li YR. (2021) Exploration of silicon functions to integrate with biotic stress tolerance and crop improvement. *Biol. Res*. 54, 19. https://doi.org/10.1186/s40659-021-00344-4

Songsri P, Jogloy S, Holbrook CC, Kesmala T, Vorasoot N, Akkasaeng C, Patanothai A. (2009) Association of root, specific leaf area and SPAD chlorophyll meter reading to water use efficiency of peanut under different available soil water. *Agric Water Manage*. 96, 790–798.

Spiteller G. (2003) The relationship between changes in the cell wall, lipid peroxidation, proliferation, senescence and cell death. *Physiol. Plant*. 119, 5–18. https://doi.org/10.1034/j.1399-3054.2003.00097.x

Sugiharto B. (2018) Biotechnology of drought-tolerant sugarcane. In: de Oliveira AB. (Ed.), *Sugarcane – Technology and Research*. InTech: London. https://doi.org/10.5772/intechopen.72436

Sussich F, Skopec C, Brady J, Cesàro A. (2001) Reversible dehydration of trehalose and anhydrobiosis: From solution state to an exotic crystal? *Carbohydr. Res*. 334, 165–176. https://doi.org/10.1016/S0008-6215(01)00189-6.

Taratima W, Ritmaha T, Jongrungklang N, Maneerattanarungroj P, Kunpratum N. (2020) Effect of stress on the leaf anatomy of sugarcane cultivars with different drought tolerance (*Saccharum officinarum*, Poaceae). *Rev. Biol. Trop*. 68(4), 1159.

Tezara W, Driscoll S, Lawlor DW. (2008) Partitioning of photosynthetic electron flow between CO_2 assimilation and O_2 reduction in sunflower plants under water deficit. *Photosynthetica* 46(1), 127–134.

Thind SK. (1996) Effect of drought on leaf epidermal structure of sugarcane. *Sugarcane* 1, 15–17.

Thirugnanasambandam PP, Hoang NV, Furtado A, Botha FC, Henry RJ. (2017) Association of variation in the sugarcane transcriptome with sugar content. *BMC Genom*. 18, 909.

Tian D, Qin L, Verma KK, Wei L, Li J, Li B, Zhou W, He Z, Wei D, Huang S, Long S, Huang Q, Li C, Wei S. (2023) Transcriptomic and metabolomic differences between banana varieties which are resistant or susceptible to *Fusarium wilt*. *PeerJ* 11, e16549. https://doi.org/10.7717/peerj.16549

Turner NC, O'Toole JC, Cruz RT, Namuco OS, Ahmad S. (1986) Responses of seven diverse rice cultivars to water deficits. I. Stress development, canopy temperature, leaf rolling and growth. *Field Crops Res.* 13, 257–271. https://doi.org/10.1016/0378-4290(86)90027-4

Vadez V, Grondinm A, Chenu K, Henry A, Laplaze L, Millet EJ, Carminati A. (2024) Crop traits and production under drought. *Nat. Rev. Earth Environ.* 5, 211–225.

Vallino M, Greppi D, Novero M, Bonfante P, Lupotto E. (2009) Rice root colonization by mycorrhizal and endophytic fungi in aerobic soil. *Ann. Appl. Biol.* 154, 195–204.

Valliyodan B, Nguyen HT. (2006) Understanding regulatory networks and engineering for enhanced drought tolerance in plants. *Curr. Opin. Plant Biol.* 9, 1–7.

Van Dillewijn C. (1952) *Botany of Sugarcane*. Chronica Botanica: Waltham.

Vanková R, Dobrá J, Štorchová H. (2012) Recovery from drought stress in tobacco: An active process associated with the reversal of senescence in some plant parts and the sacrifice of others. *Plant Signal Behav.* 7, 19–21. https://doi.org/10.4161/psb.7.1.18375

Vantini JS, Dedemo GC, Jovino Gimenez DF, Fonseca LF, Tezza RID, Mutton MA, Ferro JA, Ferro MIT. (2015) Differential gene expression in drought-tolerant sugarcane roots. *Genet. Mol. Res.* 14, 7196–7207. https://doi.org/10.4238/2015.June.29.13

Vendruscolo ECG, Schuster I, Pileggi M, Scapim CA, Molinari HBC, Marur CJ, Vieira LGE. (2007) Stress-induced synthesis of proline confers tolerance to water deficit in transgenic wheat. *J. Plant Physiol.* 164(10), 1367–1376.

Verbruggen N, Hermans C. (2008) Proline accumulation in plants: A review. *Amino Acids* 35(4), 753–759.

Verma KK, Joshi A, Song X-P, Liang Q, Xu L, Huang H-r, Wu K-C, Seth CS, Arora J and Li Y-R. (2024a) Regulatory mechanisms of plant rhizobacteria on plants to the adaptation of adverse agroclimatic variables. *Front. Plant Sci.* 15, 1377793. https://doi.org/10.3389/fpls.2024.1377793

Verma KK, Joshi A, Song X-P, Singh S, Kumari A, Arora J, Singh SK, Solanki MK, Seth CS, Li Y-R. (2024b) Synergistic interactions of nanoparticles and plant growth promoting rhizobacteria enhancing soil-plant systems: A multigenerational perspective. *Front. Plant Sci.* 15, 1376214. https://doi.org/10.3389/fpls.2024.1376214

Verma KK, Li DM, Singh M, Rajput VD, Malviya MK, Minkina T, Singh RK, Singh P, Song XP, Li YR. (2020a) Interactive role of silicon and plant-rhizobacteria mitigating abiotic stresses: A new approach for sustainable agriculture and climate change. *Plants* 9, 1055. https://doi.org/10.3390/plants9091055.

Verma KK, Rajput VD, Singh M, Verma CL, Minkina T. (2022a) Influence of waterlogging on plant physiology and molecular biology. In: *Understanding Abiotic Stresses*. Nova Science Publishers, Inc.: New York.

Verma KK, Siddiqui N, Singh M. (2022b) Understanding the plant physiological and molecular adaptation strategies to soil salinity. In: *Understanding Abiotic Stresses*. Nova Science Publishers, Inc.: New York. https://doi.org/10.52305/UBUJ4024

Verma KK, Singh P, Song X-P, Malviya MK, Singh RK, Chen G-L, Solomon S, Li YR. (2020b) Mitigating climate change for sugarcane improvement: Role of silicon in alleviating abiotic stresses. *Sugar Tech.* 22(5), 741–749. https://doi.org/10.1007/s12355-020-00831-0.

Verma KK, Singh RK, Song QQ, Singh P, Zhang B-Q, Song X-P, Chen G-L, Li YR. (2019a) Silicon alleviates drought stress of sugarcane plants by improving antioxidant responses. *Biomed. J. Sci. Tech. Res.* 002957. https://doi.org/10.26717/BJSTR.2019.17.002957

Verma KK, Song X, Degu HD, Guo D, Joshi A, Huang H, Xu L, Singh M, Huang D, Rajput VD, Li YR. (2023a) Recent advances in nitrogen and nano-nitrogen fertilizers for sustainable crop production: A mini-review. *Chem. Biol. Technol. Agric.* 10, 1–14. https://doi.org/10.1186/s40538-023-00488-3
Verma KK, Song X, Verma CL, Huang H, Singh M, Xu L, Wu J, Khan ZH, Zhang B, Li Y. (2023b) Mathematical modeling of climate and fluoride effects on sugarcane photosynthesis with silicon nanoparticles. *Plant Physiol. Biochem.* 204, 108089. https://doi.org/10.1016/j.plaphy.2023.108089
Verma KK, Song XP, Budeguer F, Nikpay A, Enrique R, Singh M, Zhang B, Wu, JM, Li YR. (2022c) Genetic engineering: An efficient approach to mitigating biotic and abiotic stresses in sugarcane cultivation. *Plant Signal. Behav.* 17(1), e2108253.
Verma KK, Song XP, Joshi A, Rajput VD, Singh M, Sharma A, Singh RK, Li DM, Arora J, Minkina T, Li YR. (2022d) Nanofertilizer possibilities for healthy soil, water and food in future: An overview. *Front. Plant Sci.* 13, 865048. https://doi.org/10.3389/fpls.2022.865048
Verma KK, Song XP, Joshi A, Tian DD, Rajput VD, Singh M, Arora J, Minkina TM, Li YR. (2022e) Recent trends in nano-fertilizer for advancing sustainable agriculture under the era of climate change to ensure food security for future generations. *Nanomaterials* 12, 173. *https://doi.org/10.3390/nano12010173.*
Verma KK, Song XP, Li DM, Singh M, Wu JM, Singh RK, Sharma A, Zhang BQ, Li YR. (2022f) Silicon and soil microorganisms improves rhizospheric soil health with bacterial community, plant growth, performance and yield. *Plant Signal. Behav.* 17(1), e2104004.
Verma KK, Song XP, Lin B, Guo DJ, Singh M, Rajput VD, Singh RK, Singh P, Sharma A, Malviya MK, Chen GL, Li YR. (2022g) Silicon induced drought tolerance in crop plants: Physiological adaptation strategies. *Silicon* 14(6), 2473–2487. https://doi.org/10.1007/s12633-021-01071-x
Verma KK, Song XP, Rajput VD, Boldyreva V, Zhang BQ, Minkina T, Li YR. (2022h) Morpho-physiological, biochemical and ultrastructural modifications on sugarcane to prolonged water deficit. In: Verma KK. et al. (Eds.), *Agro-industrial Perspectives on Sugarcane Production Under Environmental Stress*. https://doi.org/10.1007/978-981-19-3955-6_11.
Verma KK, Song XP, Rajput VD, Solomon S, Li YR, Rao GP. (2022i) *Agro-industrial Perspectives on Sugarcane Production Under Environmental Stress*. Springer Nature: Singapore, https://doi.org/10.1007/978-981-19-3955-6
Verma KK, Song XP, Singh M, Tian DD, Rajput VD, Minkina T, Li YR. (2023c) Association of silicon and soil microorganisms induces stress mitigation, increasing plant productivity. In: de Mello Prado R. (Ed.), *Benefits of Silicon in the Nutrition of Plants*, pp. 299–328. https://doi.org/10.1007/978-3-031-26673-7_17
Verma KK, Song XP, Tian DD, Guo DJ, Chen ZL, Zhong CS, Nikpay A, Singh M, Rajput VD, Singh RK, Minkina T, Li YR. (2021a) Influence of silicon on biocontrol strategies to manage biotic stress for crop protection, performance and improvement. *Plants* 10, 2163. https://doi.org/10.3390/plants10102163
Verma KK, Song XP, Tian DD, Singh M, Verma CL, Rajput VD, Singh RK, Sharma A, Singh P, Malviya MK, Li YR. (2021b) Investigation of defensive role of silicon during drought stress induced by irrigation capacity in sugarcane: Physiological and biochemical characteristics. *ACS Omega* 6, 19811–19821.

Verma KK, Song XP, Verma CL, Chen ZL, Rajput VD, Wu KC, Liao F, Chen GL, Li YR. (2021c) Functional relationship between photosynthetic leaf gas exchange in response to silicon application and water stress mitigation in sugarcane. *Biol. Res.* 54, 15. https://doi.org/10.1186/s40659-021-00338-2

Verma KK, Song XP, Verma CL, Malviya MK, Guo DJ, Rajput VD, Sharma A, Wei KJ, Chen GL, Solomon S, Li YR. (2021d) Predication of photosynthetic leaf gas exchange of sugarcane (*Saccharum* spp.) leaves in response to leaf positions to foliar spray of potassium salt of active phosphorus under limited water irrigation. *ACS Omega* 6, 2396–2409.

Verma KK, Song X-P, Xu L, Huang H-R, Liang Q, Seth CS, Li Y-R. (2023d) Nano-microplastic and agro-ecosystems: A mini-review. *Front. Plant Sci.* 14, 1283852. https://doi.org/10.3389/fpls.2023.1283852

Verma KK, Song XP, Yadav G, Degu HD, Parvaiz A, Singh M, Huang HR, Mustafa G, Xu L, Li YR. (2022j) Impact of agroclimatic variables on proteo-genomics in sugarcane (*Saccharum* spp.) plant productivity. *ACS Omega* 7, 22997–23008.

Verma KK, Song XP, Zeng Y, Guo DJ, Sing M, Rajput VD, Malviya MK, Wei KJ, Sharma A, Li DP, Chen GL, Li YR. (2021e) Foliar application of silicon boosts growth, photosynthetic leaf gas exchange, antioxidative response and resistance to limited water irrigation in sugarcane (*Saccharum officinarum* L.). *Plant Physiol. Biochem.* 166, 582–592.

Verma KK, Song XP,Zeng Y, Li DM, Guo DJ, Rajput VD, Chen GL, Barakhov A, Minkina TM, Li YR. (2020c) Characteristics and correlation of leaf stomata and its relationship with photosynthesis on *Saccharum* spp. under different irrigation and silicon application. *ACS Omega* 5, 24145–24153.

Verma KK, Vatsal S, Singh M, Rajput VD, Minkina T. (2022k) Coping with water stress: Effects and adaptive mechanisms to plants. In: *Understanding Abiotic Stresses*. Nova Science Publishers, Inc.: New York.

Verma KK, Wu KC, Verma CL, Li DM, Malviya MK, Singh RK, Singh P, Chen GL, Song XP, Li YR. (2020d) Developing mathematical model for diurnal dynamics on photosynthetic responses in sugarcane responsive to different irrigation and silicon application. *PeerJ* 8, e10154. https://doi.org/10.7717/peerj.10154.

Verma KK, Liu X-H, Wu K-C, Singh RK, Song QQ, Malviya MK, Song X-P, Singh P, Verma CL, Li YR. (2020e) The impact of silicon on photosynthetic and biochemical responses of sugarcane under different soil moisture levels. *Silicon* 12, 1355–1367.

Verma KK, Wu K-C, Singh P, Malviya MK, Singh RK, Song X-P, Li YR. (2019b) The protective role of silicon in sugarcane under water stress: Photosynthesis and antioxidant enzymes. *Biomed J. Sci. Tech. Res.* 002685. https://doi.org/0.26717/BJSTR.2019.15.002685

Vinocur B, Altman A. (2005) Recent advances in engineering plant tolerance to abiotic stress: Achievements and limitations. *Curr. Opinion Biotechnol.* 16, 1–10.

Wahid A. (2004) Analysis of toxic and osmotic effects of sodium chloride on leaf growth and economic yield of sugarcane. *Bot. Bull. Acad. Sin.* 45, 133–141.

Wang GP, Li F, Zhang J, Zhao MR, Hui Z, Wang W. (2010) Overaccumulation of glycine betaine enhances tolerance of the photosynthetic apparatus to drought and heat stress in wheat. *Photosynthetica* 48, 30–41.

Wassem M, Ali A, Tahir M, Nadeem MA, Ayub M, Tanveer A, Ahmad R, Hussain M. (2011) Mechanism of drought tolerance in plant and its management through different methods. *Continental J. Agric. Sci.* 5(1), 10–25.

Wasson AP, Richards RA, Chatrath R, Misra SC, Prasad SV, Rebetzke GJ, et al. (2012) Traits and selection strategies to improve root systems and water uptake in water-limited wheat crops. *J. Exp. Bot.* 63, 3485–3498. https://doi.org/10.1093/jxb/eru250

Welbaum GE, Meinzer FC. (1990) Compartmentation of solutes and water in developing sugarcane stalk tissue. *Plant Physiol.* 93, 1147–1153. https://doi.org/10.1104/pp.93.3.1147

Wingler A. (2002) The function of trehalose biosynthesis in plants. *Phytochemistry* 60(5), 437–440.

Wingler A, Fritzius T, Wiemken A, Boller T, Aeschbacher RA. (2000) Trehalose induces the ADP-glucose pyrophosphorylase gene, ApL3, and starch synthesis in Arabidopsis. *Plant Physiol.* 124, 105–114.

Wu K-C, Huang C-M, Verma KK, Deng Z-N, Huang H-R, Pang T, Cao H-Q, Luo H-B, Jiang S-L, Xu L. (2022) Transcriptomic responses of *Saccharum spontaneum* roots in response to polyethylene glycol–6000 stimulated drought stress. *Front. Plant Sci.* 13, 992755. https://doi.org/10.3389/fpls.2022.992755

Xu L, Deng Z-N, Wu K-C, Malviya MK, Solanki MK, Verma KK, Pang T, Li Y-J, Liu X-Y, Kashyap BK, Dessoky ES, Wang W-Z and Huang H-R. (2022) Transcriptome analysis reveals a gene expression pattern that contributes to sugarcane bud propagation induced by Indole-3-Butyric acid. *Front. Plant Sci.* 13, 852886. https://doi.org/10.3389/fpls.2022.852886

Yadav UP, Ivakov A, Feil R, Duan GY, Walther D, Giavalisco P, Piques M, Carillo P, Hubberten HM, Stitt M, Lunn JE. (2014) The sucrose-trehalose 6-phosphate (Tre6P) nexus: Specificity and mechanisms of sucrose signalling by Tre6P. *J. Exp. Bot.* 65(4), 1051–1068.

Youzhong L, Zujian Z, Xingyou G, Jianchang Y, Qingsen Z. (2004) Physiological and ecological effects of crimpy leaf character in rice (*Oryza sativa* L.) II. Photosynthetic character, dry mass production, and yield forming. *Zuo Wuxue Bao* 30(9), 883.

Yuan S, Li Y, Peng S. (2015) Leaf lateral asymmetry in morphological and physiological traits of rice plant. *PLoS One* 10(6), 0129832.

Zang U, Goisser M, Häberle KH, Matyssek R, Matzner E, Borken W. (2014) Effects of drought stress on photosynthesis, rhizosphere respiration, and fine-root characteristics of beech saplings: A rhizotron field study. *J. Plant Nutr. Soil Sci.* 177(2), 168–177.

Zhang SZ, Yang BP, Feng CL, Chen RK, Luo JP, Cai WW, Liu FH. (2006) Expression of the Grifola frondose trehalose synthase gene and improvement of drought tolerance in sugarcane (*Saccharum officinarum* L.). *J. Integrat. Plant Biol.* 48(4), 453–459.

Zhao D, Barry G, Comstock JC. (2010) Sugarcane response to water-deficit stress during early growth on organic and sand soils. *Amer. J. Agric. Biol. Sci.* 5(3), 403–414.

Zheng F-J, Lin B, Yang Y-X, Fang X-C, Verma KK, Chen G-L. (2023) Efficacy and functionality of sugarcane original vinegar on mice. *Front. Microbiol.* 14, 1224666. https://doi.org/10.3389/fmicb.2023.1224666

Zhu JK. (2001) Plant salt tolerance. *Trends Plant Sci.* 6, 66–71. 10.1016/S1360-1385(00)01838-0

Zivcak M, Brestic M, Balatova Z, Drevenakova P, Olsovska K, Kalaji KM, Yang X, Allakhverdiev SI. (2013) Photosynthetic electron transport and specific photoprotective responses in wheat leaves under drought stress. *Photosynth. Res.* 117, 529–546.

Zou LP, Sun XH, Zhang ZG, Liu P, Wu JX, Tian CJ, Qiu JL, Lu TG. (2011) Leaf rolling controlled by the homeodomain leucine zipper class IV gene Roc5 in rice. *Plant Physiol.* 156(3), 1602.

CHAPTER 4

Integration of Sugarcane Production Strategies for Betterment of Sugarcane Production System in India

LALAN SHARMA*, V. P. JAISWAL, S. K. SHUKLA, and DINESH SINGH

ICAR—Indian Institute of Sugarcane Research, Lucknow, Uttar Pradesh, India

*Corresponding author

ABSTRACT

Sugarcane is a commercial crop that grows in tropical and subtropical regions around the world. Over 100 countries produce sugarcane as a major crop for sugar production, accounting for around 80% of the world's sugar production. Sugarcane not only grows best in hot and humid conditions, ideal climate for better sprouting and germination but also needs fairly dry sunshine and hot weather for higher tiller formation. Advanced technologies such as improved sugarcane varieties, trench method of planting, optimized crop geometry, use drip irrigation, integrated nutrients management, integrated weed management, and trash mulching in ratoon crop, crop diversification, and mechanization have been found effective in improving crop growth and development and sugar recovery percentage. In this book chapter, we will discuss these advanced technologies/methods for increasing sugarcane yield and sugar recovery, reducing cultivation costs, improving productivity, and improving soil health. The adoption of these advanced sugarcane production technologies during sugarcane cultivation will undoubtedly boost sugarcane production and productivity and sustainable sugarcane production as well.

Sugarcane Production: Environmental Impacts and Socioeconomic Issues.
Krishan K. Verma, Xiu-Peng Song, Lin Xu, Rajan Bhatt, and Yang-Rui Li (Eds.)

DOI: 10.1201/9781779641137-4

4.1 INTRODUCTION

Sugarcane (*Sachharum officinarum*) is a major commercial crop in India, grown over an area of approximately 5.0 Mha with an average yield of 82 t/ha, producing around 400 MT. Recent advancements in sugarcane productivity have been made possible through the development and adoption of improved varieties, crop production and protection technologies, and effective dissemination through research and development agencies (Verma et al., 2024a, 2024b). Uttar Pradesh and Maharashtra are the leading states contributing to over 69% of the country's sugarcane production. They occupy about 56% of the sugarcane area at the national level (Shukla et al., 2017, 2019). The achievable possible yield for sugarcane is about 280 t/ha/year. Experimental plots in Brazil have demonstrated yields of 236–280 t of cane/ha (Bogden, 1977). In some cases, sugarcane yield of 250–300 t/ha has been recorded on fields of progressive farmers in tropical states like Maharashtra (Katoch, 2017). However, to improve sugarcane productivity beyond 200 t/ha, selection of variety, availability of irrigation water, plant population management, tillage practices, and nutrient and weed management strategies are key components (Katoch, 2017; Verma et al., 2019a, 2019b; Song et al., 2019, 2021, 2022). The sugarcane yield can be increased by adopting improved varieties developed for the region (Ram et al., 2017; Verma et al., Verma et al., 2021a, 2021b). The genetic makeup of the crop/variety governs the physiological and biochemical enhancements in biomass accumulation, trench method of planting (Yadav and Kumar, 2005; Verma et al., 2021c, 2021d), organic content integrated nutrient management, and other sugarcane production technologies. These have been identified and validated for improving quantitative (yield) as well as qualitative (sugar content) traits. In this chapter, we will discuss integration of advanced technologies/methods for increasing sugarcane yield and sugar recovery, reducing cultivation costs, improving productivity, and improving soil health.

4.2 DEVELOPMENT OF IMPROVED SUGARCANE VARIETIES

Improved sugarcane varieties play a crucial role in enhancing sugarcane production and productivity. These varieties can produce higher yields and are more resistant/tolerant to biotic and abiotic factors (Verma et al., 2021e, 2022a, 2023a, 2023b). In India, numerous research organizations, institutes, universities, and research stations are contributing to the development of high-yielding and high-sugar varieties that can withstand biotic and abiotic stresses (Krishna et al., 2022; Huang et al., 2024). The All India Coordinated Research Project (AICRP) on sugarcane operates under the umbrella of the Indian Institute of Sugarcane Research (ICAR), which coordinates sugarcane research at the national level. With 22 regular and 15 voluntary centers in different sugarcane-growing states of India, AICRP on Sugarcane coordinates sugarcane research on crop improvement, production, and protection technologies across the country. A number of sugarcane varieties have been identified for high-yield potential, high sugar recovery, resistance to biotic stresses, and tolerance to drought and flood conditions. Some of the most prominent sugarcane varieties in India are as follows: Co 86032 (contributes >70% sugarcane area in the Peninsular zone), Co 0238 (covered >60% sugarcane area in the North Western Zone and Central Zone), and CoLk 94184 (covered around 28%), BO 91 (around 20%), Co 6907 (around 20%), and Co Bln 9104 (around 18%) sugarcane area in the Eastern Zone of the country (Shukla et al., 2017). These varieties are widely adopted and occupy a significant portion of the sugarcane area and production in the country (Table 4.1).

4.2.1 *TRENCH METHOD OF PLANTING*

The trench method of planting sugarcane is typically suggested in coastal regions that are susceptible to climatic conditions. However, it is now also being recommended and adopted in subtropical regions of India. During the rainy season, strong winds can cause the sugarcane crop to lodge. To safeguard the crop from these winds, trenches are dug using a ridger-mounted tractor at a distance of 90–120 cm. The trenches should be about 30 cm deep, and the setts in the trenches are planted end-to-end. Once planted, the trenches are filled with loose soil. Paired Row Trench Planting (30:150 cm) is a modified version of the trench planting method, and it is being used in various parts of subtropical India. This method requires about 12–15 t/ha seed cane through the planting of three bud setts in paired row planting in trenches. Sugarcane yields >100 t/ha can be easily obtained through the adoption of this planting method (Shukla et al., 2017, 2021). However, in Maharashtra and other tropical regions where individual cane weight is

TABLE 4.1 List of Sugarcane Varieties Identified and Released for Sugarcane Cultivation in India

Sugarcane Cultivation Zone	Sugarcane Varieties Identified and Released
Peninsular Zone	Co 6907, CoC 671, MS 7455, Co 7219, Co 7318, Co 62175, CoM 7125, Co 1295, MS 7110, CoJN 86141, Co 7527, Co 8021, Co 85004, Co 87044, Co 86032, Co 87025, CoM 88121 (CoM 7714), Co 8371, Co 91010, Co 94008, Co 99004, Co 99006, Co 2001–13, Co 2001–15, Co 0314, Co 0218, CoM 0265, Co 0403, CoSnk 05103, CoSnk 05104, Co 06022, CoN 05071, Co 06027, PI 07131, Co 09004, Co 10026, Co 12009, VSI 12121, Co 13013, MS 13081, Co 14012, MS 14082, Co 14005, and Co 11015
North West Zone	Co 7717, CoJ 64, CoJ 75, Co 6304, CoLk 7701, CoS 771, CoS 802, CoS 767, CoLk 7901, CoLk 8001, CoLk 8102, CoPant 84211, CoPant 90223, CoH 92201, CoS 91230, CoS 95255, CoPant 93227, CoS 94270, CoPant 97222, CoS 96268, Co 98014, CoH 119, CoJ 20193 CoJ 89), CoS 96275, Co 0118, Co 0238, Co 0239, Co 0124, Co 0237, CoH 128, CoPK 05191, Co 05011, Co 05009, CoPant 05224, CoLk 07201, Co 06034, CoPb 08212, C0 09022, CoLk 09204, CoLk 11206, CoLk 11203, Co 12029, CoS 12232, Co 13035, CoPb14185, CoLk 14204, Co 15023, CoLk 15201, CoLk 15207, CoLk 14201 CoLk 15206, Co 16030, Co 17018, CoPb 17215, CoLk 16202, CoS 16233, Co 17018, CoPb 17215, and CoLk 16202
North Central & North Eastern Zones	BO 90, Co 7201, S101/72, BO 100, BO 91, BO 109, CoS 7918, Co 87263, Co 87268, BO 120, CoBln 9605, CoP 9103, BO 128, Co 89029, CoSe 95422, CoSe 92423, CoSe 96234, CoSe 96436, CoLk 94184, Co 0232, Co 0233, CoSe 01421, BO 146, CoSe 05451, CoBln 04174, CoP 06436 (CoP 2061), CoP 09437, UP 09453, CoLk 12207, CoLk 12209, CoSe 11453, CoLk 15466, CoP 11438, CoLk 16466, CoLk 16470, CoP 17437, CoP 16437, CoLk 16470, CoS 16233, CoP 17437, and CoP 16437
East Coast Zone	Co 7508, CoC 771, Co 86249, CoC 01061, CoOr 03151 CoA 03081, CoA 05322, CoA 05323, Co 06030, CoA 08323, CoC 08336, CoA 11321, CoC 13339, and CoA 17321

higher, the spacing may be increased up to 150 cm in single-row planting. Trench planting prevents the lodging of cane in both plant and ratoon crops. Paired row planting is the best practice for surface/subsurface drip irrigation systems and involves maximum resource use efficiency and higher cane productivity. In subtropical India, this method can increase cane yield by 25–30%. In addition, 55–60% water savings have been reported through the adoption of drip irrigation systems in trenches. Machines for trench planting and laying of drippers/laterals under subsurface drip have been developed at ICAR–Indian Institute of Spices Research (IISR), Lucknow. Adopting the trench method of planting and subsurface drip system can help farmers take 3–4 ratoon crops easily. This method can increase multiple ratooning in the sugarcane system (Figure 4.1).

FIGURE 4.1 Trench and paired row trench method of cane planting.

4.2.2 DRIP IRRIGATION SYSTEM IN SUGARCANE

Sugarcane is a water-loving crop, but water scarcity can limit its growth and development. To address this issue, sugarcane growers in water-deficient areas are recommended to use drip irrigation. In this method, water is supplied directly to the root zone using tubes and emitters placed along the delivery line. This allows for precise control and manipulation of soil moisture, resulting in improved water economy, growth, and crop yield (Chen et al., 2020; Singh et al., 2020; Verma et al., 2020a, 2020b, 2022b, 2023c, 2023d). The uniformity of water application in drip irrigation can reach as high as 95% if the system is well-designed and operated. Another recommended irrigation method is subsurface irrigation, which requires proper care and monitoring of the infrastructure developed (Deng et al., 2023; Guo et al., 2023). This system also reduces water wastage and promotes better crop growth and development. By adopting the drip method, sugarcane growers can save up to 50–60% of irrigation water, fertilizers, and improve nutrient use efficiency (Shukla et al., 2017, 2019, 2020; Joshi et al., 2023). The cost of a drip system for sugarcane is around Rs. 1.25 to 1.5 lakhs/ha. In surface drip systems, laterals can be removed for inter-culture operations. In the subsurface system, the drip system can be used for one plant crop and a minimum of three ratoons without removing dippers and laterals. The government encourages the use of drip irrigation, offering an 80–90% subsidy to farmers to minimize production costs and maximize natural resource use efficiency at farmer's field.

4.2.3 TRASH-MULCHING APPLICATION IN SUGARCANE

Trash mulching is a natural method of managing resources in sugarcane production system. Sugarcane trash contains a significant amount of cellulose (33.3–36.1%), hemicellulose (18.4–28.9%), and lignin (26.1–40.7%), and other components. Using trash mulch not only saves irrigation water but also improves cane yield potential by controlling weed growth. Trash mulching is widely recommended to conserve soil moisture and control weed growth. The sugarcane main crop produces approximately 8–10 tons of trash per hectare. Utilizing trash can help manage agricultural waste, improve soil health, and increase microbial population (Verma et al., 2022c, 2022d, 2022e; Chaudhary et al., 2023; Chen et al., 2023). For better trash utilization, *Trichoderma* can be used at 10 kg/ha (cfu 10^7/g) to speed up the decomposition process (Shukla et al., 2017, 2020; Solanki et al., 2023). In addition to increasing carbon content, trash leachate can add approximately 30 kg of potassium per hectare after the first irrigation in the ratoon crop. Trash mulching can also improve soil organic carbon in the ratoon crop and restrict unproductive tiller growth while encouraging millable cane growth. *Trichoderma* spp. are excellent natural decomposers that influence the rate of organic matter decomposition (Figure 4.2).

FIGURE 4.2 Trash mulching in sugarcane ratoon crop.

4.2.4 INTEGRATED NUTRIENT MANAGEMENT IN SUGARCANE

The importance of microbes in modern agriculture is increasing due to the declining trend in factor productivity and soil organic carbon content. Sustainability in agriculture has always been considered a critical issue to feed the growing population of the world (Kumari et al., 2022, 2023a, 2023b; Verma et al., 2022f, 2022g, 2022h; Singh et al., 2023; Li et al., 2023; Jagadesh et al., 2024). Nowadays, high cropping intensity, minimum use of organics due to their unavailability, and higher use of chemical fertilizers in intensive agriculture are causing imbalanced nutrient use and resulting in deficiencies of several macro and micronutrients. To sustain soil health and improve the response of nutrient application and sugarcane yield (Anas et al., 2021a, 2021b; Verma et al., 2022i, 2022j, 2022k; Guo et al., 2021, 2022), it is recommended to apply well-decomposed farm yard manure (FYM)/compost at a rate of 10–15 t/ha. Soil application of *Azotobacter* or *Azospirillum* at a rate of 10 kg/ha cfu 10^7–10^9 in two equal splits at 30 and 60 days after planting can save up to 25% of nitrogen fertilizers. However, in subtropical India, *Glucoacetobactor diazotrophicus* has been found beneficial. Bio-fertilizers should be mixed with well-decomposed manure or compost and applied as a basal dose. The application of organic manures along with bio-fertilizers not only improves sugarcane production but also sustains soil health (Singh et al., 2021; Huang et al., 2022; Joshi et al., 2022; Tian et al., 2023; Zheng et al., 2023). This application is a low-cost, easy-to-multiply, and eco-friendly input that can cut down the expenditures incurred on inorganic fertilizers, which pollute soil and the environment. Intercropped green manuring of Sesbania in sugarcane has also been found beneficial in improving soil quality and ratoon yield (Malviya et al., 2021, 2022). In areas where sulfur deficiency is reported, the application of 40–60 kg/ha sulfur through gypsum or elemental sulfur has been found beneficial. A foliar spray of 1% ferrous sulfate ($FeSo_4$) 2–3 times during the tillering stage can rectify iron chlorosis in the sugarcane ratoon crop (Shukla et al., 2017, 2019, 2021). The integration of biofertilizers could reduce the doses of chemical fertilizers by up to 25% in sugarcane production in the country.

4.2.5 WEED MANAGEMENT

The growth of sugarcane crop is often hindered by the presence of several weeds. Since the crop is an annual one, it is recommended to perform three manual hoeings during the tillering stage, specifically at 45, 90, and 120 days after planting, to manage the weeds economically. However, due to factors such as lack of manpower and high labor costs, pre-emergence application of weedicides like atrazine @ 2.0 kg ai/ha or metribuzin @ 1.0 kg ai/ha (mixed with 800–1000 l of water per hectare) is also applicable. For sugarcane ratoon crop, manual hoeings at 30, 60, and 90 days after ratooning can be successfully practiced. Additionally, controlling weed growth through trash mulching in between sugarcane rows can also be effective. Applying trash and using 2, 4 D @ 1 kg ai/ha at 30 days can also improve the growth and yield of ratoon crop by controlling broad-leaved weeds (Shukla et al., 2017, 2019; Oliveira et al., 2022; Sharma et al., 2021, 2022; Bhatt et al., 2023a, 2023b). Binding weed is a perpetual problem in sugarcane ratoon particularly under water-logged conditions. Some chemicals have been identified for the management of binding weeds (*Convolvulus arvensis*, *Ipomoea* spp., *Cardiospermum hispidum*, *Coccinia grandis*, *Rhyncosia minima*, and *Merremim arginita*). Application of Atrazine @ 2 kg ai/ha or metribuzine @ 1.25 kg ai/ha as pre-emergence followed by Dicamba @ 350 g ai/ha at 75 DAP is effective for controlling binding weeds in sugarcane (Figure 4.3).

FIGURE 4.3 Binding weeds infested sugarcane crop.

4.2.6 DROUGHT MANAGEMENT IN SUGARCANE

Sugarcane crop thrives well in areas with high moisture and humidity levels, but unfortunately, there has been a decrease in annual rainfall in most sugarcane-growing areas. Although sugarcane is an irrigated crop, it is often grown in a moisture-stressed environment in states such as Maharashtra, Karnataka, Andhra Pradesh, Tamil Nadu, and western Uttar Pradesh. The annual irrigation water requirement for subtropical India is about 1500–1600 mm, while it ranges from 2000 to 3500 mm in tropical India. As a result, sugarcane in the northern region requires 6–8 irrigations, while in the southern region, it requires 30–40 irrigations. To conserve water in sugarcane cultivation, key components include planting methods, trash mulching, weed management, organic application, K nutrition, and the use of drip irrigation. However, under prevailing drought conditions, it is recommended to apply FYM @ 10 t/ha in furrows before planting, soak setts in saturated lime water, mulch trash 60 days after planting, foliar spray urea and KCl @ 2.5% at 90, 105 & 120 days after planting, and apply an additional 60 kg K_2O/ha at 150–180 DAP to improve cane productivity (Shukla et al., 2017, 2019, 2020). In addition, sugarcane planting following the pit/trench method may be adopted in light- and medium-textured soils to minimize adverse effects in drought conditions.

4.2.7 INSECT PESTS AND DISEASE MANAGEMENT IN SUGARCANE

Sugarcane is a crop that is grown annually, and it is prone to being attacked by various insect pests and diseases. With the changing climate, even minor diseases and insect pests can cause significant economic damage to the crop (Verma et al., 2020c, 2020d, 2020e; Wu et al., 2022; Xu et al., 2022). Thus, effective management is essential to improve cane yield and productivity, especially when dealing with biotic stresses. Strategic management technologies need to be employed to ensure a favorable benefit-cost ratio. Management strategies include selecting healthy seed cane, treating the sett with insecticide/fungicides, monitoring the crop fields regularly, and applying cultural operations and protection technologies as needed. Bio-control agents and botanicals are being widely recommended as sustainable means of managing adverse situations while preserving the environment. At the national level, sugarcane varieties are identified based on their tolerance to red rot, wilt, and smut. However, the continuous use of seeds from a single variety increases the risk of occurrence of insect pests and diseases in selected pockets. Therefore, varietal diversification is crucial to control these pests and diseases. Borer pests in sugarcane can be managed, and several pesticides are available in the market to control these natural enemies. Pesticides' efficiency can be improved and environmental pollution minimized by applying them at the right time and dose. Application of fungicide azoxystrobin + difenconazole registered with CIB for sugarcane disease management is recommended for the management of fungal diseases of sugarcane. Pheromone traps technology is advocated for monitoring of early shoot borer, top borer, stalk borer, and internode borer in sugarcane crop. Soil application of chlorantraniliprole 0.4 G @ 22.5 kg/ha at the time of planting and 60 days after planting or chlorantraniliprole 18.5 suspension concentrate @ 375 ml/ha applied as spray after 30 and 60 days of planting was found most effective for the control of sugarcane early shoot borer. It increased cane yield, total cane height, sucrose (% in juice), and CCS (%) in North Western region of the country (Figure 4.4).

FIGURE 4.4 Pheromone traps technology and infected cane by red rot disease.

4.2.8 RATOON MANAGEMENT

Ratooning is a beneficial technique in sugarcane cultivation. It enables early maturation of the crop in just 10 months. To improve the growth of new shoots from the stubble, it is recommended to harvest the plant crop close to the ground level or shave the stubble if harvesting is not possible. After harvesting the plant crop, it is important to dismantle the ridges. Deep cultivation (off-barring) and irrigation in the inter-row spacing can improve the physical condition of the soil and prune the stubble roots. Fertilizer nitrogen should be split into at least three portions and applied at the time of initiation, 45 days, and 75 days of ratoon initiation. For the ratoon crop, it is recommended to use 25% more nitrogen than the plant crop. The ratoon crop should be fertilized with K at the rate of 80 kg K_2O/ha to prevent lodging and improve sugar recovery. The amount of K_2O and phosphorus required can be determined through soil testing. If the gap between setts in the field exceeds 10–15%, gap filling using pregerminated setts is essential. To avoid excessive tillers and lodging in the crop, earthing up should be done in July–August and wrapping/propping should be used.

4.2.9 CROP DIVERSIFICATION: THE INTERCROPPING

Intercropping can be a profitable option for sugarcane growers in India. In subtropical regions, autumn sugarcane planting can be intercropped with mustard, potato, or rajma (French bean). Intercropping pulses such as green gram and black gram, as well as vegetables like onion, with spring sugarcane is also recommended to increase income per hectare. In tropical regions, preseasonal planting of vegetables like onion, knol khol, and cabbage can be profitable. Simultaneous planting of wheat and sugarcane in December, particularly in western Uttar Pradesh, can increase sugarcane productivity. Delayed sugarcane planting until May can still result in increased crop productivity and farmer income with the use of technology. Planting sugarcane and wheat in the Furrow Raised Bed System (FIRBs) can also increase crop yields (Shukla et al., 2017, 2019). Additionally, planting high-value crops like gladiolus, marigold, and others with sugarcane can be profitable, depending on the location and market demand. Selecting short-duration crops and varieties compatible with sugarcane is crucial (Figure 4.5). Common irrigation and field management practices for both the component crops and sugarcane can help reduce the cost of production. The management practices revolve around the component crop until the time of its harvest. After harvesting the component crop, sugarcane management practices and input applications are adopted.

FIGURE 4.5 Sugarcane intercropped with potato and cauliflower crops.

4.2.10 MECHANIZATION IN SUGARCANE

Advancements in sugarcane cultivation have been made possible through the adoption of farm machinery. Sugarcane is an annual crop that requires 10–12 months to mature, and its growth and productivity are directly related to soil tilth. Therefore, preparing good soil tilth for sugarcane cultivation is crucial, and farm machinery plays a vital role in this process. The government and private organizations have developed various farm machines, ranging from simple to advanced labor-saving devices. The ICAR–Indian Institute of Sugarcane Research, Lucknow, has designed and developed several machines that are suitable for different soil and agro-climatic conditions and prevailing agronomic practices. These ICAR–IISR machines include the sugarcane cutter planter, ratoon management device, planter for intercrops, trench planter, crop residue shredder, and sugarcane trash mulchar (Shukla et al., 2017). Farmers can use these machines in their fields, and they can be purchased on a custom-hiring basis through collaborations with sugar mills or cooperative societies.

4.3 CONCLUSION

Sugarcane is a crop that is propagated vegetatively and requires 10–12 months to mature. To improve crop yields of sugarcane plant and ratoon crops, high-yielding, high-sugar varieties have been developed at the national level, in addition to advanced crop production and protection technologies. It is recommended to use the trench method of planting, optimize crop geometry, use drip irrigation, integrated nutrient management, integrated weed management, and trash mulching in ratoon crop production technologies. Crop management issues such as the time of ratoon initiation (February and March), gap filling if it exceeds 10–15%, amelioration of Fe chlorosis, and correcting micronutrients and S deficiency also need to be addressed to increase crop productivity. Integrated schedules for insect pests and disease management have also been developed to reduce the cost of production. The country has achieved cane productivity of around 82 t/ha, but there is still room for improvement through the integration of advanced production and protection technologies.

KEYWORDS

- **sugarcane**
- **production strategies**
- **advanced techniques**
- **management**
- **mechanization**

REFRENCES

Anas M, Verma KK, Haq IU, Naeem M, Li Q, Liao F, Xiang YX, Li YR. (2021a) Characterization of exotic, native and wild-type genotypes of sugarcane (*Saccharum* spp. hybrids) for internal nitrogen use efficiency under different nitrogen levels. *Sugar. Tech*. 23, 1258–1267. https://doi.org/10.1007/s12355-021-01014-1

Anas M, Verma KK, Riaz M, Qiang L, Liao F, Liu Y, Li YR. (2021b) Physio-morphological and biochemical mechanism of nitrogen use efficiency in sugarcane (*Saccharum* spp.) genotypes under different growth stages and nitrogen levels. *J. Plant. Interact.* 16, 332–343.

Bhatt R, Verma KK, Jamwal NS. (2023a) Can paddy-compost eliminate N fertilization for climate-smart sugarcane? *Sugar. Tech.* 25, 1036–1046.https://doi.org/10.1007/s12355-023-01270-3

Bhatt R, Verma KK, Kumar R, Sanghera GS. (2023b) Foliar application of potassium salt of active phosphorus (PSAP) mitigates insect pests and improves yield along with sugarcane quality in response to agroclimatic conditions of Punjab. *Sugar. Tech.* 25(3), 660–669.

Bogden AV. (1977) *Tropical Pasture and Fodder Plants (Tropical Agriculture)*. Longman Group (Far East, Limited). ISBN 978-0582466760.

Chaudhary P, Xu M, Ahamad L, Chaudhary A, Kumar G, Adeleke BS, Verma KK, Hu D-M, Siric I, Kumar P, Popescu SM, Fayssal SA. (2023) Application of synthetic consortia for improvement of soil fertility, pollution remediation, and agricultural productivity: A review. *Agronomy* 13, 643. https://doi.org/10.3390/agronomy13030643.

Chen G-L, Zheng FJ, Lin B, Lao SB, He J, Huang Z, Zeng Y, Sun J, Verma KK. (2020) Phenolic and volatile compounds in the production of sugarcane vinegar. *ACS Omega* 5, 30587–30595.

Chen G-L, Zheng F-J, Lin B, Yang Y-X, Fang X-C, Verma KK, Yang L-F. (2023) Vinegar: A potential source of healthy and functional food with special reference to sugarcane vinegar. *Front. Nutr.* 10, 1145862. https://doi.org/10.3389/fnut.2023.1145862

Deng ZN, Huang HR, Huang CM, Verma KK, Luo HB, Jiang SL, Li YJ, Cao HQ, Wu KC, Xu L. (2023) Short-term limited water irrigation influences the photosynthetic pigments, enzymatic and non-enzymatic activities in *Saccharum spontaneum* L. at vegetative stage. *Sugar. Tech.* 25, 110–118. https://doi.org/10.1007/s12355-022-01175-7

Guo D-J, Li D-P, Chen Z-D, Yang B, Verma KK, Singh RK, Singh P, Khan Q, Sharma A, Qin Y, Zhang B-Q, Song X-P, Li Y-R. (2023) Effect of endophytic diazotroph *Enterobacter roggenkampii* ED5 on nitrogen-metabolism-related microecology in the sugarcane rhizosphere at different nitrogen levels. *Front. Microbiol.* 14, 1132016. https://doi.org/10.3389/fmicb.2023.1132016

Guo D-J, Li D-P, Singh RK, Singh P, Sharma A, Verma KK, Qin Y, Khan Q, Lu Z, Malviya MK, Song X-P, Xing Y-X and Li Y-R. (2021) Differential protein expression analysis of two sugarcane varieties in response to diazotrophic plant growth-promoting endophyte *Enterobacter roggenkampii* ED5. *Front. Plant Sci.* 12, 727741. https://doi.org/10.3389/fpls.2021.727741

Guo DJ, Li DP, Singh RK, Singh P, Verma KK, Sharma A, Qin Y, Khan Q, Song XP, Malviya MK, Xing YX, Li YR. (2022) Comparative transcriptome analysis of two sugarcane varieties in response to diazitrophic plant growth promoting endophyte *Enterobacter roggenkampii* ED5. *J. Plant. Interact.* 17(1), 75–84.

Huang CM, Wu KC, Verma KK, Luo HB, Deng ZN, Shi GY, Cao HQ, Wu XJ, Ye LP, Wei YW, Xu L. (2024) Isolation and expression analysis of cellulose synthase 3 (*Ces3*) genes from sugarcane (*Saccharum officinarum* L.). *Chem. Biol. Technol. Agric.* 11, 71. https://doi.org/10.1186/s40538-024-00597-7

Huang HR, Gao YJ, Malviya MK, Verma KK, Solanki MK, Huang YX, Li X, Deng Y, Yan J, Tang SY, Wang LW, Xu L. (2022) Genetic diversity analysis of sugarcane (*Saccharum* spp. Hybrids) among high-sucrose clones of GT series and commonly used parents by using microsatellite markers in Guangxi, China. *Sugar. Tech.* 24, 397–407. https://doi.org/10.1007/s12355-021-01033-y

Jagadesh M, Dash M, Kumari A, Singh SK, Verma KK, Kumar P, Bhatt R, Sharma SK. (2024) Revealing the hidden world of soil microbes: Metagenomic insights into plant, bacteria, and fungi interactions for sustainable agriculture and ecosystem restoration. *Microbiol. Res.* 285, 127764. https://doi.org/10.1016/j.micres.2024.127764

Joshi A, Verma KK, Rajput VD, Arora J. (2023) Metabolic engineering of microorganisms in advancing biofuel production. In: *Advances in Lignocellulosic Biofuel Production Systems*. pp. 181–202. https://doi.org/10.1016/B978-0-323-91192-4.00016-X

Joshi A, Verma KK, Rajput VD, Minkina T, Arora J. (2022) Recent advances in metabolic engineering of microorganisms for advancing lignocellulose-derived biofuels. *Bioengineered* 13(4), 8135–8163. https://doi.org/10.1080/21655979.2022.2051856.

Katoch M. (2017) https://www.thebetterindia.com/124175/sureshkabade-100-tonnes-sugarcane-in-one-acre/.
Krishna D, Sachan HK, Verma KK. (2022) Abiotic stresses in plants: An overview. In: *Understanding Abiotic Stresses*. Nova Science Publishers, Inc.: New York.
Kumari A, Dash M, Singh SK, Jagadesh M, Mathpal B, Mishra PK, Pandey SK, Verma KK. (2023a) Soil microbes: A natural solution for mitigating the impact of climate change. *Environ. Monit. Assess.* 195, 1436. https://doi.org/10.1007/s10661-023-11988-y
Kumari A, Lakshmi GA, Krishna GK, Patni B, Prakash S, Bhattacharyya M, Singh SK, Verma KK. (2022) Climate change and its impact on crops: A comprehensive investigation for sustainable agriculture. *Agronomy* 12, 3008. https://doi.org/10.3390/agronomy12123008
Kumari A, Singh SK, Mathpal B, Verma KK, Garg VK, Bhattacharyya M, Bhatt R. (2023b) The biosynthesis, mechanism of action, and physiological functions of melatonin in horticultural plants: A review. *Horticulturae* 9, 913. https://doi.org/10.3390/horticulturae9080913
Li A-M, Liao F, Wang M, Chen Z-L, Qin C-X, Huang R-Q, Verma KK, Li Y-R, Que Y-X, Pan Y-Q, et al. (2023) Transcriptomic and proteomic landscape of sugarcane response to biotic and abiotic stressors. *Int. J. Mol. Sci.* 24, 8913. https://doi.org/10.3390/ijms24108913
Malviya MK, Li C-N, Lakshmanan P, Solanki MK, Wang Z, Solanki AC, Nong Q, Verma KK, Singh RK, Singh P, Sharma A, Guo D-J, Dessoky ES, Song X-P, Li Y-R. (2022) High-throughput sequencing-based analysis of rhizosphere and diazotrophic bacterial diversity among wild progenitor and closely related species of sugarcane (*Saccharum* spp. inter-specific hybrids). *Front. Plant Sci.* 13, 829337. *https://doi.org/10.3389/fpls.2022.829337.*
Malviya MK, Solanki MK, Li CN, Wang Z, Zeng Y, Verma KK, Singh RK, Singh P, Huang HR, Yang LT, Song XP, Li YR. (2021) Sugarcane-legume intercropping can enrich the soil microbiome and plant growth. *Front. Sustain. Food Syst.* 5, 606595. https://doi.org/10.3389/fsufs.2021.606595.
Oliveira MA, Verma KK, Bhatt R, Oliveira TBA. (2022) Impact of green and organic fertilizers on soil fertility and sugarcane productivity. In: Verma KK. et al. (Eds.), *Agro-Industrial Perspectives on Sugarcane Production Under Environmental Stress*. https://doi.org/10.1007/978-981-19-3955-6_11.
Ram B, Hemaprabha G, Murali P. (2017) Importance of new varieties of sugarcane suitable for tropical India. *Indian Farming* 67, 16–18.
Sharma A, Singh RK, Singh P, Vaishnav A, Guo DJ, Verma KK, Li DP, Song XP, Malviya MK, Khan N, Lakshmanan P, Li YR. (2021) Insights into the bacterial and nitric oxide-induced salt tolerance in sugarcane and their growth-promoting abilities. *Microorganisms* 9, 2203. https://doi.org/10.3390/microorganisms9112203.
Sharma A, Song X-P, Singh RK, Vaishnav A, Gupta S, Singh P, Guo D-J, Verma KK, Li Y-R. (2022) Impact of carbendazim on cellular growth, defence system and plant growth promoting traits of *Priestia megaterium* ANCB-12 isolated from sugarcane rhizosphere. *Front. Microbiol.* 13, 1005942. https://doi.org/10.3389/fmicb.2022.1005942
Shukla SK, Jaiswal VP, Sharma L, Pathak AD, Singh AK, Gupta C, Awasthi SK, Gaur A, Zubair A, Tiwari R. (2020) Sugarcane yield using minimum tillage technology through

subsoiling: Beneficial impact on soil compaction, carbon conservation and activity of soil enzymes. *Sugar Tech.* 22, 987–1006.

Shukla SK, Sharma L, Awasthi SK, Pathak AD. (2017) *Sugarcane in India: Package of Practices for Different Agroclimatic Zones.* Indian Institute of Sugarcane Research: Lucknow, India, pp. 1–64.

Shukla SK, Sharma L, Jaiswal VP, Pathak AD, Awasthi SK, Zubsir A, Yadav SK. (2021) Identification of appropriate agri-technologies minimizing yield gaps in different sugarcane-growing states of India. *Sugar Tech.* 23, 580–595. https://doi.org/10.1007/s12355-020-00907-x

Shukla SK, Sharma L, Jaiswal VP, Pathak AD, Tiwari R, Awasthi SK, Gaur A. (2020) Soil quality parameters vis-a-vis growth and yield attributes of sugarcane as influenced by integration of microbial consortium with NPK fertilizers. *Sci. Rep.* 10, 19180. https://doi.org/10.1038/s41598-020-75829-5

Shukla SK, Solomon S, Sharma L, Jaiswal VP, Pathak AD, Singh P. (2019) Green technologies for improving cane sugar productivity and sustaining soil fertility in sugarcane-based cropping system. *Sugar Tech.* 21(2), 186–196.

Singh P, Singh RK, Li HB, Guo DJ, Sharma A, Verma KK, Solanki MK, Upadhyay SK, Lakshmanan P, Yang LT, Li YR. (2023) Nitrogen fixation and phytohormone stimulation of sugarcane plant through plant growth promoting diazotrophic *Pseudomonas. Biotechnol. Genetic Eng. Rev.* 40(1), 15–35. https://doi.org/10.1080/02648725.2023.2177814

Singh P, Singh RK, Song QQ, Li HB, Guo DJ, Malviya MK, Verma KK, Song XP, Lakshmanan P, Yang LT, Li YR. (2021) Comparative analysis of protein and differential responses of defense-related gene and enzyme activity reveals the long-term molecular responses of sugarcane inoculated with *Sporisorium scitamineum. J. Plant Interactions* 16(1), 12–29.

Singh RK, Singh P, Li HB, Song QQ, Guo DJ, Solanki MK, Verma KK, Malviya MK, Song XP, Lakshmanan P, Yang LT, Li YR. (2020) Diversity of nitrogen-fixing rhizobacteria associated with sugarcane: A comprehensive study of plant-microbe interactions for growth enhancement in *Saccharum* spp. *BMC Plant Biol.* 20, 220. https://doi.org/10.1186/s12870-020-02400-9.

Solanki MK, Verma KK, Dastogeer KMG, Mora-Poblete F, Mundra S. (2023) Microbial resilience in plant nutrient management towards sustainable farming. *Front. Microbiol.* 14, 1280811. https://doi.org/10.3389/fmicb.2023.1280811

Song X-P, Mo F, Verma KK, Wei J, Zhang X, Yang L, Li YR. (2019) Effect of sugarcane smut (*Ustilago scitaminea* Syd.) on ultrastructure and biochemical indices of sugarcane. *Biomed J. Sci. Tech. Res.* 17(1), 002950. https://doi.org/10.26717/BJSTR.2019.17.002950.

Song XP, Mo F, Yan M, Zhang X, Zhang B, Huang X, Huang D, Pan Y, Verma KK, Li YR. (2022) Effect of smut infection on the photosynthetic physiological characteristics and related defense enzymes of sugarcane. *Life* 12, 1201. https://doi.org/10.3390/life12081201

Song XP, Verma KK, Tian DD, Zhang XQ, Liang YJ, Huang X, Li CN, Li YR. (2021) Exploration of silicon functions to integrate with biotic stress tolerance and crop improvement. *Biol. Res.* 54, 19. https://doi.org/10.1186/s40659-021-00344-4

Tian D, Qin L, Verma KK, Wei L, Li J, Li B, Zhou W, He Z, Wei D, Huang S, Long S, Huang Q, Li C, Wei S. (2023) Transcriptomic and metabolomic differences between banana varieties which are resistant or susceptible to *Fusarium wilt. PeerJ.* 11, e16549. https://doi.org/10.7717/peerj.16549

Verma KK, Anas M, Chen Z, Rajput VD, Malviya MK, Verma CL, Singh RK, Singh P, Song XP, Li YR. (2020) Silicon supply improves leaf gas exchange, antioxidant defense system and growth in sugarcane responsive to water limitation. *Plants* 9, 1032, https://doi.org/10.3390/plants9081032.

Verma KK, Joshi A, Song X-P, Liang Q, Xu L, Huang H-r, Wu K-C, Seth CS, Arora J and Li Y-R. (2024a) Regulatory mechanisms of plant rhizobacteria on plants to the adaptation of adverse agroclimatic variables. *Front. Plant Sci.* 15, 1377793. https://doi.org/10.3389/fpls.2024.1377793

Verma KK, Joshi A, Song X-P, Singh S, Kumari A, Arora J, Singh SK, Solanki MK, Seth CS, Li Y-R. (2024b) Synergistic interactions of nanoparticles and plant growth promoting rhizobacteria enhancing soil-plant systems: A multigenerational perspective. *Front. Plant Sci.* 15, 1376214. https://doi.org/10.3389/fpls.2024.1376214

Verma KK, Li DM, Singh M, Rajput VD, Malviya MK, Minkina T, Singh RK, Singh P, Song XP, Li YR. (2020a) Interactive role of silicon and plant-rhizobacteria mitigating abiotic stresses: A new approach for sustainable agriculture and climate change. *Plants* 9, 1055. https://doi.org/10.3390/plants9091055.

Verma KK, Rajput VD, Singh M, Verma CL, Minkina T. (2022a) Influence of waterlogging on plant physiology and molecular biology. In: *Understanding Abiotic Stresses*. Nova Science Publishers, Inc.: New York.

Verma KK, Siddiqui N, Singh M. (2022b) Understanding the plant physiological and molecular adaptation strategies to soil salinity. In: *Understanding Abiotic Stresses.* Nova Science Publishers, Inc.: New York. https://doi.org/10.52305/UBUJ4024

Verma KK, Singh P, Song X-P, Malviya MK, Singh RK, Chen G-L, Solomon S, Li YR. (2020b) Mitigating climate change for sugarcane improvement: Role of silicon in alleviating abiotic stresses. *Sugar Tech.* 22(5), 741–749. https://doi.org/10.1007/s12355-020-00831-0.

Verma KK, Singh RK, Song QQ, Singh P, Zhang B-Q, Song X-P, Chen G-L, Li YR. (2019a) Silicon alleviates drought stress of sugarcane plants by improving antioxidant responses. *Biomed. J. Sci. Tech. Res.* 002957. https://doi.org/10.26717/BJSTR.2019.17.002957

Verma KK, Song X, Degu HD, Guo D, Joshi A, Huang H, Xu L, Singh M, Huang D, Rajput VD, Li YR. (2023a) Recent advances in nitrogen and nano-nitrogen fertilizers for sustainable crop production: A mini-review. *Chem. Biol. Technol. Agric.* 10, 1–14. https://doi.org/10.1186/s40538-023-00488-3

Verma KK, Song X, Verma CL, Huang H, Singh M, Xu L, Wu J, Khan ZH, Zhang B, Li Y. (2023b) Mathematical modeling of climate and fluoride effects on sugarcane photosynthesis with silicon nanoparticles. *Plant Physiol. Biochem.* 204, 108089. https://doi.org/10.1016/j.plaphy.2023.108089

Verma KK, Song XP, Budeguer F, Nikpay A, Enrique R, Singh M, Zhang B, Wu, JM, Li YR. (2022c) Genetic engineering: An efficient approach to mitigating biotic and abiotic stresses in sugarcane cultivation. *Plant Signal. Behav.* 17(1), e2108253.

Verma KK, Song XP, Joshi A, Rajput VD, Singh M, Sharma A, Singh RK, Li DM, Arora J, Minkina T, Li YR. (2022d) Nanofertilizer possibilities for healthy soil, water and food in future: An overview. Front. Plant Sci. 13, 865048. https://doi.org/10.3389/fpls.2022.865048

Verma KK, Song XP, Joshi A, Tian DD, Rajput VD, Singh M, Arora J, Minkina TM, Li YR. (2022e) Recent trends in nano-fertilizer for advancing sustainable agriculture under the era of climate change to ensure food security for future generations. *Nanomaterials 12*, 173. https://doi.org/10.3390/nano12010173.

Verma KK, Song XP, Li DM, Singh M, Wu JM, Singh RK, Sharma A, Zhang BQ, LI YR. (2022f) Silicon and soil microorganisms improves rhizospheric soil health with bacterial community, plant growth, performance and yield. *Plant Signal. Behav.* 17(1), e2104004.

Verma KK, Song XP, Lin B, Guo DJ, Singh M, Rajput VD, Singh RK, Singh P, Sharma A, Malviya MK, Chen GL, Li YR. (2022g) Silicon induced drought tolerance in crop plants: Physiological adaptation strategies. *Silicon* 14(6), 2473–2487. https://doi.org/10.1007/s12633-021-01071-x

Verma KK, Song XP, Rajput VD, Boldyreva V, Zhang BQ, Minkina T, Li YR. (2022h) Morpho-physiological, biochemical and ultrastructural modifications on sugarcane to prolonged water deficit. In: Verma KK. et al. (Eds.), *Agro-Industrial Perspectives on Sugarcane Production Under Environmental Stress*. https://doi.org/10.1007/978-981-19-3955-6_11.

Verma KK, Song XP, Rajput VD, Solomon S, Li YR, Rao GP. (2022i) *Agro-Industrial Perspectives on Sugarcane Production Under Environmental Stress*. Springer Nature Singapore, https://doi.org/10.1007/978-981-19-3955-6

Verma KK, Song XP, Singh M, Tian DD, Rajput VD, Minkina T, Li YR. (2023c) Association of silicon and soil microorganisms induces stress mitigation, increasing plant productivity. In: de Mello Prado R. (ed.), *Benefits of Silicon in the Nutrition of Plants*. pp. 299–328. https://doi.org/10.1007/978-3-031-26673-7_17

Verma KK, Song XP, Tian DD, Guo DJ, Chen ZL, Zhong CS, Nikpay A, Singh M, Rajput VD, Singh RK, Minkina T, Li YR. (2021a) Influence of silicon on biocontrol strategies to manage biotic stress for crop protection, performance and improvement. *Plants* 10, 2163. https://doi.org/10.3390/plants10102163

Verma KK, Song XP, Tian DD, Singh M, Verma CL, Rajput VD, Singh RK, Sharma A, Singh P, Malviya MK, Li YR. (2021b) Investigation of defensive role of silicon during drought stress induced by irrigation capacity in sugarcane: Physiological and biochemical characteristics. *ACS Omega* 6, 19811–19821.

Verma KK, Song XP, Verma CL, Chen ZL, Rajput VD, Wu KC, Liao F, Chen GL, Li YR. (2021c) Functional relationship between photosynthetic leaf gas exchange in response to silicon application and water stress mitigation in sugarcane. *Biol. Res.* 54, 15. https://doi.org/10.1186/s40659-021-00338-2

Verma KK, Song XP, Verma CL, Malviya MK, Guo DJ, Rajput VD, Sharma A, Wei KJ, Chen GL, Solomon S, Li YR. (2021d) Predication of photosynthetic leaf gas exchange of sugarcane (*Saccharum* spp.) leaves in response to leaf positions to foliar spray of potassium salt of active phosphorus under limited water irrigation. *ACS Omega* 6, 2396–2409.

Verma KK, Song X-P, Xu L, Huang H-R, Liang Q, Seth CS, Li Y-R. (2023d) Nano-microplastic and agro-ecosystems: A mini-review. *Front. Plant Sci.* 14, 1283852. https://doi.org/10.3389/fpls.2023.1283852

Verma KK, Song XP, Yadav G, Degu HD, Parvaiz A, Singh M, Huang HR, Mustafa G, Xu L, Li YR. (2022j) Impact of agroclimatic variables on proteo-genomics in sugarcane (*Saccharum* spp.) plant productivity. *ACS Omega* 7, 22997–23008.

Verma KK, Song XP, Zeng Y, Guo DJ, Sing M, Rajput VD, Malviya MK, Wei KJ, Sharma A, Li DP, Chen GL, Li YR. (2021e) Foliar application of silicon boosts growth, photosynthetic leaf gas exchange, antioxidative response and resistance to limited water irrigation in sugarcane (*Saccharum officinarum* L.). *Plant Physiol. Biochem.* 166, 582–592.

Verma KK, Song XP,Zeng Y, Li DM, Guo DJ, Rajput VD, Chen GL, Barakhov A, Minkina TM, Li YR. (2020c) Characteristics and correlation of leaf stomata and its relationship with photosynthesis on *Saccharum* spp. under different irrigation and silicon application. *ACS Omega* 5, 24145–24153.

Verma KK, Vatsal S, Singh M, Rajput VD, Minkina T. (2022k) Coping with water stress: Effects and adaptive mechanisms to plants. In: *Understanding Abiotic Stresses*. Nova Science Publishers, Inc.: New York.

Verma KK, Wu KC, Verma CL, Li DM, Malviya MK, Singh RK, Singh P, Chen GL, Song XP, Li YR. (2020d) Developing mathematical model for diurnal dynamics on photosynthetic responses in sugarcane responsive to different irrigation and silicon application. *PeerJ.* 8, e10154. https://doi.org/10.7717/peerj.10154.

Verma KK, Liu X-H, Wu K-C, Singh RK, Song QQ, Malviya MK, Song X-P, Singh P, Verma CL, Li YR. (2020e) The impact of silicon on photosynthetic and biochemical responses of sugarcane under different soil moisture levels. *Silicon* 12, 1355–1367.

Verma KK, Wu K-C, Singh P, Malviya MK, Singh RK, Song X-P, Li YR. (2019b) The protective role of silicon in sugarcane under water stress: Photosynthesis and antioxidant enzymes. *Biomed. J. Sci. Tech. Res.* 002685. https://doi.org/0.26717/BJSTR.2019.15.002685

Wu K-C, Huang C-M, Verma KK, Deng Z-N, Huang H-R, Pang T, Cao H-Q, Luo H-B, Jiang S-L, Xu L. (2022) Transcriptomic responses of *Saccharum spontaneum* roots in response to polyethylene glycol–6000 stimulated drought stress. *Front. Plant Sci.* 13, 992755. https://doi.org/10.3389/fpls.2022.992755

Xu L, Deng Z-N, Wu K-C, Malviya MK, Solanki MK, Verma KK, Pang T, Li Y-J, Liu X-Y, Kashyap BK, Dessoky ES, Wang W-Z and Huang H-R. (2022) Transcriptome analysis reveals a gene expression pattern that contributes to sugarcane bud propagation induced by Indole-3-Butyric acid. *Front. Plant Sci.* 13, 852886. https://doi.org/10.3389/fpls.2022.852886

Yadav R, Kumar R. (2005) On-farm comparison of ring-pit and conventional flat planting methods for yield and quality of sugarcane in Northwest India. *Ind. J. Agric. Sci.* 75, 605–607.

Zheng F-J, Lin B, Yang Y-X, Fang X-C, Verma KK, Chen G-L. (2023) Efficacy and functionality of sugarcane original vinegar on mice. *Front. Microbiol.* 14, 1224666. https://doi.org/10.3389/fmicb.2023.1224666

CHAPTER 5

Physiological and Molecular Insights into Heavy Metal Stress Resilience Efficiency and Sustainability

PRASANN KUMAR*, JAISURIYAN K., and BHUPENDRA MATHPAL

Department of Agronomy, School of Agriculture, Lovely Professional University, Phagwara, Punjab, India

*Corresponding author

ABSTRACT

The extensive physiological and molecular investigations are necessary to address the increasing influence of heavy metals on plant systems. This study elucidates the intricate resilience mechanisms, enhancing efficiency and sustainability. Environmental concerns, such as high levels of heavy metals threaten agricultural productivity and ecosystems. The presence of diverse metal transporters enables specific absorption, while the process of vacuolar sequestration effectively prevents the harmful effects of metal toxicity. The complexity of plant resilience is demonstrated by the contribution of metabolic changes in primary and secondary metabolism to adaptive responses. The genome-wide expression profiling uncovers the intricate network of genes that respond to metal and their distinct reactions to stress. Transcriptional regulatory networks, which involve transcription factors and epigenetic modifications, are considered for molecular orchestration and resilience. Reactive oxygen species play a crucial role as signaling molecules in the plant's response to heavy metal stress. Analyze the coordination of antioxidant defence systems, encompassing enzymatic and nonenzymatic elements, to mitigate oxidative stress caused by

Sugarcane Production: Environmental Impacts and Socioeconomic Issues.
Krishan K. Verma, Xiu-Peng Song, Lin Xu, Rajan Bhatt, and Yang-Rui Li (Eds.)

DOI: 10.1201/9781779641137-5

heavy metals. Advanced imaging techniques enable the visualization of metal distribution in plant tissues *in vivo*. Integrating omics technologies, such as metabolomics, proteomics, and genomics, allows for a comprehensive comprehension of the molecular landscape of mental stress.

5.1 INTRODUCTION

The dynamic relationship between heavy metal stress in soil ecosystems and the dynamics of metal accumulation and translocation in plants is an essential aspect of environmental science that has far-reaching implications for the health and sustainability of ecosystems (Chakraborti et al., 2022; Ijaz et al., 2023; Huang et al., 2024). This in-depth analysis delves into the myriad facets of heavy metal stress in soil ecosystems. It investigates the origins of this stress, its effects on the health of the soil, its influence on the growth of plants, and the available remediation strategies. In addition, it investigates the complex processes that govern the accumulation and translocation of metals within plants, shedding light on the mechanisms that plants employ to defend themselves against the effects of heavy metal stress (Verma et al., 2019a, 2019b; 2022a; Jagadesh et al., 2024). A complex interaction between natural and anthropogenic sources introduces heavy metals into soil ecosystems for the first time. The elevated concentrations of heavy metals in soil result from various factors, including industrial activities, agricultural practices, and atmospheric deposition. Agricultural practices, including using metal-containing fertilizers and pesticides, worsen soil contamination. Industrial discharges release metals such as lead, cadmium, and mercury into the environment (Krishna et al., 2022; Verma et al., 2024a, 2024b). The heavy metal burden in soils is further increased by atmospheric deposition caused by emissions from various sources, including industrial facilities and roadway traffic. It is important to note that heavy metal stress significantly impacts the soil's physicochemical properties, affecting the soil's overall health and fertility (Kumari et al., 2022, 2023a, 2023b). Metal stress can cause significant changes in the soil's pH, organic matter content, and cation exchange capacity. These three factors are especially susceptible to change (Maqsood et al., 2023; Mohapatra et al., 2023).

Acidic heavy metals can bring the pH of the soil down, which can affect the availability of nutrients and the activity of microbiomes. In addition, high metal concentrations can potentially upset the delicate balance of the microbial communities in the soil, which can put the soil's health at risk. Heavy metal stress is a more severe stress to the plants because they are the primary components of soil ecosystems (Guo et al., 2022; Bhatt et al., 2023a, 2023b). Toxic metals affect plants' growth and development, resulting in chlorosis, inhibited growth, and a diminished capacity for reproduction. There is a correlation between the disruption of essential physiological and biochemical processes, such as photosynthesis and nutrient uptake, and the overall impairment of plant health. Certain plants can tolerate certain metals through adaptive mechanisms; however, other plants cannot withstand the toxic effects of these metals, which lead to a reduction in biodiversity and ecological resilience. Therefore, efficient soil remediation strategies are required to address the issue of heavy metal stress. Several approaches are utilized to mitigate the negative effects of heavy metals on soil health, each of which has advantages and limitations. The cultivation of plants that are capable of accumulating metals to extract metals from the soil is a process known as phytoextraction, which has demonstrated that it has the potential to reduce the concentrations of toxic ions. Using plants to immobilize metals in the soil to prevent their migration to water sources is an example of phytostabilization. Bioaugmentation, which uses microorganisms to improve metal remediation, is another advanced approach. Several factors should be considered when selecting an appropriate remediation method. These include the type and concentration of metals, the characteristics of the soil, and the desired land use application (Ali et al., 2022; Khalid et al., 2023; Mohan et al., 2023).

The first step in the journey of heavy metals within plants is when the roots take them up from the soil solution. For example, the availability of metals in the soil, the pH of the soil, and the presence of competing ions are all factors that can affect this process. Plants use specialized transporters on the root cell membranes to selectively absorb nutrients. Different plant species have different levels of efficiency when it comes to the uptake of metals, and this efficiency is highly dependent on the bioavailability of metals in the rhizosphere. Heavy metals are transported throughout the plant by the xylem, moving from the roots to the shoots after the plant has absorbed them. Several factors can affect this translocation process, including the chemical properties of the metal, the plant species involved, and the conditions of the environment. During translocation, metal ions may undergo complexation with organic ligands, impacting their mobility and toxicity. A solid understanding of the factors that govern metal translocation is necessary to accurately forecast the potential effects of metal contamination on plants' health and ecosystems' dynamics.

The patterns of metal accumulation within the organs and tissues of plants exhibit a remarkable degree of diversity across the categories of plant species. Metal hyperaccumulation is a phenomenon that occurs when certain plants prefer to accumulate particular metals in particular organs compared to other crops. Because of their one-of-a-kind physiological adaptations, hyperaccumulator plants can accumulate extremely high concentrations of metals in their aerial parts without experiencing significant toxicity. By investigating the mechanisms responsible for the accumulation of metals, one can gain insights into the potential applications of particular plant species for phytoremediation. When exposed to heavy metals, plants undergo a wide range of physiological responses to handle the stress they are experiencing. Plants employ various strategies to mitigate the toxic effects of metals, including activating antioxidant defense systems, sequestrating metals in vacuoles, and chelating metals by phytochelatins (Dilnawaz et al., 2023; Rai et al., 2023; Chaudhary et al., 2023).

The process of metal sequestration in vacuoles functions as a protective mechanism, preventing metals from interfering with the processes essential to the cell mechanisms. Metal ions are chelated by phytochelatins, small peptides produced in response to metal stress. This reduces the toxicity of the metal ions. To combat the oxidative stress that heavy metals cause, plants can benefit from activating antioxidant enzymatic activities like superoxide dismutase (SOD) and catalase (CAT). To summarize, the intricate relationship between heavy metal stress in soil ecosystems and the dynamics of metal accumulation and translocation in plants exemplifies the complexity of environmental processes (Chen et al., 2023; Malviya et al., 2021, 2022). Identifying the causes and consequences of metal stress in soil is essential for developing efficient remediation strategies and promoting environmentally responsible land management practices. Concurrently, deciphering the mechanisms that govern the uptake, translocation, and accumulation of metals in plants contributes to a better understanding of how plants adapt to environmental threats (Huang et al., 2022; Verma et al., 2021a, 2021b; Tian et al., 2023). It is essential to conduct research that spans multiple disciplines and work together to address the challenges posed by heavy metal contamination and its influence on the interactions between plants and soil as we work toward achieving ecological sustainability (Figure 5.1).

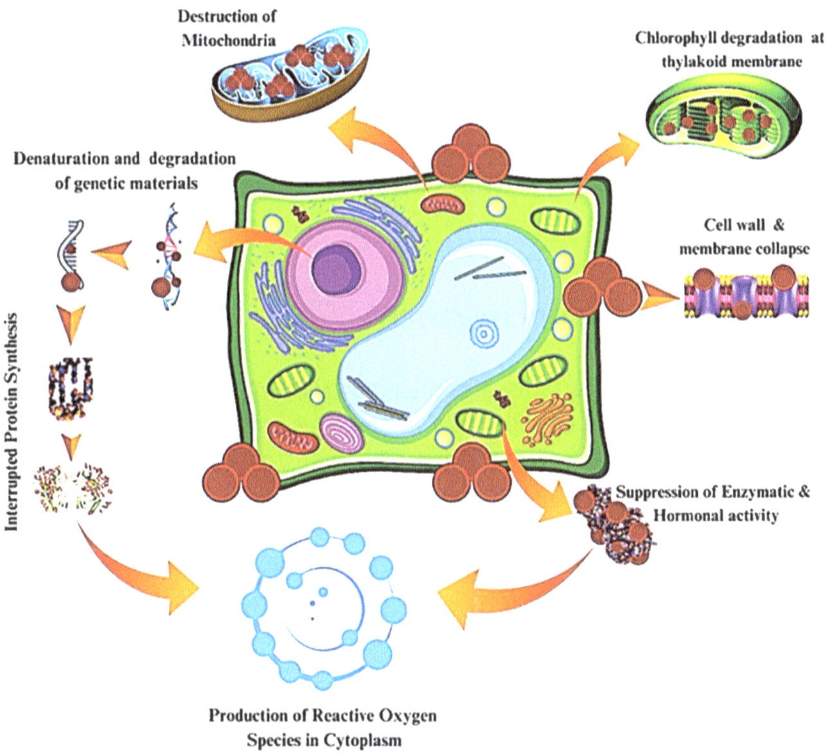

FIGURE 5.1 Stress-induced cellular damage of plants.

5.2 PHYSIOLOGICAL RESPONSES TO HEAVY METAL STRESS

Several dynamic processes determine the fate of metals within plants, and plants' physiological responses to heavy metal stress are intricate and multifunctional. This discussion aims to shed light on how plants navigate the challenges of heavy metal stress by delving into the mechanisms underlying metal uptake dynamics, the diversity and specificity of metal transporters, and the dynamics of root-shoot metal translocation (Sheri et al., 2023; Wahab et al., 2024).

5.2.1 MECHANISMS OF ENERGY ABSORPTION

Plant roots are responsible for the first step in the physiological response to heavy metal stress, which is the uptake of metals. Several factors govern this process, including the concentration of metals in the soil solution, the pH of the soil, and the presence of competing ions. Plants use passive and active absorption mechanisms to remove metals from the rhizospheric soil. Passive absorption takes place through channels that are not selective. In contrast, active absorption involves specialized transport proteins that selectively facilitate the entry of particular metal ions into root cells.

5.2.2 SELECTIVITY AND SPECIFICITY

The selectivity and specificity of metal uptake are determined by the variety of transporters in the root cells' membranes. It is possible to observe selectivity in metal uptake because different metal transporters have varying affinities for particular metal ions. One example is the iron-regulated transporter (IRT) family, which plays a role in the uptake of divalent metal cations such as iron, zinc, and manganese. This exemplifies the complex regulatory mechanisms that govern the uptake of metals in response to the plant's nutritional requirements (Choudhary and Muthamilarasan, 2022; Etesami and Glick, 2024).

5.2.3 RHIZOSPHERIC INTERACTIONS

Within the context of the dynamics of metal uptake, the rhizosphere, which is the soil region surrounding the roots, plays an essential role. Root exudates that contain organic acids, phenolic compounds, and proteins affect the solubility of metals in the rhizosphere. Roots secrete organic acids, which help to mobilize metals in the soil and make them more accessible for uptake. Roots also release organic acids (Guo et al., 2021; Sharma et al., 2021; Song et al., 2019, 2021; Solanki et al., 2023). Furthermore, root exudates can form complexes with metal ions, which can affect the speciation and bioavailability of the metal ions. It is essential to have a solid understanding of these rhizosphere interactions to decipher the intricate relationship between plants and the immediate soil environment in which they are located when metal stress conditions are present (Singh et al., 2021; Zhao et al., 2024; Zheng et al., 2023a).

5.3 DIVERSIFICATION OF TRANSPORTERS AND THEIR SPECIFICITY

5.3.1 FAMILIES OF METAL TRANSPORT CONTAINERS

Plant cells contain various metal transporter families, highlighting the complexity of metal homeostasis. Transporters belonging to several different families are involved in the process of heavy metal uptake and translocation. Each of these transporters possesses distinctive structural and functional characteristics. For example, the zinc–iron–permease transporter family, the natural resistance-associated macrophage protein transporter family, and the ATP-binding cassette transporter family are all components of metal transport processes. Some of these families are responsible for the specificity observed in metal uptake, which enables plants to differentiate between essential metals and not crucial (Sharma et al., 2022a; Bera et al., 2024).

5.3.2 UPTAKE AND ACCUMULATION OF METALS

Metal transporters are an essential component in the process by which plant roots selectively take in metals. As an illustration, members of the IRT family play a significant role in the uptake of essential divalent metal cations, such as zinc and iron. Transporters, such as those belonging to the heavy metal ATPase family, play a role in removing toxic metals from root cells, which serves as a mechanism for this process. Plants can maintain metal homeostasis and protect themselves from the poisonous effects of heavy metal accumulation thanks to the intricate coordination of these transporter families (Iqbal et al., 2023; Shinde et al., 2024).

5.3.3 REGULATION OF TRANSPORTER

The expression of metal transporters is subject to stringent regulation in response to shifting environmental conditions, including metal stress. To adjust the plant's capacity to take in metals by its physiological requirements, transcription factors and regulatory proteins are responsible for modulating the expression of metal transporters (Joshi et al., 2022; Sharma et al., 2022b; Song et al., 2022). Metal stress-responsive elements

(MREs) located in the promoter regions of transporter genes play a pivotal role in orchestrating the plant's adaptive responses to fluctuations in metal availability. Unraveling the regulatory networks that govern transporter expression is necessary to better understand the molecular mechanisms underlying plant responses to heavy metal stress (Oliveira et al., 2022; Sharma et al., 2022b; Kaya and Ugurlar, 2023) (Figure 5.2).

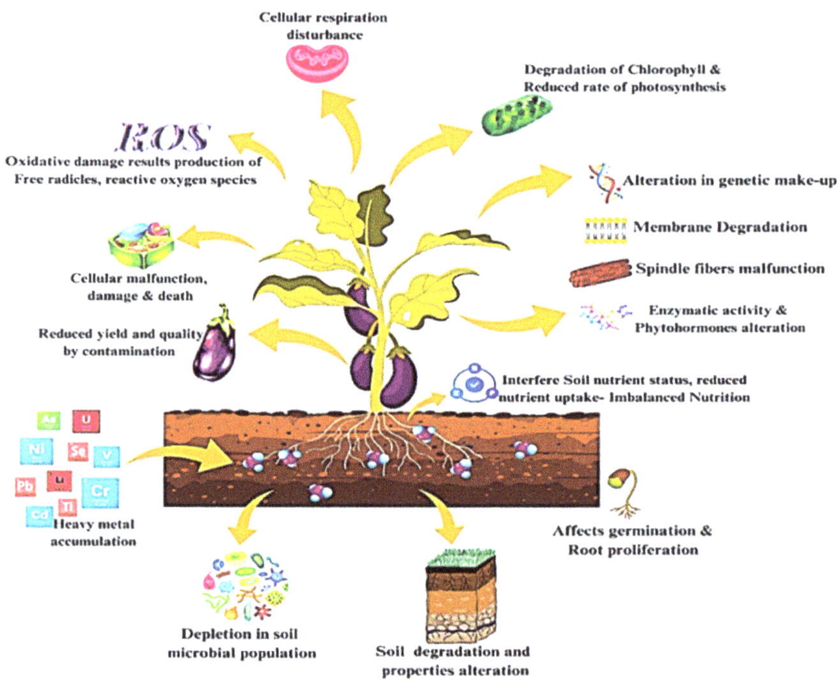

FIGURE 5.2 Heavy metal exposure and responses of plants.

5.4 TRANSFORMATIONS IN THE METALS LOCATION: ROOT-SHOOT

5.4.1 TRANSLOCATION OF IONS IN PLANT

Following the uptake of metals by the roots, the xylem is responsible for the dynamic process of metal translocation within the crop. The xylem is the primary vascular tissue responsible for the upward movement of water and dissolved minerals, including metals, from the roots to the shoots of the plant. It is possible to improve the solubility and mobility of metal ions by loading them into the xylem vessels in the root, which then form complexes with organic ligands. Factors such as the metal's chemical properties and the plant's physiological status are among the factors that influence the efficiency of metal translocation, which varies from plant species to plant species (Rani et al., 2023; Balooch et al., 2024).

5.4.2 PATTERNS OF SHOOT ACCUMULATION

Following their arrival in the shoot, metals go through the process of complexation with a variety of ligands, such as organic acids and proteins, which in turn affect their distribution throughout the organs and tissues of the plant. The accumulation patterns of shoots are highly variable between different plant species and types of metals. Metal compartmentalization is a phenomenon that occurs in certain plants and is characterized by the preferential accumulation of particular metals in particular organs. To evaluate the potential of various plant species for phytoremediation and develop an understanding of the impact that metal stress has on plant health, it is essential to have a solid understanding of the factors that govern shoot accumulation (Ebrahimi-Zarandi et al., 2023; Joel et al., 2023).

5.4.3 DETOXIFICATION FUNCTIONS OF METALS

Plants utilize a variety of mechanisms to detoxify metals in the shoot, thereby preventing the adverse effects of metals on the processes that occur within the cell. Among the methods plants use to reduce the toxicity of accumulated metals, some of the strategies they employ include the sequestration of metals in vacuoles, the chelation of metals by phytochelatins, and the formation of metal-binding proteins. A process known as vacuolar sequestration involves the storage of metals within vacuoles to prevent them from interfering with the cell's essential functions. The synthesis of phytochelatins, which occur in response to metal stress, chelates metal ions, thereby reducing the toxicity of the metal ions (Siddique and Kumar, 2018; Naik and Kumar, 2020; Chen et al., 2020; Singh et al., 2020).

The formation of metal-binding proteins assists in the sequestration of metals in particular cellular compartments, thereby reducing metals' impact on the processes within the cell. The physiological responses of plants to heavy metal stress involve a complex set of processes, ranging from the dynamics of metal uptake at the root level to the diversity and specificity of metal transporters, as well as the dynamics of metal translocation between the root and the shoot (Zheng et al., 2023b; Verma et al., 2023a, 2023b; Li et al., 2023). Developing effective strategies to mitigate the adverse effects of heavy metal stress on plant health and ecosystem dynamics requires a thorough understanding of these processes, which is essential for developing such strategies. To lay the groundwork for developing crop varieties that have increased metal tolerance and phytoremediation potential, it is necessary to unravel the molecular mechanisms that govern the transport and accumulation of metal by plants. Continued research in this area is essential for addressing the challenges posed by heavy metal contamination and ensuring the resilience of plant–soil interactions in the face of environmental stressors. This is especially important as we achieve sustainable environmental management (Sharma and Kumar, 1999; Paul et al., 2005; Singh et al., 2023) (Figure 5.3).

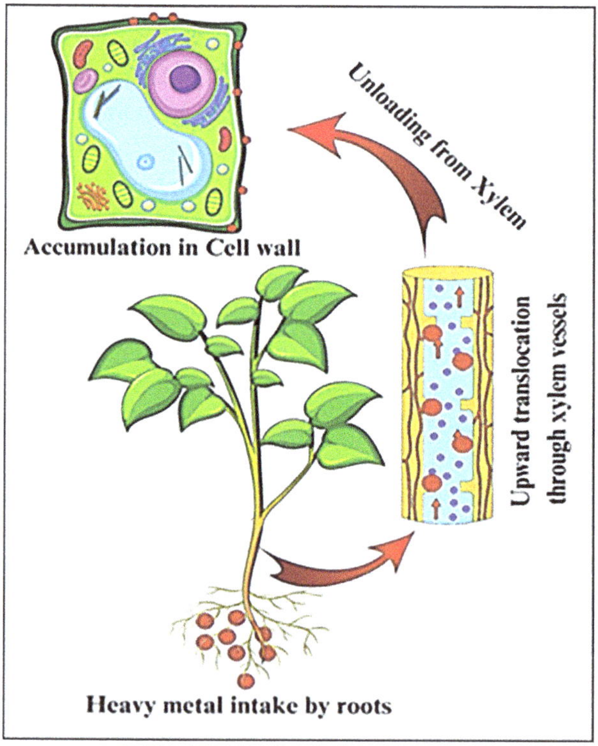

FIGURE 5.3 Translocation of metals in plants.

5.5 CELLULAR COMPARTMENTALIZATION STRATEGIES

Cellular compartmentalization of heavy metals is crucial for maintaining plant metal homeostasis, protecting cellular functions, and facilitating essential physiological processes. The vacuole and the Golgi apparatus are critical cellular compartments that have vital functions in controlling the storage and regulation of heavy metals in plant cells. This discussion explores the methods plants use to store substances in vacuoles and the role of the Golgi apparatus in regulating the balance of metals within the plant (Kumar et al., 2011a, 2011b).

5.5.1 MECHANISMS OF VACUOLE SEQUESTRATION

5.5.1.1 THE VACUOLE SERVES AS A LOCATION FOR STORAGE

The vacuole, a substantial organelle enclosed by a membrane in plant cells, is a primary repository for diverse substances, including heavy metals. Vacuole sequestration is an important cellular mechanism that removes metals from the cell and prevents them from disrupting vital cellular functions. Plants can counteract the harmful consequences of metal buildup and uphold cellular balance by segregating metals in the vacuole (Pankaj et al., 2012a; Srivastav et al., 2023).

5.5.1.2 TRANSPORTERS ENABLING VACUOLE SEQUESTRATION

Vacuole sequestration entails the active transportation of metal ions into the vacuole. The process is facilitated by specific transporters, such as the Vacuole Iron Transporter (VIT) family and the cation/H^+ exchanger (CAX) family. VIT transporters facilitate the transport of divalent metal cations, such as iron, zinc, and manganese, into the vacuole. CAX transporters, however, aid in transporting different metal cations into the vacuole by exchanging them for protons, thereby assisting in the metal detoxification processes.

5.5.1.3 COMPLEXATION AND BINDING PROCESS OF METALS

When metal ions enter the vacuole, they frequently form complexes with different ligands, including organic acids and peptides. This process helps to decrease their toxicity. Metals can be more soluble and accessible in vacuoles by forming complexes with organic acids such as citrate and malate. In addition, intracellularly synthesized metal-binding proteins can sequester metals in the vacuole, inhibiting their interaction with cellular components. Creating these metal complexes enhances the overall effectiveness of vacuole sequestration and is a crucial component of cellular metal detoxification strategies (Kumar et al., 2019a, 2019b; Goud et al., 2022; Dey et al., 2023).

5.5.1.4 pH REGULATION OF VACUOLE

The control of vacuole pH is another crucial element of vacuole sequestration mechanisms. The vacuole regulates an acidic pH, which increases the solubility of metal ions and aids in their sequestration. Proton-pumping ATPases in the vacuole membrane actively transport protons into the vacuole, creating its acidic environment. The regulation of pH is crucial for maximizing the efficiency of metal sequestration within the vacuole, guaranteeing the effective isolation of toxic metal ions from the cellular environment.

5.6 ROLE OF THE GOLGI APPARATUS IN MAINTAINING MENTAL BALANCE

5.6.1 GOLGI APPARATUS IS A CENTER FOR SORTING AND MODIFYING CELLULAR COMPONENTS

The Golgi apparatus is essential for maintaining metal balance by acting as a hub for sorting and modifying proteins responsible for transporting and storing metals. The Golgi apparatus alters and assembles proteins, such as metal transporters and chelating proteins, before directing them to specific compartments within the cell.

5.6.2 CHANGES OF METAL TRANSPORTERS

Metal transporters produced in the endoplasmic reticulum are transferred to the Golgi apparatus, where they undergo posttranslational modifications. Modifications such as glycosylation and phosphorylation influence metal transporters' stability, activity, and cellular localization. Golgi-mediated modifications enhance the efficiency of metal transporters by facilitating the transport of metals across cellular membranes, thereby ensuring their proper functioning (Kumar et al., 2018a–2018g; Yomso et al., 2021; Kumari et al., 2022).

5.6.3 PROTEIN TRAFFICKING THROUGH CHELATION

Proteins involved in chelation, such as metallothioneins and phytochelatins, are produced and altered in the Golgi apparatus before being transported to their ultimate locations, where they play crucial roles in removing metals from the body. Metallothioneins, compact proteins rich in cysteine, play a vital role in the binding and removal of metals from the body. Phytochelatins, conversely, are peptides produced as a response to metal stress and play a role in binding metal ions. The Golgi apparatus ensures the appropriate modification and transportation of these chelating proteins to specific cellular compartments where the removal of metals is necessary.

5.6.4 GOLGI-MEDIATED VESICLE TRANSPORT

The Golgi apparatus is responsible for transporting vesicles, which aid in moving metal-related proteins to different parts of the cell. Membrane-bound structures called vesicles, which contain metal transporters, chelating proteins, or other components related to metal balance in the body, are formed from the Golgi apparatus and are targeted to specific areas within the cell. Vesicle-mediated transport guarantees the accurate transportation of metal-related proteins to their exact locations, thereby playing a role in regulating cellular metal homeostasis (Reddy et al., 2022; Sharma et al., 2023; Upadhyay et al., 2023).

Plants demonstrate advanced responses to heavy metal stress through cellular compartmentalization strategies, specifically vacuolar sequestration mechanisms and the participation of the Golgi apparatus in metal homeostasis. The vacuole serves as a storage and detoxification organelle, effectively sequestering and isolating heavy metals to prevent their harmful impact on cellular mechanisms. The Golgi apparatus is crucial for sorting, modifying, transporting metal transport, and detoxifying proteins. This ensures the accurate regulation of metal homeostasis in plant cells. Knowing these cellular strategies offers a valuable understanding of the molecular mechanisms that drive plant responses to heavy metal stress. Additional investigation in this area shows potential for developing biotechnological methods to increase plant resistance to heavy metals and enhance the effectiveness of phytoremediation techniques. As we aim to achieve sustainable environmental management, it is crucial to understand the complex strategies cells use to compartmentalize substances. This understanding is essential for reducing the adverse effects of heavy metal contamination on plant health and the dynamics of ecosystems (Kandpal et al., 2018; Kumar et al., 2018a–2018g; Devi et al., 2023a, 2023b, 2023c; Hidangmayum et al., 2023).

5.7 METABOLIC BALANCE

Metabolism is a complex network of biochemical processes in living organisms that adapt to internal and external stimuli. Metabolic adaptations are crucial in preserving cellular balance and adjusting to the era of climate change. This chapter examines the complex interaction between metabolic adaptations, their influence on both primary and secondary metabolism, and the essential role of metabolites in providing stress resistance.

5.7.1 METABOLIC ADAPTATIONS AND FUNDAMENTAL BIOCHEMICAL PROCESSES

Primary metabolism refers to the essential biochemical pathways producing energy and necessary cellular function components. Cells adapt their metabolic processes to efficiently create energy and utilize resources in response to environmental signals. The transition from aerobic to anaerobic metabolism in response to oxygen deprivation is a well-known illustration of such adaptations. Under conditions of stress, the metabolic profile of cells experiences substantial changes. Increasing glycolysis and initiating alternative energy-generating pathways guarantee a prompt and effective reaction to energy requirements. Moreover, the activation of stress-responsive transcription factors, such as hypoxia-inducible factor-1, coordinates metabolic adaptations to improve cell viability in unfavorable circumstances (Pathak et al., 2017; Kandpal et al., 2018; Kumar and Dwivedi, 2020; Kumar et al., 2022, 2023; Deng et al., 2023).

5.7.2 SECONDARY METABOLISM AND ITS SUSCEPTIBILITY TO ENVIRONMENTAL STIMULI

Secondary metabolism encompasses the production of specialized metabolites that do not directly participate in vital cellular processes but have significant functions in ecological interactions, defense mechanisms, and stress adaptation. These metabolites are frequently produced in reaction to particular environmental stimuli, acting as signaling molecules or aiding the organism's resistance to stresses. Environmental stressors, including temperature fluctuations, nutrient availability, and pathogen attacks, may trigger secondary metabolic responses. Plants produce secondary metabolites, such as flavonoids and alkaloids, to protect themselves from herbivores and pathogens. The responsiveness of secondary metabolism to environmental stimuli emphasizes its adaptive significance and underscores the crucial role of metabolic adaptations in dealing with stress (Kumar and Dwivedi, 2018; Guo et al., 2023; Devi et al., 2023a, 2023b, 2023c; Joshi et al., 2023).

The role of metabolites in stress resilience is essential. Metabolites, small molecules generated during metabolism, play a critical role in determining an individual's ability to withstand and recover from stress. These molecules function as signaling entities, agents that influence cellular processes, and controllers of gene expression. Comprehending the function of metabolites in stress resilience offers knowledge of the molecular mechanisms that drive adaptive responses. Phytohormones in plants are a class of metabolites that have received considerable attention about stress resilience. One example is the role of abscisic acid (ABA) in controlling plant reactions to different stress factors, such as drought and salinity (Anas et al., 2021a, 2021b; Verma et al., 2023c, 2023d). ABA induces stomatal closure, decreases transpiration, and stimulates stress-responsive genes, thereby augmenting the plant's resilience to unfavorable circumstances. Likewise, in animals and microorganisms, particular metabolites serve as stress signals, regulating cellular processes to enhance survival (Wu et al., 2022; Xu et al., 2022).

5.7.3 CROSS-TALK AND INTEGRATION OF METABOLIC FUNCTIONS

Metabolic adaptations in reaction to stress are not independent occurrences but entail complex communication between metabolic pathways. The integration of primary and secondary metabolism enables cells to allocate resources effectively, striking a balance between energy generation and the synthesis of metabolites related to stress. Activating the pentose phosphate pathway (PPP) during oxidative stress serves as it reduces the equivalents of antioxidant defense and generates ribose-5-phosphate, a precursor for nucleotide biosynthesis. This demonstrates how metabolic adaptations are precisely regulated to fulfill various cellular requirements simultaneously (Dubey et al., 2002; Verma et al., 2020a, 2020b; Kumar et al., 2021a, 2021b; Sharma et al., 2023).

Metabolic adaptations are crucial for the survival and adjustment of living organisms in response to environmental challenges. The effects of these modifications have a broader reach than just primary metabolism, as they also impact secondary metabolism. This, in turn, affects the synthesis of specific metabolites essential for enhancing the ability to withstand stress (Verma et al., 2021c, 2021d, 2021e). Metabolites coordinate adaptive responses to stressors by acting as signaling molecules and effectors. As our comprehension of metabolic regulation becomes more profound, it unveils novel opportunities for devising approaches to augment stress tolerance in crops, enhance animal disease resistance, and investigate therapeutic interventions in human health. The complex process of metabolic adjustments reveals the remarkable ability embedded within the cellular machinery, emphasizing life's intricate and beautiful biochemical foundations.

5.8 MOLECULAR MECHANISMS OF TOLERANCE

Resilience is a crucial component of an organism's strategy for survival, which involves complex molecular mechanisms that allow it to adapt and flourish in response to various stressors. The study of these mechanisms has become more detailed and sophisticated due to the introduction of advanced technologies, and one potent tool is genome-wide expression profiling. By adopting this comprehensive approach, researchers can thoroughly examine an organism's genomic landscape, uncovering the intricate molecular details that determine its resilience. This discussion explores the molecular processes involved in resilience, specifically genome-wide expression profiling, identifying metal-responsive genes, and examining differential expression patterns during stress. Transcriptome analysis across the entire genome: genome-wide expression profiling is an innovative method that offers a comprehensive overview of an organism's gene expression patterns across various conditions (Dwivedi and Kumar, 2011; Kumar et al., 2018a–2018g, 2019a, 2019b; Dey et al., 2023).

It entails the concurrent examination of the expression levels of all genes in an organism, providing a comprehensive comprehension of its molecular activities. The importance of this approach resides in its capacity to reveal the genes implicated in particular biological processes, including those linked to resilience (Verma et al., 2020c, 2020d, 2020e; Guo et al., 2022). Genome-wide expression profiling enables researchers to observe alterations in gene expression patterns in response to stress within the resilience framework. Various stressors, including external environmental challenges and internal disturbances, induce dynamic changes in the expression of proteins. By examining these alterations at a genomic level, scientists can pinpoint crucial participants in the intricate molecular mechanisms of resilience. This comprehensive analysis clarifies the pathways and networks involved in adaptive responses, which form the basis for specific interventions and a more profound comprehension of an organism's capacity to endure challenges (Figure 5.4).

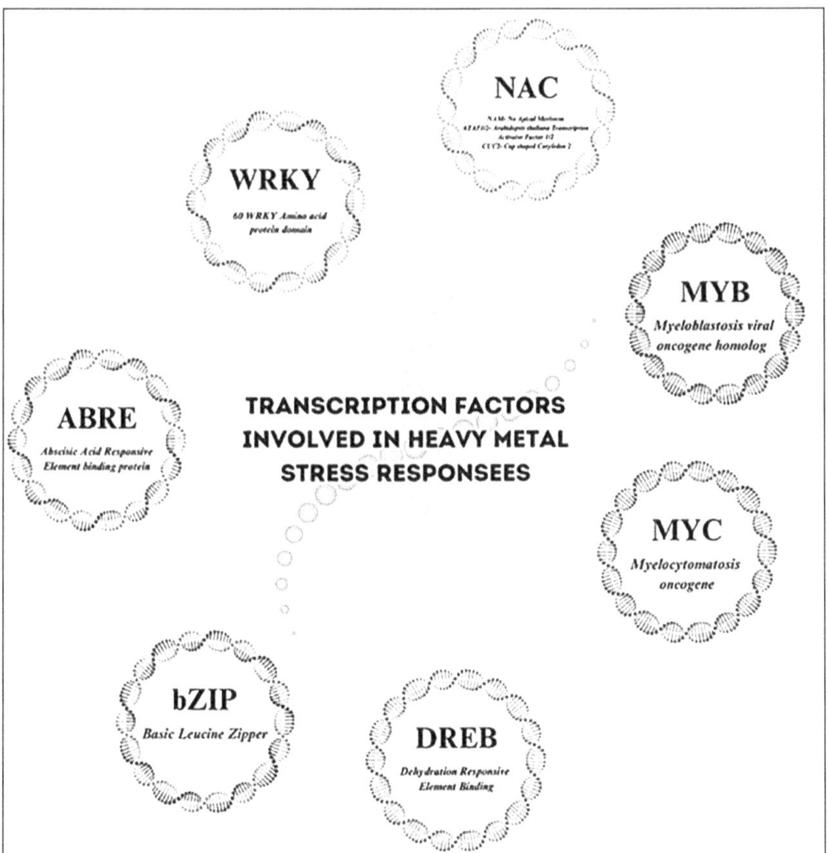

FIGURE 5.4 Transcription factors involved in heavy metal stress responses.

5.8.1 RESPONSIVE GENES OF METALS

Metals, which can be both necessary micronutrients and harmful pollutants, significantly impact the balance of cellular functions. Identifying metal-responsive genes is essential for comprehending an organism's ability to withstand stress caused by metals. Genome-wide expression profiling is crucial for revealing the complex genetic mechanisms that control metal tolerance. Researchers investigate metal-responsive genes by exposing organisms to controlled metal levels and carefully analyzing the subsequent alterations in gene expression. Organisms that can survive and grow in polluted environments often show distinct genetic characteristics related to the presence of heavy metals. Whole-genome expression analysis Profiling enables the detection of selectively activated or suppressed genes in reaction to metal exposure, offering an understanding of the adaptive tactics employed by these organisms. The knowledge acquired from identifying metal-responsive genes surpasses mere academic curiosity. It has practical ramifications in diverse domains, such as environmental science, agriculture, and medicine (Verma et al., 2022b, 2022c, 2022d). Ecological science relies on comprehending how organisms adapt to metal pollutants to develop practical bioremediation and environmental restoration approaches. Manipulating the expression of identified genes allows for the practical development of crops with improved metal tolerance in agriculture. Furthermore, in medicine, knowledge about metal-responsive genes enhances our comprehension of diseases associated with metal toxicity, thus enabling the development of precise therapeutic interventions (Kotia et al., 2021; Kumar et al., 2018a–2018g, 2021a, 2021b; Kumar and Pathak, 2019; Verma et al., 2022e, 2022f; Sharma et al., 2023).

5.8.2 DIFFERENTIAL EXPRESSION PATTERNS UNDER STRESS CONDITIONS

Stress, in its various manifestations, is a ubiquitous element of existence. From microbes to humans, organisms have developed complex mechanisms to deal with multiple stressors. Studying differential expression patterns in stressful conditions offers a detailed understanding of how genes coordinate responses to preserve cellular integrity and adverse functionality. Genome-wide expression profiling enables researchers to analyze and contrast gene expression patterns in normal and stressful conditions (Kumar et al., 2018a–2018g; Kumar and Naik, 2020; Verma et al., 2022g, 2022h). The emerging differential expression patterns provide valuable insights into the genes implicated in stress responses. For instance, when a plant experiences drought stress, it may increase the expression of genes associated with water conservation and decrease the expression of genes related to growth. Likewise, when bacteria are subjected to the pressure of antibiotics, they may activate genes that confer resistance to ensure their survival. The variability of these distinct expression patterns demonstrates the ability and durability of organisms to adjust and recover from challenges. Comprehending the molecular foundation of resilience in stressful conditions has extensive implications in diverse fields. Deciphering the distinct gene expression patterns in crops subjected to environmental stressors is beneficial for developing stress-tolerant crop varieties in agriculture (Kumar et al., 2018a–2018g; Siddique et al., 2018a, 2018b; Verma et al., 2022i, 2022j, 2022k).

This is especially pertinent in climate change, as unpredictable weather patterns substantially hinder crop production. Identifying distinct expression patterns in response to physiological stressors in medicine enhances our comprehension of stress-related diseases. It provides opportunities for the development of diagnostic markers and therapeutic strategies to improve resilience in individuals experiencing chronic stress. Integrating various molecular techniques, such as genome-wide expression profiling and identification of metal-responsive genes, along with examining differential expression patterns in stress conditions, provides a comprehensive understanding of the molecular mechanisms underlying resilience. These approaches are not mutually exclusive but complementary, offering additional layers of information that enhance our understanding of how organisms navigate complex environments. For example, a research study on a plant species that can grow well in soil contaminated with metals may use genome-wide expression profiling to identify genes that are affected by exposure to metals. Concurrently, researchers can examine the distinct gene expression patterns in stressful situations, revealing how these genes that respond to metals interact with more extensive stress response networks. Synthesis of these discoveries illuminates the molecular pathways determining the plant's ability to withstand stress caused by metals and other environmental adversities (Devi et al., 2023a, 2023b, 2023c; Kumar and Mistri, 2020).

5.8.3 OBSTACLES AND PROSPECTS

Despite the significant progress in understanding the molecular mechanisms of resilience through advanced techniques, there are still ongoing challenges. Researchers in the field face continuing challenges due to the intricate nature of biological systems, the interaction of multiple stressors, and the ever-changing process of gene regulation. A significant obstacle is integrating multi-omics data to build extensive molecular regulation. In addition to genome-wide expression profiling, advancements in technologies such as proteomics, metabolomics, and epigenomics offer supplementary layers of information. Combining these datasets poses a challenging task but holds the potential to reveal complex regulatory networks and interactions between various molecular entities. Another potential area for future investigation involves utilizing machine learning and computational modeling to interpret the extensive datasets produced by genome-wide expression profiling (Aley et al., 2022; Das et al., 2022).

Using predictive models that can anticipate an organism's response to specific stressors based on its genomic profile presents intriguing prospects for customized interventions across various domains. In addition, comprehending the resilience of microbial communities, which have crucial functions in maintaining ecological equilibrium and biogeochemical processes, necessitates using advanced methodologies. Metagenomics, which involves analyzing genetic material obtained directly from environmental samples and sophisticated analytical tools, shows potential in understanding how microbial communities collectively respond to stressors. The investigation of molecular mechanisms of resilience, which includes genome-wide expression profiling, the identification of metal-responsive genes, and the analysis of differential expression patterns in stress conditions, is a constantly changing and developing area of research. These methodologies enhance our comprehension of how organisms navigate obstacles and provide practical implications across various fields. By combining perspectives from these approaches, a comprehensive understanding of the molecular aspects of resilience can be achieved (Siddique et al., 2018a, 2018b; Kumar et al., 2022). The repercussions have wide-ranging effects, encompassing the creation of stress-resistant crops and the advancement of specific treatments for stress-related illnesses. The field's ongoing progress in technology is driving it forward. At the same time, the intricate molecular mechanisms of resilience present an intriguing frontier that has the potential to revolutionize our approach to agriculture, medicine, environmental conservation, and more.

5.9 TRANSCRIPTIONAL REGULATORY MECHANISMS

Transcriptional regulation is an intricate and carefully coordinated process that controls the expression of genes, enabling cells to adapt flexibly to different stresses. Transcriptional regulation involves two important aspects: the influence of transcription factors on metal stress responses and the impact of epigenetic modifications on chromatin remodeling. These factors determine how cells respond to environmental challenges (Kumar et al., 2013, 2018a–2018g). This exploration focuses on the complex realm of transcriptional regulatory networks, revealing how cells respond to metal stress and the significant influence of epigenetic modifications on chromatin dynamics. Cellular homeostasis is significantly compromised by metal stress, necessitating the activation of strong defense mechanisms by organisms. Transcription factors play a crucial role in these stress responses by coordinating the activation of genes involved in metal detoxification, transport, and cellular repair. The metal-responsive element-binding transcription factor (MTF) is a well-known family crucial in responding to metal stress. MTFs identify precise DNA sequences called MREs within the promoters of target genes. When exposed to metal stress, MTFs undergo structural changes, enabling them to attach to MREs and initiate the transcription of genes essential for maintaining body balance (Kumar et al., 2017, 2021a, 2021b).

This encompasses genes that encode metallothioneins, transporters, and enzymes involved in detoxification processes. Furthermore, the particularity of metal stress responses is frequently determined by additional categories of transcription factors. For instance, zinc finger transcription factors are essential for coordinating to zinc-stressed plants. These factors control the activation of zinc absorption and storage genes and impact the communication between various pathways related to metal stress. The metal stress responses underscore the interdependent nature of transcriptional regulatory networks. Interactions between pathways guarantee a thorough and flexible cellular reaction. Cells detect and react to variations in metal levels, and the complex interaction of transcription factors influences the gene expression pattern, allowing cells to adjust their metal tolerance mechanisms dynamically. Gaining a comprehensive understanding of the subtle distinctions of transcription factors in metal-stress responses is essential for deciphering the molecular foundation of metal toxicity and has practical implications. Controlling the function of these transcription factors shows potential for modifying crops to withstand higher levels of metals, reducing the negative effects of metal pollution on natural ecosystems, and improving treatment methods for metal-related diseases in humans. In addition to direct DNA sequence recognition by transcription factors, chromatin's epigenetic landscape substantially impacts gene expression (Kumar, Koul, et al., 2022; Saini et al., 2023).

Epigenetic modifications, such as DNA methylation, histone modifications, and chromatin remodeling, are complex mechanisms that regulate gene expression at the transcriptional level. DNA methylation adds methyl groups to cytosine residues, typically in CpG dinucleotides found in gene promoters. This alteration typically suppresses the activity of genes by obstructing the interaction between transcription factors or enlisting proteins that promote a repressive state of chromatin. Within the framework of metal stress, DNA methylation arises as a governing mechanism that influences the transcriptional response. Research has demonstrated changes in DNA methylation patterns due to exposure to metals. This affects the activation of genes that eliminate metals from the body and the body's response to stress (Pankaj et al., 2012b; Chakraborty et al., 2021).

The complex relationship between DNA methylation and metal stress highlights the intricate process of epigenetic regulation in cellular adaptation. Histones, the protein constituents of chromatin, undergo diverse post-translational alterations that regulate chromatin organization and gene expression. The modifications of histones, such as acetylation, methylation, phosphorylation, and ubiquitination, have been extensively studied and are known to affect transcriptional activity. Within the framework of metal stress responses, histone modifications function as dynamic regulators of gene expression. Acetylation of histones is commonly linked to the activation of transcription. Research has shown that there is a higher level of histone acetylation at the specific regions of DNA called promoters, which regulate the expression of genes that respond to metal stress. This increased histone acetylation helps to facilitate the expression of these genes when the organism is exposed to metal stress. On the other hand, histone deacetylation is associated with inhibiting gene expression, creating a sensitive equilibrium in the control of metal stress-responsive genes. Chromatin remodeling entails the relocation or removal of nucleosomes, which are the fundamental structural components of chromatin. This dynamic process enables transcription factors to access their target DNA sequences and promotes the creation of chromatin environments that either promote or inhibit transcription. Chromatin remodeling caused by metal stress is an intriguing aspect of controlling gene expression. Research has uncovered changes in the arrangement of nucleosomes and the accessibility of chromatin at gene locations that respond to metals (Kumar et al., 2020).

The interaction between chromatin remodelers and metal-responsive transcription factors influences the timing and location of the transcriptional response to metal stress, thereby enhancing the ability of cells to adapt to difficult environments. The amalgamation of transcriptional regulation and epigenetic modifications constitutes an intricate regulatory network that controls cellular reactions to metal stress. Transcription factors and epigenetic changes do not function independently; they work together and mutually affect each other in a dynamic interaction. Transcription factors implicated in metal stress responses can enlist chromatin-modifying enzymes to specific genes, affecting their epigenetic state (Pathak et al., 2017; Kumar et al., 2021a, 2021b). On the other hand, epigenetic modifications can establish a chromatin environment that is either favorable or inhibitory to the attachment of transcription factors. This complex interaction guarantees the accurate and context-specific control of gene expression in reaction to metal stress. Gaining insight into the interaction between the control of gene expression and chemical modifications of DNA is essential for unraveling the precise molecular processes that enable organisms to withstand the harmful effects of metal exposure (Kumar et al., 2017, 2021a, 2021b; Pathak et al., 2017; Siddique et al., 2018a, 2018b; Kumar and Mistri, 2020; Chakraborty et al., 2021).

Furthermore, this understanding has significant implications for creating specific interventions, such as epigenetic treatments, that can control how cells react to metal toxicity in various biological systems. Despite notable advancements, there are still ongoing difficulties in understanding the intricate transcriptional regulatory networks and epigenetic mechanisms involved in metal stress. Researchers face ongoing challenges due to the dynamic nature of regulatory processes, the diversity of metal stressors, and the complexity of cellular aspects. An obstacle arises in understanding the complexities of context-dependent reactions to various metals. Although there are similarities in the transcriptional and epigenetic reactions to metal stress, subtle differences depend on the specific metal, its concentration, and the length of exposure (Ali et al., 2022; Etesami and Glick, 2024; Goud et al., 2022; Mohapatra et al., 2023; Rai et al., 2023; Sharma and Kumar, 1999; Srivastav et al., 2023).

A comprehensive understanding of these subtle distinctions is essential for formulating precise strategies in various environmental and biomedical scenarios. Another potential area for future investigation involves utilizing high-throughput technologies and systems biology methods to understand the intricate nature of transcriptional and epigenetic landscapes. Integrating multi-omics data, encompassing genomics, transcriptomics, and epigenomics, enables a thorough comprehension of the molecular mechanisms involved. Moreover, sophisticated computational modeling and machine learning methods can forecast and control transcriptional and epigenetic reactions in different metal stress situations. Studying the transcriptional regulatory networks, particularly metal stress responses and their interaction with epigenetic modifications, is an exciting area of research in plant science. This comprehensive viewpoint enhances our comprehension of how cells adapt in difficult conditions by examining the functions of metal-responsive transcription factors and investigating the complexities of epigenetic regulation and chromatin remodeling. Transcriptional regulation and epigenetic modifications are interconnected elements within a complex regulatory network that controls how cells respond to metal stress. Incorporating these regulatory layers guarantees cells' accuracy, flexibility, and robustness when confronted with environmental challenges. As scientific research advances, the knowledge acquired by understanding these molecular mechanisms shows potential for various practical uses, such as ecological bioremediation and the development of specific treatments for metal-related diseases. Ultimately, this contributes to the wider field of sustainable and resilient biology (Kumar et al., 2011b; Paul et al., 2005; Chakraborti et al., 2022; Choudhary and Muthamilarasan, 2022; Dilnawaz et al., 2023).

5.10 DYNAMICS OF REACTIVE OXYGEN SPECIES

Reactive oxygen species (ROS) are extremely reactive molecules generated within cells due to regular cellular metabolism. ROS have crucial functions in multiple physiological processes. However, an imbalance in their production and removal can result in oxidative stress, a condition associated with numerous diseases. Gaining insight into the mechanisms of ROS, their signaling networks, and the complex interaction with antioxidant defense systems is crucial for deciphering cellular responses to stress and preserving redox homeostasis. ROS, including molecules like superoxide ($O_2^{\cdot -}$), H_2O_2, and hydroxyl radicals ($\cdot OH$), serve as signaling molecules in various cellular processes (Kandpal et al., 2018; Kumar et al., 2023; Devi et al., 2023a, 2023b, 2023c).

The ROS signaling network is a complex system that combines environmental cues, cellular stress, and developmental signals to regulate cellular responses. ROS are vital in important physiological processes, such as regulating cell growth, specialization, and immune reactions. Enzymes such as NADPH oxidases, mitochondria, and peroxisomes are important components of the ROS signaling network. These cellular compartments generate ROS as part of regular metabolic processes and, importantly, in response to specific triggers. NADPH oxidases, for example, produce superoxide as a component of the immune response to fight against pathogens, emphasizing the dual characteristic of ROS as both advantageous and potentially detrimental signaling molecules. ROS, or reactive oxygen species, are secondary messengers in stress responses. They transmit signals from different stressors to trigger specific cellular pathways. Exposure to environmental stressors, such as ultraviolet radiation or pollutants, increases ROS levels. The increase in ROS serves as a signal, triggering the activation of genes and pathways that respond to stress. In plants, ROS signaling is integral to defense mechanisms against pathogens and environmental stressors, while in animals, it plays a role in immune responses and cellular adaptation to stress (Kumar et al., 2018a–2018g; Kotia et al., 2021; Das et al., 2022).

The versatility of ROS as secondary messengers is apparent in their participation in abiotic stress responses, such as drought, extreme temperatures, and exposure to heavy metals. The delicate equilibrium between producing and removing ROS is meticulously regulated to guarantee a suitable and targeted cellular reaction to external stimuli. The interaction between ROS and other signaling pathways increases the intricacy of cellular responses. ROS can regulate the function of different signaling molecules, such as kinases, phosphatases, and transcription factors, affecting various cellular processes. An exemplary instance is the interplay between ROS and mitogen-activated protein kinase (MAPK)-signaling pathways (Chakraborty et al., 2021; Kumar et al., 2021a, 2021b; Aley et al.,

2022; Saini et al., 2023). ROS can initiate MAPK cascades, resulting in the activation of stress-responsive genes. In contrast, MAPKs can control the functioning of enzymes that produce ROS, thus creating a feedback mechanism that precisely adjusts cellular reactions to stress. Furthermore, ROS engages with signaling pathways implicated in inflammation, apoptosis, and autophagy. Incorporating ROS signaling into these pathways emphasizes the diverse function of ROS in coordinating cellular responses. It emphasizes their importance as central participants in the coordination of stress-related occurrences. Cells have developed intricate antioxidant defense systems to effectively regulate levels of ROS and uphold redox homeostasis, thus mitigating potential damage caused by ROS.

The defense systems consist of enzymatic and nonenzymatic antioxidants that work together to counteract ROS and hinder oxidative harm. Enzymatic antioxidants play a vital role in protecting against ROS. SOD, CAT, and peroxidases are crucial enzymes that eliminate ROS by transforming them into molecules that are less reactive or harmless. SOD facilitates the conversion of superoxide radicals into hydrogen peroxide and oxygen through a process called dismutation. Catalase catalyzes the decomposition of hydrogen peroxide into water and oxygen, inhibiting ROS buildup. Peroxidases, such as glutathione peroxidases and peroxiredoxins, play a role in removing different types of peroxides and reducing oxidative stress. Vitamins C and E, glutathione, and flavonoids are nonenzymatic antioxidants (Kandpal et al., 2018; Kumar et al., 2018a–2018g, 2019a, 2019b, 2021; Kumar and Dwivedi, 2018). They work alongside enzymatic defenses by removing ROS and restoring enzymatic antioxidants. Vitamin C functions as a powerful antioxidant that dissolves in water, effectively neutralizing ROS within the watery environment of cells. Vitamin E, a type of antioxidant that dissolves in lipids, shields cell membranes from damage caused by the oxidation of lipids, which is triggered by ROS.

Glutathione, a three-amino acid antioxidant, engages in redox reactions and acts as a cofactor for specific antioxidant enzymes. The wide range of nonenzymatic antioxidants demonstrates the ability of cells to adapt to various oxidative conditions and emphasizes the significance of maintaining a strong antioxidant defense system. The interaction between ROS signaling networks and antioxidant defense systems forms a highly regulated and coordinated defense mechanism against oxidative stress. This coordination entails the adaptive modulation of gene expression, enzyme activity, and the presence of nonenzymatic antioxidants in response to the cellular redox state. During periods of stress, cells activate particular transcription factors, such as nuclear factor erythroid 2-related factor 2, which coordinate the increase in expression of genes that encode antioxidant and phase II detoxification enzymes. This transcriptional response increases the ability of cells to counteract ROS and adjust to oxidative stress (Sharma et al., 2022c; Shinde et al., 2024; Bera et al., 2024). Moreover, cells maintain an intricate equilibrium between generating and removing ROS. Mitochondria, acting as both generators and recipients of ROS, have a pivotal function in maintaining this equilibrium. Mitochondrial ROS can signal molecules in cellular processes. However, an overabundance of mitochondrial ROS production is linked to cellular dysfunction and aging. Cells can control ROS levels, which affect autophagy regulation. Autophagy is a cellular process that involves breaking down and recycling damaged components.

ROS can impact autophagic activity, aiding in maintaining cellular balance by eliminating structures damaged by oxidation and reducing oxidative stress. ROS dynamics, including ROS signaling networks, their function as secondary messengers in stress responses, interaction with other signaling pathways, and coordination of antioxidant defense systems, constitute a complex and interconnected framework in cellular biology. ROS play a crucial role in maintaining normal cellular function and are involved in important signaling processes when carefully controlled. Nevertheless, equilibrium in ROS concentrations can result in oxidative stress associated with various pathological conditions. Gaining a comprehensive understanding of the intricate nature of ROS dynamics and the cellular mechanisms developed to mitigate oxidative stress is essential for furthering our comprehension of how cells respond to environmental stressors, aging, and diseases linked to an imbalance in the redox state (Dubey et al., 2002; Dwivedi and Kumar, 2011; Kandpal et al., 2018; Kumar and Dwivedi, 2020). Studying the complex molecular processes of ROS signaling and antioxidant defenses provides opportunities for developing therapeutic interventions that target conditions related to oxidative stress. This highlights the crucial role of ROS in maintaining cellular health by maintaining a delicate balance.

5.11 INTERACTION OF PLANT MICROBES

Plant–microbe interactions are complex networks that significantly impact plants' growth, development, and ability to withstand stress. Two significant components of these interactions are the mutualistic relationship between mycorrhizal fungi and plants and the protective effects mediated by rhizobacteria. This exploration focuses on the intricate dynamics of mycorrhizal associations, particularly their impact on the absorption and tolerance of metals, as well as the signaling pathways that contribute to the enhanced ability of mycorrhizal-assisted stress resilience. In addition, we investigate the function of beneficial rhizobacteria in the metal detoxification process and the developing field of manipulating plant–microbe collaborations to improve resilience (Kumar et al., 2018a–2018g; Kumar et al., 2021a, 2021b, 2022; Das et al., 2022). Mycorrhizal symbiosis refers to a mutually beneficial relationship between plants and fungi. The fungus establishes itself in the plant's roots, creating arbuscules and vesicles. The mutualistic association between the plant and the fungus improves the plant's ability to absorb nutrients, particularly phosphorus, while the fungus obtains carbohydrates from the plant. In addition to nutrient exchange, mycorrhizal associations play a crucial role in stress resilience, particularly in metal tolerance. Mycorrhizal associations' impact on plant metal absorption and resilience is intriguing.

Mycorrhizal fungi can improve the uptake of vital metals, such as phosphorus, and have a critical function in reducing the harmful impacts of metals, including heavy metals like cadmium and zinc. Mycorrhizal hyphae expand the range of the plant root systems, enabling access to areas abundant in nutrients and facilitating the absorption of metals. Additionally, mycorrhizal fungi create metal-binding complexes, decreasing the amount of toxic metals the plant can absorb. The importance of mycorrhizae in plant adaptation to challenging environments is highlighted by their ability to both acquire nutrients and detoxify ions. Mycorrhizal-assisted stress resilience is not solely a physical interaction but encompasses complex signaling pathways regulating the plant's stress reactions. When exposed to stressors like heavy metal contamination or drought, mycorrhizal associations initiate signaling cascades that activate plant stress-responsive genes. The jasmonic acid (JA) pathway is a significant signaling pathway that plays a role in stress resilience with the assistance of mycorrhizal. JA is a phytohormone that exhibits a response to diverse environmental stresses, such as herbivory and pathogen assaults. Mycorrhizal associations stimulate the synthesis of JA in plants, preparing them for a proactive response to stressors (Kumar et al., 2013, 2017; Chakraborty et al., 2021; Das et al., 2022; Kumar et al., 2022).

The involvement of the salicylic acid (SA) pathway is also suggested in stress responses mediated by mycorrhizal interactions. SA is a plant hormone that plays a role in protecting against pathogens. The activation of the SA pathway by mycorrhizae indicates that mycorrhizae play a complex role in enhancing plant immunity and resilience to stress. Gaining knowledge about these signaling pathways offers a deeper understanding of the molecular mechanisms that contribute to the improved ability to withstand stress through mycorrhizal associations. This knowledge can be used to develop specific approaches to genetically modifying crops to be more tolerant of stress. Aside from mycorrhizal symbiosis, another aspect of plant–microbe interactions pertains to the impact of beneficial rhizobacteria on plant health and their ability to withstand stress. Rhizobacteria, which inhabit the rhizosphere (the soil area surrounding plant roots), provide protective benefits to plants through various mechanisms (Figure 5.5).

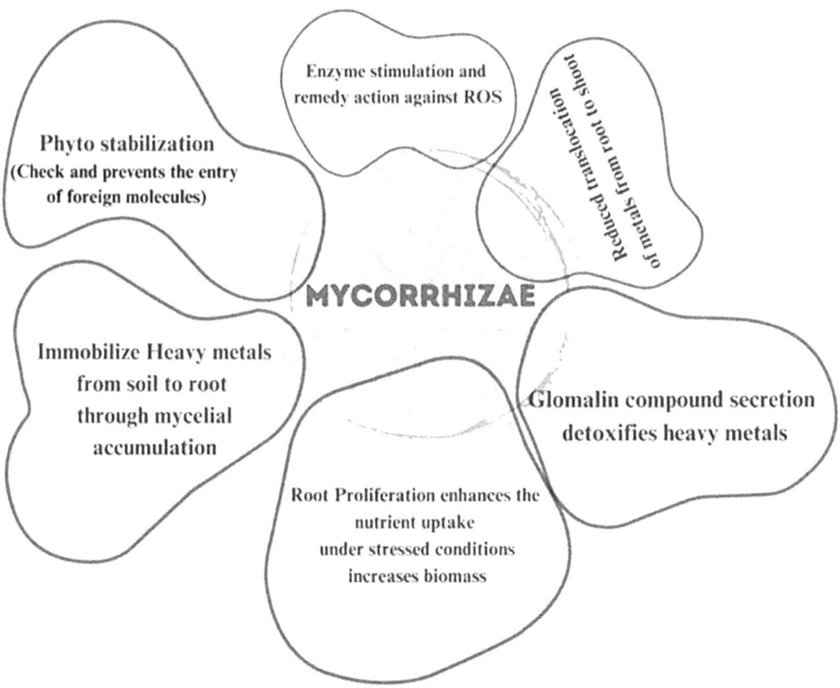

FIGURE 5.5 Interaction of plant microbes.

Rhizobacteria that are beneficial play a significant role in the process of metal detoxification. These bacteria can alter the availability of metals in the rhizosphere, thereby increasing or decreasing their accessibility to plants. Rhizobacteria play a crucial role in mitigating the absorption of harmful metals by plants through mechanisms like metal sequestration and precipitation. Furthermore, specific rhizobacteria can chemically convert toxic metal ions into less detrimental forms. For example, certain bacteria synthesize metal-chelating compounds and organic acids that attach to metals, inhibiting plant roots' absorption. Rhizobacteria detoxification mechanisms demonstrate their potential for sustainable methods to alleviate plant stress caused by metals. With the advancement of our knowledge regarding plant–microbe interactions, there is a growing desire to manipulate these relationships to improve the ability of plants to withstand environmental stressors. This entails strategically using plant and microbial elements to maximize advantageous results. In the context of mycorrhizal associations, genetic engineering methods strive to increase the expression of genes involved in stress response pathways activated by mycorrhizae (Dubey et al., 2002; Kumar et al., 2018a–2018g, 2019a, 2019b; Sharma et al., 2023).

This involves altering the expression of genes associated with JA and SA signaling pathways and genes involved in nutrient transportation and metal detoxification. The objective is to develop plants that demonstrate enhanced resistance to stress by optimizing their mycorrhizal interactions. In the same way, engineering plant–rhizobacteria partnerships entails carefully selecting or modifying rhizobacterial strains with improved ability to mobilize nutrients, detoxify metals, and produce substances that promote plant growth. By tailoring the rhizobacterial component to specific stressors or environmental conditions, it is possible to create customized plant–microbe partnerships that enhance stress resilience. Plant–microbe interactions, including mycorrhizal symbiosis and rhizobacteria-mediated protection, are complex and dynamic networks that significantly impact plant health, growth, and ability to withstand stress. Mycorrhizal associations improve the absorption of nutrients and have a vital function in removing metals from the system, with signaling pathways playing a role in the ability to withstand stress. Conversely, advantageous rhizobacteria actively remove metals from plants and enhance their health and well-being. As our understanding of these interactions increases, there is a growing focus on utilizing this knowledge for sustainable agriculture.

The collaboration between engineering plants and microbes presents promising opportunities for developing resilient crops, mitigating the environmental consequences of metal pollution, and enhancing plant nutrient absorption efficiency. The complex interaction between plants and microbes gradually reveals its mysteries, offering potential novel approaches to tackle the difficulties posed by an evolving and stress-ridden environment.

5.12 ADVANCED TECHNOLOGIES FOR ASSESSMENT AND MONITORING

Advanced technologies are required to assess and monitor the distribution and dynamics of metals in environmental sciences. These technologies provide detailed insights necessary for effective management and mitigation approaches. Researchers are obtaining unprecedented access to the complex mechanisms regulating metal interactions in ecosystems by utilizing state-of-the-art methods, such as imaging, advanced spectroscopy, and omics integration. Imaging technologies have significantly transformed our ability to visualize the distribution of metals in living organisms and ecosystems (Siddique et al., 2018a, 2018b; Kumar et al., 2021a, 2021b; Saini et al., 2023). Methods such as X-ray microtomography and magnetic resonance imaging facilitate nonintrusive, detailed imaging, enabling researchers to observe the precise arrangement of metals within organisms and environmental matrices. Honoring metal distribution in living organisms allows for a dynamic comprehension of the movement and accumulation of metals within plants, animals, and ecosystems as time progresses. This insight is essential for evaluating the influence of metal pollutants, comprehending bioaccumulation pathways, and formulating specific approaches for environmental remediation. Real-time monitoring of metal dynamics is crucial for capturing ecological systems' dynamic processes and responses. Advanced spectroscopic techniques, such as laser-induced breakdown spectroscopy (LIBS) and synchrotron radiation-based methods, provide exceptional abilities for monitoring metal concentrations and speciation in real-time and on-site. LIBS allows for quick and precise analysis of chemical elements with high sensitivity. This technology is invaluable for conducting on-site assessments, enabling prompt decision-making in metal contamination emergencies (Kumar et al., 2017, 2018a–2018g, 2019a, 2019b; Kumar and Mistri, 2020; Kumar and Naik, 2020; Chakraborty et al., 2021; Aley et al., 2022; Sharma et al., 2023).

In contrast, synchrotron-based techniques offer highly detailed information about the specific chemical forms of metals, allowing for a better understanding of their availability in biological systems and their potential effects on the environment. Omics technologies, including metabolomics, proteomics, and genomics, have become influential tools in understanding the intricate relationship between organisms and metal contaminants. Integrating these omics approaches enables a comprehensive comprehension of the molecular mechanisms that control metal responses in biological systems. Metabolomics enables the thorough analysis of small molecules that play a role in cellular processes. By studying the metabolome, scientists can detect and determine the metabolic pathways affected by exposure to metals, thus capturing a momentary representation of the cellular reaction. Proteomics explores the complete set of proteins present in a cell or organism. This method uncovers changes in protein expression and modifications that occur after translation, which are caused by exposure to metal stress. Comprehending the proteome offers an understanding of the molecular mechanisms utilized for metal detoxification and cellular defense. Genomics, the scientific discipline that investigates the entirety of an organism's genetic material, aids in the identification of genetic elements that impact the tolerance and reactions to metals.

Scientists can discover the genetic foundation of adaptive characteristics by identifying metal-responsive genes and potentially manipulating plants or microorganisms to have improved metal tolerance. Data integration is essential for extracting comprehensive insights from various datasets produced by imaging, spectroscopy, and omics technologies. Integrating advanced computational methods and bioinformatics tools allows for synthesizing information from diverse sources, enabling a comprehensive comprehension of metal dynamics in intricate environmental aspects. Data integration allows scientists to establish a connection between spatial information obtained from imaging techniques and the molecular information obtained from omics analyses. This collaboration allows for developing comprehensive models that accurately depict the complex connections between metal exposure, biological reactions, and ecological results. State-of-the-art environmental science assessment and monitoring technologies provide a complete method for comprehending the distribution and behavior of metals. Imaging methods offer spatial understanding, sophisticated spectroscopy allows for immediate monitoring, and integration of omics reveals the molecular mechanisms. Combining these strategies and advanced data integration techniques enables scientists to understand the intricacies of metal interactions in ecosystems, facilitating informed decision-making for sustainable environmental management.

5.13 INTERVENTIONS FOR SUSTAINABLE AGRICULTURE

Technological interventions have become catalysts for transformative change in pursuing agriculture sustainability. These interventions address climate change, resource limitations, and food security challenges. Two crucial areas driving this revolution are advancements in genetic engineering, specifically in clustered regularly interspaced short palindromic repeats (CRISPR)/Cas-based precision modifications, and strategies in precision agriculture, utilizing sensor-based precision management and data-driven decision-making. These advanced technologies have the potential to enhance agricultural productivity and promote environmental stewardship and resilience in changing rural landscapes. The advent of genetic engineering, specifically the groundbreaking CRISPR/Cas technology, has introduced a new period characterized by accurate alterations in agriculture. CRISPR/Cas facilitates precise and targeted modifications in the genetic code of crops, enabling scientists to enhance desirable characteristics with exceptional accuracy (Paul et al., 2005; Kumar et al., 2011a; Siddique and Kumar, 2018; Joel et al., 2023).

This technology allows for the manipulation of specific genes, facilitating the creation of crops with excellent resistance to environmental stressors, improved nutritional value, and heightened immunity against pests and diseases. CRISPR/Cas-based genetic engineering provides high accuracy, reducing unintended genetic changes. This effectively addresses the concerns that were associated with previous genetic modification strategies. Attaining such a high degree of precision is crucial for cultivating crops that can flourish in various and difficult agroecological circumstances. This, in turn, supports sustainable agricultural methods that emphasize effectiveness, conservation of resources, and minimizing environmental harm.

Genetic engineering surpasses the constraints of conventional crop breeding, allowing for the development of crops with customized characteristics well-suited to modern agricultural obstacles. Genetically modified crops can demonstrate an improved ability to withstand drought, increased resistance to particular diseases, and enhanced nutrient utilization efficiency. These alterations enhance crop productivity and decrease reliance on chemical inputs, promoting environmentally sustainable farming methods. Genetically modified crops with improved resistance help encourage climate-smart agriculture, enabling farmers to adjust to shifting climatic conditions and alleviate the effects of severe weather occurrences. Through genetic engineering, agriculture gains the ability to adapt and withstand challenges, thus establishing the basis for sustainable food production in the presence of a changing and unpredictable climate. Precision agriculture is a transformative change in farming methods that utilizes technology to maximize resource utilization and improve productivity (Paul et al., 2005; Kumar et al., 2011a; Siddique and Kumar, 2018; Dilnawaz et al., 2023; Joel et al., 2023; Mohan et al., 2023; Sheri et al., 2023; Zhao et al., 2024).

The foundation of this approach lies in using sensor-based precision management, which offers up-to-the-minute information on a range of factors such as soil moisture, nutrient levels, and crop health. These sensors, commonly used on drones or ground-based vehicles, allow farmers to observe and react to changing field conditions with exceptional accuracy. Utilizing sensor-based precision management enables the precise and efficient utilization of resources, thereby minimizing both waste and environmental impact. By utilizing real-time data, farmers can customize irrigation, fertilization, and pest control methods, resulting in optimized resource allocation, reduced expenses, and a minimized ecological impact of agriculture. By utilizing data, this approach enhances sustainability by harmonizing agricultural methods with the unique requirements of individual crops and fields, thereby enhancing the overall effectiveness and adaptability of the farm system. Integrating data-driven decision-making into farming practices is crucial for the success of precision agriculture. The extensive quantities of data gathered from sensors, satellites, and on-farm monitoring devices are analyzed using sophisticated analytics, machine learning algorithms, and broad data applications.

The abundance of data enables farmers to make well-informed choices regarding crop management, resource distribution, and risk reduction. Data-driven decision-making enables the detection of trends, patterns, and correlations that may go unnoticed through traditional observation methods. It allows farmers to optimize planting schedules, efficiently manage irrigation, and promptly address emerging pest or disease threats. Through data utilization, agriculture transforms into a flexible and adaptable system capable of responding to evolving circumstances and contributing to producing sustainable and resilient food. Implementing technological advancements in sustainable agriculture, such as genetic engineering breakthroughs and precision agriculture techniques, profoundly alters the trajectory of food production. These technologies provide accurate and flexible solutions that improve agricultural productivity and help protect the environment while adapting to changing challenges. These advancements have the potential to establish a sustainable and efficient farming system that can fulfill the needs of a rapidly increasing global population while also protecting the planet's well-being.

5.14 CHALLENGES AND OPPORTUNITIES

As we enter a new era of agricultural advancement, it is crucial that we actively explore new possibilities and tackle obstacles to guarantee the long-term viability and adaptability of global food systems. Nanotechnology and synthetic biology have great potential for reducing crop stress. Simultaneously, global collaborations and policy frameworks are essential for cultivating international cooperation and advancing sustainable practices. Nevertheless, the emergence of promising technologies necessitates incorporating ethical considerations and the establishment of comprehensive policy frameworks. These measures are crucial to effectively tackle potential challenges and guarantee these technologies' fair and responsible implementation. The capacity of nanotechnology to manipulate matter at the nanoscale presents revolutionary opportunities for alleviating stress in agriculture. Nanoparticles can be engineered to transport nutrients, optimize water utilization, and mitigate the effects of environmental stressors on crops. Nanotechnology allows for precisely manipulating plant stress-responsive pathways, offering opportunities to develop highly effective solutions with minimal environmental impact. Nanoparticle-mediated delivery systems for fertilizers and pesticides facilitate precise and regulated release, decreasing inefficiency and mitigating ecological consequences. In addition, nanosensors can offer immediate and ongoing monitoring of plant health and the environment, enabling prompt interventions. Prudent evaluation of safety, environmental consequences, and ethical considerations is crucial to guarantee responsible implementation and utilization of nanotechnology in agriculture despite its significant potential advantages.

Synthetic biology, which involves manipulating biological systems to achieve specific objectives, provides a potent set of tools for customizing method to agricultural issues. Synthetic biology facilitates the development of crops with augmented characteristics, enhanced resistance to stress, and heightened nutritional content through the design and construction of novel physical components, devices, and systems. An exemplary use case involves the creation of stress-tolerant crops that are genetically modified to endure challenging environmental circumstances, such as drought or excess soil salinity. Synthetic biology also generates bioengineered microbes that augment soil nutrient accessibility, contributing to sustainable farming practices. As we explore the boundaries of synthetic biology in agriculture, it is crucial to carefully address ethical considerations, ecological impacts, and regulatory frameworks to guarantee responsible innovation. The agricultural challenges surpass the boundaries of individual countries, requiring international collaborations to develop practical solutions. Global collaboration is essential for exchanging knowledge, resources, and technologies to tackle shared agricultural challenges, specifically those about the resilience of crops under heavy metal stress. Collaborative research initiatives, joint projects, and knowledge exchange platforms have the potential to expedite progress and strengthen the combined ability of nations to tackle agricultural challenges. Furthermore, it is imperative to expand global partnerships to establish international norms and principles for the ethical and conscientious utilization of emerging approaches in agriculture.

Standardizing regulations and policies internationally can establish a favorable atmosphere for innovation while protecting environmental and societal concerns. Strong policy frameworks are necessary to direct the implementation of new technologies and guarantee the use of sustainable agricultural methods. To effectively address resilience to heavy metal stress, policies should include directives for managing soil, monitoring water quality, and the responsible utilization of lands contaminated with metals. It is important to incorporate incentives for adopting precision farming practices, which can reduce the environmental effects of metal pollutants, into agricultural policies. Moreover, policies should prioritize the significance of research and development in sustainable agriculture, promoting investments in technologies that improve crop stress tolerance. Endorsing endeavors that advocate for agroecological practices, organic farming, and agroforestry can aid in developing robust agricultural systems that are more resistant to the negative effects of heavy metal exposure.

Nevertheless, the formulation of policies should follow a systematic and comprehensive approach involving contributions from scientists, farmers, policymakers, and other relevant parties. Ensuring a sustainable future for agriculture requires policy frameworks that effectively balance innovation with ethical considerations, environmental conservation, and social equity. Future advancements in agriculture are closely connected with developing new technologies, international partnerships, and strong policy structures. The utilization of nanotechnology and synthetic biology holds promising potential for mitigating stresses. However, it is crucial to approach their implementation with ethical considerations and responsible policies to ensure responsible deployment. International collaboration is crucial for harnessing shared knowledge and resources, while policy frameworks are essential in determining the direction of sustainable agriculture. While exploring these new areas, creating technologically advanced, socially equitable, environmentally conscious, and ethically sound solutions is crucial.

5.15 CONCLUSION AND FUTURE RECOMMENDATIONS

Investigating the physiological and molecular aspects of heavy metal stress resilience is crucial for improving agriculture's effectiveness and long-term viability in response to environmental difficulties. Comprehending the complex relationship between plants and severe metal stress requires a multifaceted understanding of the fundamental mechanisms involved, which involves integrating physiological responses and molecular adaptations. Physiological studies have revealed the various mechanisms plants employ to deal with the challenges of heavy metal stress. Plants demonstrate remarkable adaptability in their responses, from containing harmful metals in vacuoles to activating antioxidant defense systems. Gaining knowledge about these physiological adaptations serves as a basis for cultivating crops with improved stress resilience, a critical factor in guaranteeing food security in areas impacted by metal-contaminated soils. In addition to providing physiological understanding, molecular research explores the complex genetic and cellular processes that regulate responses to heavy metal stress. Studying the communication pathways, control of gene activity, and gene expression patterns related to metals improves our capacity to modify crops with a specific resistance.

The advent of sophisticated molecular methodologies, such as transcriptomics, genomics, and proteomics, has expedited our comprehension of the molecular framework that regulates the reactions to heavy metal stress. Incorporating physiological and molecular knowledge is crucial for creating comprehensive approaches to enhance resistance to severe mental stress. Identifying essential genes and pathways implicated in metal detoxification, nutrient transport, and stress signaling enables precise genetic modifications using techniques such as CRISPR/Cas. Customizing crops with improved resistance helps reduce the adverse effects of heavy metal stress on agricultural output and promotes sustainable farming methods. The ability to withstand heavy metal stress effectively leads to higher agricultural productivity, more excellent economic stability for farmers, and improved food security for expanding populations. By unraveling the complex molecular details, scientists and researchers create opportunities for precision agriculture, in which crops possess the genetic mechanisms to flourish in environments contaminated with metals. This innovative strategy tackles the immediate difficulties caused by heavy metal stress and propels agriculture toward long-term sustainability.

Sustainability, a fundamental aspect of future agricultural practices, encompasses the maintenance of ecological equilibrium, the responsible management of the environment, and the fulfillment of social obligations. Understanding the physiological and molecular mechanisms underlying the ability to withstand heavy metal stress is crucial for developing long-lasting solutions. As we deal with the challenges of severe metal stress, it is clear that combining our understanding of the body's functions with precise knowledge of molecules is crucial for guiding agriculture toward a robust, efficient, and environmentally friendly future. The interdependence between science and agriculture strengthens our capacity to address current obstacles and establish the foundation for a more sustainable and resilient global food system.

KEYWORDS

- **environmental pollution**
- **cadmium toxicity**
- **plant microbiome**
- **plant tissues**
- **metabolomics**
- **proteomics**
- **genomics**
- **mitigation strategies**
- **metal toxicity**
- **physiology**
- **molecular activities**

REFERENCES

Aley P, Singh J, Kumar P. (2022) Adapting the changing environment: Microbial way of life. In: Kumar A, Singh J, Ferreira LFRBT-MUCC. (Eds.), *Microbiome Under Changing Climate* Woodhead Publishing, pp. 507–525. https://doi.org/10.1016/B978-0-323-90571-8.00023-7

Ali S, Tyagi A, Mushtaq M, Al-Mahmoudi H, Bae H. (2022) Harnessing plant microbiome for mitigating arsenic toxicity in sustainable agriculture. *Environ. Poll.* 300, 118940. https://doi.org/10.1016/j.envpol.2022.118940

Anas M, Verma KK, Haq IU, Naeem M, Li Q, Liao F, Xiang YX, Li YR. (2021a) Characterization of exotic, native and wild-type genotypes of sugarcane (*Saccharum* spp. Hybrids) for internal nitrogen use efficiency under different nitrogen levels. *Sugar Tech.* 23, 1258–1267. https://doi.org/10.1007/s12355-021-01014-1

Anas M, Verma KK, Riaz M, Qiang L, Liao F, Liu Y, Li YR. (2021b) Physio-morphological and biochemical mechanism of nitrogen use efficiency in sugarcane (*Saccharum* spp.) genotypes under different growth stages and nitrogen levels. *J. Plant Interact.* 16, 332–343.

Balooch S, Noreen S, Mahmood S, Zahra N, Azeem A, Altaf MM, Salim Akhter M, Abbas A. (2024) Unlocking Triticum aestivum L. resilience: Exogenous proline's remarkable role in mitigating lead stress, and delving into the secrets of TaGSK1 gene expression. *South African J. Bot.* 166, 79–87. https://doi.org/10.1016/j.sajb.2023.12.022

Bera K, Sadhukhan S, Gunjal A, Choudhury A, Dutta P. (2024) Phytomicrobiome in modulating plant growth and stress resilience: An insight into the functions and emerging perspectives in agriculture. In: Parray JA, Shameem N, D.B.T.-M.D. of Egamberdieva

EF. (Eds.), *Microbiome Research in Plants and Soil*. Academic Press, pp. 31–59. https://doi.org/10.1016/B978-0-443-19121-3.00010-7

Bhatt R, Verma KK, Jamwal NS. (2023a) Can paddy-compost eliminate N fertilization for climate-smart sugarcane? *Sugar Tech*. 25, 1036–1046.https://doi.org/10.1007/s12355-023-01270-3

Bhatt R, Verma KK, Kumar R, Sanghera GS. (2023b) Foliar application of potassium salt of active phosphorus (PSAP) mitigates insect pests and improves yield along with sugarcane quality in response to agroclimatic conditions of Punjab. *Sugar Tech*. 25(3), 660–669.

Chakraborti S, Bera K, Sadhukhan S, Dutta P. (2022) Bio-priming of seeds: Plant stress management and its underlying cellular, biochemical and molecular mechanisms. *Plant Stress* 3, 100052. https://doi.org/10.1016/j.stress.2021.100052

Chakraborty S, Kumar P, Sanyal R, Mane AB, Arvind Prasanth D, Patil M, Dey A. (2021) Unravelling the regulatory role of miRNAs in secondary metabolite production in medicinal crops. *Plant Gene* 27, 100303. https://doi.org/10.1016/j.plgene.2021.100303

Chaudhary P, Xu M, Ahamad L, Chaudhary A, Kumar G, Adeleke BS, Verma KK, Hu D-M, Siric I, Kumar P, Popescu SM, Fayssal SA. (2023) Application of synthetic consortia for improvement of soil fertility, pollution remediation, and agricultural productivity: A review. *Agronomy* 13, 643. https://doi.org/10.3390/agronomy13030643.

Chen GL, Zheng FJ, Lin B, Lao SB, He J, Huang Z, Zeng Y, Sun J, Verma KK. (2020) Phenolic and volatile compounds in the production of sugarcane vinegar. *ACS Omega* 5, 30587−30595.

Chen G-L, Zheng F-J, Lin B, Yang Y-X, Fang X-C, Verma KK, Yang L-F. (2023) Vinegar: A potential source of healthy and functional food with special reference to sugarcane vinegar. *Front. Nutr*. 10, 1145862. https://doi.org/10.3389/fnut.2023.1145862

Choudhary P, Muthamilarasan M. (2022) Modulating physiological and transcriptional regulatory mechanisms for enhanced climate resilience in cereal crops. *J. Plant Physiol*. 278, 153815. https://doi.org/10.1016/j.jplph.2022.153815

Das T, Saha SC, Sunita K, Majumder M, Ghorai M, Mane AB, Prasanth DA, Kumar P, Pandey DK, Al-Tawaha AR, Batiha GE-S, Shekhawat MS, Ghosh A, Sharifi-Rad J, Dey A. (2022) Promising botanical-derived monoamine oxidase (MAO) inhibitors: Pharmacological aspects and structure-activity studies. *South African J. Bot*. 146, 127–145. https://doi.org/10.1016/j.sajb.2021.09.019

Deng ZN, Huang HR, Huang CM, Verma KK, Luo HB, Jiang SL, Li YJ, Cao HQ, Wu KC, Xu L. (2023) Short-term limited water irrigation influences the photosynthetic pigments, enzymatic and non-enzymatic activities in *Saccharum spontaneum* L. at vegetative stage. *Sugar Tech*. 25, 110–118. https://doi.org/10.1007/s12355-022-01175-7

Devi P, Dey SR, Kumar P. (2023a) Integration of water resources management in rural areas. In: *Water Resources Management for Rural Development: Challenges and Mitigation*. Elsevier, pp. 139–152. https://doi.org/10.1016/B978-0-443-18778-0.00013-1

Devi P, Dey SR, Saini L, Kumar P, Panigrahi S, Dwivedi P. (2023b) Toward sustainable agriculture: Strategies involving phytoprotectants against reactive oxygen species. In: *Reactive Oxygen Species: Prospects in Plant Metabolism*. Springer Nature, pp. 229–247. https://doi.org/10.1007/978-981-19-9794-5_13

Devi P, Kumar P, Sharma K, Singh J. (2023c) Is there a relationship between myconanotechnology and sustainable development? In: *Myconanotechnology and Application of Nanoparticles in Biology: Fundamental Concepts, Mechanism and Industrial Applications*. Elsevier, pp. 175–207. https://doi.org/10.1016/B978-0-443-15262-7.00009-7

Dey SR, Devi P, Kumar P. (2023a) Organic contaminants in aquatic environments: Sources and impact assessment. In: *Metal Organic Frameworks for Wastewater Contaminant Removal*. Wiley, pp. 299–317.

Dey SR, Sharma M, Kumar P. (2023b) Effects and responses of chromium on plants. In: *Environmental Science and Engineering*. Springer Deutschland, *Vol. F1975*, pp. 385–427. https://doi.org/10.1007/978-3-031-44029-8_14

Dilnawaz F, Misra AN, Apostolova E. (2023) Involvement of nanoparticles in mitigating plant's abiotic stress. *Plant Stress* 10, 100280. https://doi.org/10.1016/j.stress.2023.100280

Dubey A, Kumar P, Yati V. (2002) Growth performance and susceptibility of Solanum melongena subjected to different concentration of opium effluent. *Poll. Res*. 21(1), 89–90.

Dwivedi P, Kumar P. (2011) Anti-diabetic medicinal plants and their conservation: Waging green war on diabetes. *Med. Plants* 3(3), 181–189. https://doi.org/10.5958/j.0975-4261.3.3.031

Ebrahimi-Zarandi M, Etesami H, Glick BR. (2023) Fostering plant resilience to drought with Actinobacteria: Unveiling perennial allies in drought stress tolerance. *Plant Stress* 10, 100242. https://doi.org/10.1016/j.stress.2023.100242

Etesami H, Glick BR. (2024) Bacterial indole-3-acetic acid: A key regulator for plant growth, plant-microbe interactions, and agricultural adaptive resilience. *Microbiol. Res*. 281, 127602. https://doi.org/10.1016/j.micres.2024.127602

Goud EL, Singh J, Kumar P. (2022) Climate change and their impact on global food production. In: *Microbiome Under Changing Climate,* Elsevier, pp. 415–436.

Guo D-J, Li D-P, Chen Z-D, Yang B, Verma KK, Singh RK, Singh P, Khan Q, Sharma A, Qin Y, Zhang B-Q, Song X-P, Li Y-R. (2023) Effect of endophytic diazotroph *Enterobacter roggenkampii* ED5 on nitrogen-metabolism-related microecology in the sugarcane rhizosphere at different nitrogen levels. *Front. Microbiol*. 14, 1132016. https://doi.org/10.3389/fmicb.2023.1132016

Guo D-J, Li D-P, Singh RK, Singh P, Sharma A, Verma KK, Qin Y, Khan Q, Lu Z, Malviya MK, Song X-P, Xing Y-X and Li Y-R. (2021) Differential protein expression analysis of two sugarcane varieties in response to diazotrophic plant growth-promoting endophyte enterobacter roggenkampii ED5. *Front. Plant Sci*. 12, 727741. https://doi.org/10.3389/fpls.2021.727741

Guo DJ, Li DP, Singh RK, Singh P, Verma KK, Sharma A, Qin Y, Khan Q, Song XP, Malviya MK, Xing YX, Li YR. (2022) Comparative transcriptome analysis of two sugarcane varieties in response to diazitrophic plant growth promoting endophyte *Enterobacter roggenkampii* ED5. *J. Plant Interact.* 17(1), 75–84.

Hidangmayum A, Dwivedi P, Kumar P, Upadhyay SK. (2023) Seed priming and foliar application of chitosan ameliorate drought stress responses in mungbean genotypes through modulation of morpho-physiological attributes and increased antioxidative defense mechanism. *J. Plant Growth Regul*. 42(10), 6137–6154. https://doi.org/10.1007/s00344-022-10792-1

Huang CM, Wu KC, Verma KK, Luo HB, Deng ZN, Shi GY, Cao HQ, Wu XJ, Ye LP, Wei YW, Xu L. (2024) Isolation and expression analysis of cellulose synthase 3 (*Ces3*) genes from sugarcane (*Saccharum officinarum* L.). *Chem. Biol. Technol. Agric*. 11, 71. https://doi.org/10.1186/s40538-024-00597-7

Huang HR, Gao YJ, Malviya MK, Verma KK, Solanki MK, Huang YX, Li X, Deng Y, Yan J, Tang SY, Wang LW, Xu L. (2022) Genetic diversity analysis of sugarcane (*Saccharum* spp. Hybrids) among high-sucrose clones of GT series and commonly used parents by using microsatellite markers in Guangxi, China. *Sugar Tech*. 24, 397–407. https://doi.org/10.1007/s12355-021-01033-y

Ijaz M, Khan F, Ahmed T, Noman M, Zulfiqar F, Rizwan M, Chen J, Siddique HM, Li B. (2023) Nanobiotechnology to advance stress resilience in plants: Current opportunities and challenges. Mater. *Today Bio*. 22, 100759. https://doi.org/10.1016/j.mtbio.2023.100759

Iqbal B, Li G, Alabbosh KF, Hussain H, Khan I, Tariq M, Javed Q, Naeem M, Ahmad N. (2023) Advancing environmental sustainability through microbial reprogramming in growth improvement, stress alleviation, and phytoremediation. *Plant Stress* 10, 100283. https://doi.org/10.1016/j.stress.2023.100283

Jagadesh M, Dash M, Kumari A, Singh SK, Verma KK, Kumar P, Bhatt R, Sharma SK. (2024) Revealing the hidden world of soil microbes: Metagenomic insights into plant, bacteria, and fungi interactions for sustainable agriculture and ecosystem restoration. *Microbiol. Res*. 127764. https://doi.org/10.1016/j.micres.2024.127764

Joel JM, Johnson R, Puthur JT. (2023) Co-application of arbuscular mycorrhizal fungi and engineered nanomaterials: A promising strategy for crop resilience against abiotic stresses. *South African J. Bot*. 162, 314–323. https://doi.org/10.1016/j.sajb.2023.09.022

Joshi A, Verma KK, Rajput VD, Arora J. (2023) Metabolic engineering of microorganisms in advancing biofuel production. In: *Advances in Lignocellulosic Biofuel Production Systems*, pp. 181–202. https://doi.org/10.1016/B978-0-323-91192-4.00016-X

Joshi A, Verma KK, Rajput VD, Minkina T, Arora J. (2022) Recent advances in metabolic engineering of microorganisms for advancing lignocellulose-derived biofuels. *Bioengineered* 13(4), 8135–8163. https://doi.org/10.1080/21655979.2022.2051856.

Kandpal G, Kumar P, Siddique A. (2018) Effect of drought and improvement mechanism in rice: A review. *Ann. Agri.Bio. Res*. 23(2), 150–155.

Kaya C, Ugurlar F. (2023) Melatonin and stress tolerance in horticultural crops: Insights into gene regulation, epigenetic modifications, and hormonal interplay. *Scientia Horticult*. 322, 112432. https://doi.org/10.1016/j.scienta.2023.112432

Khalid F, Asif K, Rasheed Y, Ashraf H, Maqsood MF, Rana S, Zulfiqar U, Naz N, Shahbaz M, Sardar R, Riaz Z. (2023) Nano priming for boosting growth and resilience in crops under abiotic stresses. *Biocatalysis Agric. Biotechnol*. 53, 102892. https://doi.org/10.1016/j.bcab.2023.102892

Kotia A, Rutu P, Singh V, Kumar A, Dhoke S, Kumar P, Singh DK. (2021) Rheological analysis of rice husk-starch suspended in water for sustainable agriculture application. In: *2nd International Conference on Functional Materials, Manufacturing and Performances, ICFMMP 2021*, *50*, 1962–1966. https://doi.org/10.1016/j.matpr.2021.09.325

Krishna D, Sachan HK, Verma KK. (2022) Abiotic stresses in plants: An overview. In: *Understanding Abiotic Stresses*. Nova Science Publishers, Inc.: New York.

Kumar D, Rameshwar SD, Kumar P. (2019a) Effect of integrated application of inorganic and organic fertilizers on the roots of chickpea. *Plant Arch*. 19(1), 857–860.

Kumar P, Devi P, Dey SR. (2021a) Fungal volatile compounds: A source of novel in plant protection agents. In: *Volatiles and Metabolites of Microbes,* Elsevier, pp. 83–104

Kumar P, Dwivedi P. (2018) Putrescine and glomus mycorrhiza moderate cadmium actuated stress reactions in *zea mays* L. By means of extraordinary reference to sugar and protein. *Vegetos* 31(3), 74–77. https://doi.org/10.5958/2229-4473.2018.00076.9

Kumar P, Dwivedi P. (2020) Lignin estimation in sorghum leaves grown under hazardous waste site. *Plant Arch.* 20, 2558–2561.

Kumar P, Goud EL, Devi P, Dey SR, Dwivedi P. (2022a) Heavy metals: Transport in plants and their physiological and toxicological effects. In: *Plant Metal and Metalloid Transporters,* Springer Nature, pp. 23–54. https://doi.org/10.1007/978-981-19-6103-8_2

Kumar P, Goud EL, Devi P, Koul B. (2022b) Metal pollutants in the environment. In: *Environmental Microbiology: Emerging Technologies.* De Gruyter, pp. 291–323. https://doi.org/10.1515/9783110727227-012

Kumar P, Harshavardhan M, Kumar PS, Yumnam J, Jyoti N, Naik M, Misao L, Purnima Kumar S. (2018a) Effect on chlorophyll a/b ratio in cadmium contaminated maize leaves treated with putrescine and mycorrhiza. *Ann. Biol.* 34(3), 281–283.

Kumar P, Koul B, Sharma M. (2022c) Phytoremediation of heavy metals. In: *Heavy Metals in Plants: Physiological to Molecular Approach.* CRC Press, pp. 369–388) https://doi.org/10.1201/9781003110576-17

Kumar P, Krishna V, Pandey AK, Pathak S, Siddique A. (2018b) Assessment of scavenging competence for cadmium, lead, chromium and nickel metals by in vivo grown Zea mays l. Using atomic absorption spectrophotometer. *Ann. Agri Bio. Res.* 23(2), 166–168.

Kumar P, Kumar S, Harshavardhan M, Naik M, Yumnam J, Kumar PS, Jyoti N, Misao L, Purnima. (2018c) Evaluation of plant height and leaf length of sorghum grown under different sources of nutrition. *Annal. Biol.* 34(3), 284–286.

Kumar P, Kumar T, Singh S, Tuteja N, Prasad R, Singh J. (2020) Potassium: A key modulator for cell homeostasis. *J. Biotechnol.* 324, 198–210.

Kumar P, Mandal B, Dwivedi P. (2011a) Heavy metal scavenging capacity of Mentha spicata and Allium cepa. *Med. Plants* 3(4), 315–318. https://doi.org/10.5958/j.0975-4261.3.4.053

Kumar P, Mandal B, Dwivedi P. (2011b) Heavy metals scavenging of soils and sludges by ornamental plants. *J. Appl. Horticult.* 13(2), 144–146.

Kumar P, Mandal B, Dwivedi P. (2013) Combating heavy metal toxicity from hazardous waste sites by harnessing scavenging activity of some vegetable plants. *Vegetos* 26(2), 416–425. https://doi.org/10.5958/j.2229-4473.26.2.106

Kumar P, Mistri TK. (2020) Transcription factors in SOX family: Potent regulators for cancer initiation and development in the human body. *Seminars Cancer Biol.* 67, 105–113. https://doi.org/10.1016/j.semcancer.2019.06.016

Kumar P, Naik M. (2020) Biotic symbiosis and plant growth regulators as a strategy against cadmium and lead stress in chickpea. *Plant Arch.* 20, 2495–2500.

Kumar P, Pandey AK, Krishna V, Pathak S, Siddique A. (2018d) Phytoextraction of lead, chromium, cadmium and nickel by tagetes plant grown at hazardous waste site. *Ann. Biol.* 34(3), 287–289.

Kumar P, Pathak S, Amarnath KS, Veerendra Brahma Teja P, Dileep B, Kumar K, Singh M, Siddique A. (2018e) Effect of growth regulator on morpho-physiological attributes of chilli: A case study. *Plant Arch.* 18(2), 1771–1776.

Kumar P, Pathak S, Kumar M, Dwivedi P. (2018f) Role of secondary metabolites for the mitigation of cadmium toxicity in sorghum grown under mycorrhizal inoculated hazardous waste site. In: *Biotechnological Approaches for Medicinal and Aromatic Plants: Conservation, Genetic Improvement and Utilization*, pp. 199–212.

Kumar P, Pathak S. (2019) Responsiveness index of sorghum (Sorghum bicolor (L.) Moench) grown under cadmium contaminated soil treated with putrescine and mycorrhiza. *Bangladesh J. Bot.* 48(1), 139–143.

Kumar P, Saini L, Sharma M. (2023) Assessment of the phytochemical constituents and metabolites in the medicinal plants and herbal medicine used in the treatment and management of respiratory diseases. In: *Reference Series in Phytochemistry, Springer,* vol. 2023. https://doi.org/10.1007/978-3-031-21973-3_19-1

Kumar P, Sharma K, Saini L, Dey SR. (2021b) Role and behavior of microbial volatile organic compounds in mitigating stress. In: Kumar A, Singh J, JBTV, Samuel M. (Eds.), Academic Press, pp. 143–161. https://doi.org/10.1016/B978-0-12-824523-1.00010-9

Kumar P, Siddique A, Thongbam S, Chopra P, Kumar S. (2019b) Cadmium induced changes in total starch, total amylose and amylopectin content in putrescine and mycorrhiza treated sorghum crop. *Nat. Environ. Poll. Technol.* 18(2), 525–530.

Kumar P, Singh BN, Dwivedi P. (2017) Plant growth regulators, plant adaptability and plant productivity: A review on abscisic acid (ABA) signaling in plants under emerging environmental stresses. In: *Sustaining Future Food Security in Changing Environments,* Nova Science Publishers, Inc.: USA, pp. 81–97.

Kumar P, Yumnam J, Kumar PS, Misao L, Jyoti N, Naik M, Purnima Kumar S, Harshavardhan M. (2018g) Cadmium induced changes in germination of maize seed treated with mycorrhiza. *Ann. Agri. Bio. Res.* 23(2), 169–170.

Kumar V, Dwivedi P, Kumar P, Singh BN, Pandey DK, Kumar V, Bose B. (2021) Mitigation of heat stress responses in crops using nitrate primed seeds. *South African J. Bot.* 140, 25–36. https://doi.org/10.1016/j.sajb.2021.03.024

Kumari A, Dash M, Singh SK, Jagadesh M, Mathpal B, Mishra PK, Pandey SK, Verma KK. (2023a) Soil microbes: A natural solution for mitigating the impact of climate change. *Environ. Monit. Assess.* 195, 1436. https://doi.org/10.1007/s10661-023-11988-y

Kumari A, Lakshmi GA, Krishna GK, Patni B, Prakash S, Bhattacharyya M, Singh SK, Verma KK. (2022a) Climate change and its impact on crops: A comprehensive investigation for sustainable agriculture. *Agronomy* 12, 3008. https://doi.org/10.3390/agronomy12123008

Kumari A, Singh SK, Mathpal B, Verma KK, Garg VK, Bhattacharyya M, Bhatt R. (2023b) The biosynthesis, mechanism of action, and physiological functions of melatonin in horticultural plants: A review. *Horticulturae* 9, 913. https://doi.org/10.3390/horticulturae9080913

Kumari P, Singh J, Kumar P. (2022b) Impact of bioenergy for the diminution of an ascending global variability and change in the climate. In: *Microbiome Under Changing Climate*. Elsevier, pp. 469–487.

Li A-M, Liao F, Wang M, Chen Z-L, Qin C-X, Huang R-Q, Verma KK, Li Y-R, Que Y-X, Pan Y-Q, et al. (2023) Transcriptomic and proteomic landscape of sugarcane response to biotic and abiotic stressors. *Int. J. Mol. Sci.* 24, 8913. https://doi.org/10.3390/ijms24108913

Malviya MK, Li C-N, Lakshmanan P, Solanki MK, Wang Z, Solanki AC, Nong Q, Verma KK, Singh RK, Singh P, Sharma A, Guo D-J, Dessoky ES, Song X-P, Li Y-R. (2022) High-throughput sequencing-based analysis of rhizosphere and diazotrophic bacterial diversity among wild progenitor and closely related species of sugarcane (*Saccharum* spp. inter-specific hybrids). *Front. Plant Sci.* 13, 829337. https://doi.org/*10.3389/fpls.2022.829337*.

Malviya MK, Solanki MK, Li CN, Wang Z, Zeng Y, Verma KK, Singh RK, Singh P, Huang HR, Yang LT, Song XP, Li YR. (2021) Sugarcane-legume intercropping can enrich the soil microbiome and plant growth. *Front. Sustain. Food Syst.* 5, 606595. https://doi.org/10.3389/fsufs.2021.606595.

Maqsood MF, Shahbaz M, Khalid F, Rasheed Y, Asif K, Naz N, Zulfiqar U, Zulfiqar F, Moosa A, Alamer KH, Attia H. (2023) Biogenic nanoparticles application in agriculture for ROS mitigation and abiotic stress tolerance: A review. *Plant Stress* 10, 100281. https://doi.org/10.1016/j.stress.2023.100281

Mohan N, Ahlawat J, Sharma L, Pal A, Prateek, Rao P, Redhu M, Singh V, Rani N, Kumari K, Kaur T, Jogender, Yadav S. (2023) Engineered nanoparticles a novel approach in alleviating abiotic and biotic stress in millets: A complete study. *Plant Stress* 10, 100223. https://doi.org/10.1016/j.stress.2023.100223

Mohapatra B, Chamoli S, Salvi P, Saxena SC. (2023) Fostering nanoscience's strategies: A new frontier in sustainable crop improvement for abiotic stress tolerance. *Plant Nano Biol.* 3, 100026. https://doi.org/10.1016/j.plana.2023.100026

Naik M, Kumar P. (2020) Role of growth regulators and microbes for metal detoxification in plants and soil. *Plant Arch.* 20, 2820–2824.

Oliveira MA, Verma KK, Bhatt R, Oliveira TBA. (2022) Impact of green and organic fertilizers on soil fertility and sugarcane productivity. In: Verma KK. et al. (Eds.), *Agro-industrial Perspectives on Sugarcane Production Under Environmental Stress*. https://doi.org/10.1007/978-981-19-3955-6_11.

Pankaj PK, Kumar P, Nigam RC, Mishra PK. (2012a) Monitoring and Surveillance of synthetic pyrethroids and organophosphate in different brands of soft drinks. *J. Chem. Pharmaceut. Res.* 4(8), 3939–3943.

Pankaj PK, Kumar P, Nigam RC, Mishra PK. (2012b) To studies organochlorine insecticides in different brands of cold drinks. *J. Chem. Pharmaceut. Res.* 4(8), 3934–3938.

Pathak S, Kumar P, Mishra PK, Kumar M. (2017) Mycorrhiza assisted approach for bioremediation with special reference to biosorption. *Poll. Res.* 36(2), 329–332.

Paul A, Sharma SR, Sresty TVS, Devi S, Bala S, Kumar PS, Saradhi PP, Frutos R, Altosaar I, Kumar PA. (2005) Transgenic cabbage (*Brassica oleracea* var. capitata) resistant to Diamondback moth (*Plutella xylostella*). *Ind. J. Biotechnol.* 4, 72–77.

Rai PK, Sonne C, Kim K-H. (2023) Heavy metals and arsenic stress in food crops: Elucidating antioxidative defense mechanisms in hyperaccumulators for food security, agricultural sustainability, and human health. *Sci. Total Environ.* 874, 162327. https://doi.org/10.1016/j.scitotenv.2023.162327

Rani A, Guleria M, Sharma Y, Sharma S, Chaudhary A, Sharma R, Kumar P. (2023) Insights into elicitor's role in augmenting secondary metabolites production and climate resilience in genus Ocimum—A globally important medicinal and aromatic crop. *Industrial Crops Prod.* 202, 117078. https://doi.org/10.1016/j.indcrop.2023.117078

Reddy SS, Kumar P, Dwivedi P. (2022) Heavy Metal Transporters, Phytoremediation Potential, and Biofortification. In: *Plant Metal and Metalloid Transporters,* Springer, pp. 387–405.

Saini L, Sharma K, Kumar P. (2023) Water efficiency in agriculture as a vital approach toward water management in rural areas. In: *Water Resources Management for Rural Development: Challenges and Mitigation.* Elsevier, pp. 127–138. https://doi.org/10.1016/B978-0-443-18778-0.00020-9

Sharma A, Singh RK, Singh P, Vaishnav A, Guo DJ, Verma KK, Li DP, Song XP, Malviya MK, Khan N, Lakshmanan P, Li YR. (2021) Insights into the bacterial and nitric oxide-induced salt tolerance in sugarcane and their growth-promoting abilities. *Microorganisms* 9, 2203. https://doi.org/10.3390/microorganisms9112203.

Sharma A, Song X-P , Singh RK, Vaishnav A, Gupta S, Singh P, Guo D-J, Verma KK, Li Y-R. (2022a) Impact of carbendazim on cellular growth, defence system and plant growth promoting traits of *Priestia megaterium* ANCB-12 isolated from sugarcane rhizosphere. *Front. Microbiol.* 13, 1005942. doi:10.3389/fmicb.2022.1005942

Sharma K, Devi P, Kumar P, Dey A, Dwivedi P. (2023a) Hazardous phytotoxic nature of reactive oxygen species in agriculture. In: *Reactive Oxygen Species: Prospects in Plant Metabolism.* Springer, pp. 135–146. https://doi.org/10.1007/978-981-19-9794-5_8

Sharma K, Devi P, Kumar P. (2023b) Metal-organic frameworks for wastewater contaminants removal. In: *Metal Organic Frameworks for Wastewater Contaminant Removal.* Wiley, pp. 95–117.

Sharma M, Dey SR, Kumar P. (2023c) Antioxidant defence: A key mechanism of chromium tolerance. In: *Environmental Science and Engineering.* Springer, F1975, pp. 91–116. https://doi.org/10.1007/978-3-031-44029-8_5

Sharma M, Kumar P, Verma V, Sharma R, Bhargava B, Irfan M. (2022b) Understanding plant stress memory response for abiotic stress resilience: Molecular insights and prospects. *Plant Physiol. Biochem.* 179, 10–24. https://doi.org/10.1016/j.plaphy.2022.03.004

Sharma PC, Kumar P. (1999) Alleviation of salinity stress during germination in Brassica juncea by pre-sowing chilling treatments to seeds. *Biologia Plant.* 42(3), 451–455. https://doi.org/10.1023/A:1002481709121

Sharma S, Shree B, Sharma A, Irfan M, Kumar P. (2022c) Nanoparticle-based toxicity in perishable vegetable crops: Molecular insights, impact on human health and mitigation strategies for sustainable cultivation. *Environ. Res.* 212, 113168. https://doi.org/10.1016/j.envres.2022.113168

Sheri V, Kumar M, Jaconis S, Zhang B. (2023) Antioxidant defense in cotton under environmental stresses: Unraveling the crucial role of a universal defense regulator for enhanced cotton sustainability. *Plant Physiol. Biochem.* 204, 108141. https://doi.org/10.1016/j.plaphy.2023.108141

Shinde NA, Kawar PG, Dalvi SG. (2024) Chitosan-based nanoconjugates: A promising solution for enhancing crop drought-stress resilience and sustainable yield in the face of climate change. *Plant Nano Biol.* 7, 100059. https://doi.org/10.1016/j.plana.2024.100059

Siddique A, Dubey AP, Kumar P. (2018a) Cadmium induced physio-chemical changes in roots of wheat. *Vegetos* 31(3), 113–118. https://doi.org/10.5958/2229-4473.2018.00081.2

Siddique A, Kandpal G, Kumar P. (2018b) Proline accumulation and its defensive role under diverse stress condition in plants: An overview. *J. Pure Appl. Microbiol.* 12(3), 1655–1659.

Siddique A, Kumar P. (2018) Physiological and biochemical basis of pre-sowing soaking seed treatment -an overview. *Plant Arch.* 18(2), 1933–1937.

Singh P, Singh RK, Li HB, Guo DJ, Sharma A, Verma KK, Solanki MK, Upadhyay SK, Lakshmanan P, Yang LT, Li YR. (2023) Nitrogen fixation and phytohormone stimulation of sugarcane plant through plant growth promoting diazotrophic *Pseudomonas. Biotechnol. Genet. Eng. Rev.* 40(1), 15–35. https://doi.org/10.1080/02648725.2023.2177814

Singh P, Singh RK, Song QQ, Li HB, Guo DJ, Malviya MK, Verma KK, Song XP, Lakshmanan P, Yang LT, Li YR. (2021) Comparative analysis of protein and differential responses of defense-related gene and enzyme activity reveals the long-term molecular responses of sugarcane inoculated with *Sporisorium scitamineum. J. Plant Interact.* 16(1), 12–29.

Singh RK, Singh P, Li HB, Song QQ, Guo DJ, Solanki MK, Verma KK, Malviya MK, Song XP, Lakshmanan P, Yang LT, Li YR. (2020) Diversity of nitrogen-fixing rhizobacteria associated with sugarcane: A comprehensive study of plant-microbe interactions for growth enhancement in *Saccharum* spp. *BMC Plant Biol.* 20, 220. https://doi.org/10.1186/s12870-020-02400-9.

Solanki MK, Verma KK, Dastogeer KMG, Mora-Poblete F, Mundra S. (2023) Microbial resilience in plant nutrient management towards sustainable farming. *Front. Microbiol.* 14, 1280811. https://doi.org/10.3389/fmicb.2023.1280811

Song X-P, Mo F, Verma KK, Wei J, Zhang X, Yang L, Li YR. (2019) Effect of sugarcane smut (*Ustilago scitaminea* Syd.) on ultrastructure and biochemical indices of sugarcane. *Biomed J. Sci. Tech. Res.* 002950. https://doi.org/10.26717/BJSTR.2019.17.002950.

Song XP, Mo F, Yan M, Zhang X, Zhang B, Huang X, Huang D, Pan Y, Verma KK, Li YR. (2022) Effect of smut infection on the photosynthetic physiological characteristics and related defense enzymes of sugarcane. *Life* 12, 1201. https://doi.org/10.3390/life12081201

Song XP, Verma KK, Tian DD, Zhang XQ, Liang YJ, Huang X, Li CN, Li YR. (2021) Exploration of silicon functions to integrate with biotic stress tolerance and crop improvement. *Biol. Res.* 54, 19. https://doi.org/10.1186/s40659-021-00344-4

Srivastav AL, Patel N, Rani L, Kumar P, Dutt I, Maddodi BS, Chaudhary VK. (2023) Sustainable options for fertilizer management in agriculture to prevent water contamination: A review. *Environ. Dev. Sustain.* 26, 8303–8327. https://doi.org/10.1007/s10668-023-03117-z

Tian D, Qin L, Verma KK, Wei L, Li J, Li B, Zhou W, He Z, Wei D, Huang S, Long S, Huang Q, Li C, Wei S. (2023) Transcriptomic and metabolomic differences between banana varieties which are resistant or susceptible to *Fusarium wilt. PeerJ* 11, e16549. https://doi.org/10.7717/peerj.16549

Upadhyay SK, Devi P, Kumar V, Pathak HK, Kumar P, Rajput VD, Dwivedi P. (2023) Efficient removal of total arsenic (As3+/5+) from contaminated water by novel strategies mediated iron and plant extract activated waste flowers of marigold. *Chemosphere* 313, 137551. https://doi.org/10.1016/j.chemosphere.2022.137551

Verma KK, Joshi A, Song X-P, Liang Q, Xu L, Huang H-r, Wu K-C, Seth CS, Arora J and Li Y-R (2024a) Regulatory mechanisms of plant rhizobacteria on plants to the adaptation of adverse agroclimatic variables. *Front. Plant Sci.* 15, 1377793. https://doi.org/10.3389/fpls.2024.1377793

Verma KK, Joshi A, Song X-P, Singh S, Kumari A, Arora J, Singh SK, Solanki MK, Seth CS, Li Y-R. (2024b) Synergistic interactions of nanoparticles and plant growth promoting rhizobacteria enhancing soil-plant systems: A multigenerational perspective. *Front. Plant Sci.* 15, 1376214. https://doi.org/10.3389/fpls.2024.1376214

Verma KK, Li DM, Singh M, Rajput VD, Malviya MK, Minkina T, Singh RK, Singh P, Song XP, Li YR. (2020a) Interactive role of silicon and plant-rhizobacteria mitigating abiotic stresses: A new approach for sustainable agriculture and climate change. *Plants* 9, 1055. https://doi.org/10.3390/plants9091055.

Verma KK, Rajput VD, Singh M, Verma CL, Minkina T. (2022a) Influence of waterlogging on plant physiology and molecular biology. In: *Understanding Abiotic Stresses*. Nova Science Publishers, Inc.: New York.

Verma KK, Siddiqui N, Singh M. (2022b) Understanding the plant physiological and molecular adaptation strategies to soil salinity. In: *Understanding Abiotic Stresses*. Nova Science Publishers, Inc.: New York. https://doi.org/: 10.52305/UBUJ4024

Verma KK, Singh P, Song X-P, Malviya MK, Singh RK, Chen G-L, Solomon S, Li YR. (2020b) Mitigating climate change for sugarcane improvement: Role of silicon in alleviating abiotic stresses. *Sugar Tech.* 22(5), 741–749. https://doi.org/10.1007/s12355-020-00831-0.

Verma KK, Singh RK, Song QQ, Singh P, Zhang B-Q, Song X-P, Chen G-L, Li YR. (2019a) Silicon alleviates drought stress of sugarcane plants by improving antioxidant responses. *Biomed. J. Sci. Tech. Res.* 002957. https://doi.org/10.26717/BJSTR.2019.17.002957

Verma KK, Song X, Degu HD, Guo D, Joshi A, Huang H, Xu L, Singh M, Huang D, Rajput VD, Li YR. (2023a) Recent advances in nitrogen and nano-nitrogen fertilizers for sustainable crop production: A mini-review. *Chem. Biol. Technol. Agric.* 10, 1–14. https://doi.org/10.1186/s40538-023-00488-3

Verma KK, Song X, Verma CL, Huang H, Singh M, Xu L, Wu J, Khan ZH, Zhang B, Li Y. (2023b) Mathematical modeling of climate and fluoride effects on sugarcane photosynthesis with silicon nanoparticles. *Plant Physiol. Biochem.* 204, 108089. https://doi.org/10.1016/j.plaphy.2023.108089

Verma KK, Song XP, Budeguer F, Nikpay A, Enrique R, Singh M, Zhang B, Wu, JM, Li YR. (2022c) Genetic engineering: An efficient approach to mitigating biotic and abiotic stresses in sugarcane cultivation. *Plant Signal. Behav.* 17(1), e2108253.

Verma KK, Song XP, Joshi A, Rajput VD, Singh M, Sharma A, Singh RK, Li DM, Arora J, Minkina T, Li YR. (2022d) Nanofertilizer possibilities for healthy soil, water and food in future: An overview. *Front. Plant Sci.* 13, 865048. https://doi.org/10.3389/fpls.2022.865048

Verma KK, Song XP, Joshi A, Tian DD, Rajput VD, Singh M, Arora J, Minkina TM, Li YR. (2022e) Recent trends in nano-fertilizer for advancing sustainable agriculture under the era of climate change to ensure food security for future generations. *Nanomaterials* 12, 173. *https://doi.org/10.3390/nano12010173.*

Verma KK, Song XP, Li DM, Singh M, Wu JM, Singh RK, Sharma A, Zhang BQ, LI YR. (2022f) Silicon and soil microorganisms improves rhizospheric soil health with bacterial community, plant growth, performance and yield. *Plant Signal. Behav.* 17(1), e2104004.

Verma KK, Song XP, Lin B, Guo DJ, Singh M, Rajput VD, Singh RK, Singh P, Sharma A, Malviya MK, Chen GL, Li YR. (2022g) Silicon induced drought tolerance in crop plants:

Physiological adaptation strategies. *Silicon* 14(6), 2473–2487. https://doi.org/10.1007/s12633-021-01071-x

Verma KK, Song XP, Rajput VD, Boldyreva V, Zhang BQ, Minkina T, Li YR. (2022h) Morpho-physiological, biochemical and ultrastructural modifications on sugarcane to prolonged water deficit. Verma KK. et al. (Eds.), *Agro-industrial Perspectives on Sugarcane Production Under Environmental Stress*. https://doi.org/10.1007/978-981-19-3955-6_11.

Verma KK, Song XP, Rajput VD, Solomon S, Li YR, Rao GP. (2022i) *Agro-industrial Perspectives on Sugarcane Production Under Environmental Stress*. Springer Nature: Singapore, https://doi.org/10.1007/978-981-19-3955-6

Verma KK, Song XP, Singh M, Tian DD, Rajput VD, Minkina T, Li YR. (2023c) Association of silicon and soil microorganisms induces stress mitigation, increasing plant productivity. In: de Mello Prado R. (Ed.), *Benefits of Silicon in the Nutrition of Plants*, pp. 299–328. https://doi.org/10.1007/978-3-031-26673-7_17

Verma KK, Song XP, Tian DD, Guo DJ, Chen ZL, Zhong CS, Nikpay A, Singh M, Rajput VD, Singh RK, Minkina T, Li YR. (2021a) Influence of silicon on biocontrol strategies to manage biotic stress for crop protection, performance and improvement. *Plants* 10, 2163. https://doi.org/10.3390/plants10102163

Verma KK, Song XP, Tian DD, Singh M, Verma CL, Rajput VD, Singh RK, Sharma A, Singh P, Malviya MK, Li YR. (2021b) Investigation of defensive role of silicon during drought stress induced by irrigation capacity in sugarcane: Physiological and biochemical characteristics. *ACS Omega* 6, 19811–19821.

Verma KK, Song XP, Verma CL, Chen ZL, Rajput VD, Wu KC, Liao F, Chen GL, Li YR. (2021c) Functional relationship between photosynthetic leaf gas exchange in response to silicon application and water stress mitigation in sugarcane. *Biol. Res.* 54, 15. https://doi.org/10.1186/s40659-021-00338-2

Verma KK, Song XP, Verma CL, Malviya MK, Guo DJ, Rajput VD, Sharma A, Wei KJ, Chen GL, Solomon S, Li YR. (2021d) Predication of photosynthetic leaf gas exchange of sugarcane (*Saccharum* spp.) leaves in response to leaf positions to foliar spray of potassium salt of active phosphorus under limited water irrigation. *ACS Omega* 6, 2396–2409.

Verma KK, Song X-P, Xu L, Huang H-R, Liang Q, Seth CS, Li Y-R. (2023d) Nano-microplastic and agro-ecosystems: A mini-review. *Front. Plant Sci.* 14, 1283852. https://doi.org/10.3389/fpls.2023.1283852

Verma KK, Song XP, Yadav G, Degu HD, Parvaiz A, Singh M, Huang HR, Mustafa G, Xu L, Li YR. (2022j) Impact of agroclimatic variables on proteo-genomics in sugarcane (*Saccharum* spp.) plant productivity. *ACS Omega* 7, 22997–23008.

Verma KK, Song XP, Zeng Y, Guo DJ, Sing M, Rajput VD, Malviya MK, Wei KJ, Sharma A, Li DP, Chen GL, Li YR. (2021e) Foliar application of silicon boosts growth, photosynthetic leaf gas exchange, antioxidative response and resistance to limited water irrigation in sugarcane (*Saccharum officinarum* L.). *Plant Physiol. Biochem.* 166, 582–592.

Verma KK, Song XP,Zeng Y, Li DM, Guo DJ, Rajput VD, Chen GL, Barakhov A, Minkina TM, Li YR. (2020c) Characteristics and correlation of leaf stomata and its relationship with photosynthesis on *Saccharum* spp. under different irrigation and silicon application. *ACS Omega* 5, 24145–24153.

Verma KK, Vatsal S, Singh M, Rajput VD, Minkina T. (2022k) Coping with water stress: Effects and adaptive mechanisms to plants. In: *Understanding Abiotic Stresses*. Nova Science Publishers, Inc.: New York.

Verma KK, Wu KC, Verma CL, Li DM, Malviya MK, Singh RK, Singh P, Chen GL, Song XP, Li YR. (2020d) Developing mathematical model for diurnal dynamics on photosynthetic responses in sugarcane responsive to different irrigation and silicon application. *PeerJ* 8, e10154. https://doi.org/10.7717/peerj.10154.

Verma KK, Liu X-H, Wu K-C, Singh RK, Song QQ, Malviya MK, Song X-P, Singh P, Verma CL, Li YR. (2020e) The impact of silicon on photosynthetic and biochemical responses of sugarcane under different soil moisture levels. *Silicon* 12, 1355–1367.

Verma KK, Wu K-C, Singh P, Malviya MK, Singh RK, Song X-P, Li YR. (2019b) The protective role of silicon in sugarcane under water stress: Photosynthesis and antioxidant enzymes. *Biomed. J. Sci. Tech. Res.* 002685. https://doi.org/0.26717/BJSTR.2019.15.002685

Wahab A, Batool F, Muhammad M, Zaman W, Mikhlef RM, Qaddoori SM, Ullah S, Abdi G, Saqib S. (2024) Unveiling the complex molecular dynamics of arbuscular mycorrhizae: A comprehensive exploration and future perspectives in harnessing phosphate-solubilizing microorganisms for sustainable progress. *Environ. Exp. Bot.* 219, 105633. https://doi.org/10.1016/j.envexpbot.2023.105633

Wu K-C, Huang C-M, Verma KK, Deng Z-N, Huang H-R, Pang T, Cao H-Q, Luo H-B, Jiang S-L, Xu L. (2022) Transcriptomic responses of *Saccharum spontaneum* roots in response to polyethylene glycol-6000 stimulated drought stress. *Front. Plant Sci.* 13, 992755. https://doi.org/10.3389/fpls.2022.992755

Xu L, Deng Z-N, Wu K-C, Malviya MK, Solanki MK, Verma KK, Pang T, Li Y-J, Liu X-Y, Kashyap BK, Dessoky ES, Wang W-Z and Huang H-R. (2022) Transcriptome analysis reveals a gene expression pattern that contributes to sugarcane bud propagation induced by Indole-3-Butyric acid. *Front. Plant Sci.* 13, 852886. https://doi.org/10.3389/fpls.2022.852886

Yomso J, Kumar P, Siddique A. (2021) Effect of cadmium induced stress on morphological growth and yield attributes of rice (*Oryza sativa* L.). *Res. Crops* 22(2), 246–250. https://doi.org/10.31830/2348-7542.2021.064

Zhao J, Yu X, Zhang C, Hou L, Wu N, Zhang W, Wang Y, Yao B, Delaplace P, Tian J. (2024) Harnessing microbial interactions with rice: Strategies for abiotic stress alleviation in the face of environmental challenges and climate change. *Sci. Total Environ.* 912, 168847. https://doi.org/10.1016/j.scitotenv.2023.168847

Zheng H, Dang Y, Diao X, Sui N. (2023b) Molecular mechanisms of stress resistance in sorghum: Implications for crop improvement strategies. *J. Integr. Agric.* 23(3), 741–768. https://doi.org/10.1016/j.jia.2023.12.023

Zheng F-J, Lin B, Yang Y-X, Fang X-C, Verma KK, Chen G-L. (2023a) Efficacy and functionality of sugarcane original vinegar on mice. *Front. Microbiol.* 14, 1224666. https://doi.org/10.3389/fmicb.2023.1224666

CHAPTER 6

Sugarcane Plant–Ratoon System with Traditional and Advanced Nutrient Approach

NAVNIT KUMAR[1,*], LALITA RANA[1], and SUMIT SOW[2]

[1]*Department of Agronomy, Sugarcane Research Institute, Dr. Rajendra Prasad Central Agricultural University, Pusa, Samastipur, Bihar, India*

[2]*Department of Agronomy, PGCA, Dr. Rajendra Prasad Central Agricultural University, Pusa, Samastipur, Bihar, India*

*Corresponding author

ABSTRACT

Sugarcane is cultivated across a wide range of agro-climatic regions, and the use of fertilizers is recommended to ensure proper development. The intrinsic fertility of soil and efficient utilization of various nutrients via fertilizers have an extensive impact on sugarcane productivity. This chapter explores global conventional and advanced nutrient management strategies in sugarcane. Ensuring equilibrium between organic and inorganic amendments is paramount to preserving the desirable soil physicochemical characteristics and supplying sustenance for its microbial community. Restoring the nutrients depleted by crops via intercropping, green manuring, appropriate recycling of crop residues and factory wastes, biofertilizers, and bio-compost constitutes sustainable sugar productivity. Soil fertility and crop nutrition must be understood to guide fertilizer application using the diagnosis and recommendation integrated system, soil test crop response, and composition nutrient diagnosis. Additionally, precise nutrient management using crop canopy sensors and unmanned

Sugarcane Production: Environmental Impacts and Socioeconomic Issues.
Krishan K. Verma, Xiu-Peng Song, Lin Xu, Rajan Bhatt, and Yang-Rui Li (Eds.)

DOI: 10.1201/9781779641137-6

aerial vehicles can enhance nutrient use efficiency in the sugarcane plant–ratoon system. Machine learning and deep learning algorithms can be employed for real-time crop nutrition management in sugarcane, reducing the need for conventional measurements of chlorophyll content. Hence, the adoption of integrated and advanced nutrient management techniques holds the potential for improving soil health, enhancing cane productivity and farmers' income.

6.1 INTRODUCTION

The sugarcane plant–ratoon system involves multiple cycles of harvesting and regrowth, with new shoots (ratoons) emerging from the stubble of previously harvested sugarcane plants. Fertilizer is one of the costliest agricultural inputs, and its application is crucial for farm profitability and environmental conservation (Song et al., 2019; Reddy et al., 2022; Verma et al., 2023a, 2023b; Huang et al., 2024). Traditional and advanced nutrient management approaches can be applied to optimize sugarcane yield and sustainability in this system (Kumar et al., 2017; Oliveira et al., 2022; Bhatt et al., 2023a, 2023b). For optimal plant growth in the early stages, it is common practice to apply a balanced fertilizer typically containing nitrogen (N), phosphorus (P), and potassium (K) during planting time (Kumar, 2012; Anas et al., 2021a, 2021b). After the initial harvest, ratoon crops rely on residual nutrients from the previous cycle. However, over time, nutrient levels may decline, leading to decreased yields (Kumar and Kumar, 2020; Verma et al., 2019a, 2019b; 2023c, 2023d; Deng et al., 2023). Therefore, additional fertilization may be necessary for subsequent ratoon crops. In traditional systems, fertilizers are often applied uniformly across the field without considering spatial variability in soil nutrient levels. This approach may result in overapplication in some areas and underapplication in others, leading to inefficiencies and potential environmental risks (Wu et al., 2022; Xu et al., 2022; Singh et al., 2020, 2023).

Dotaniya et al. (2016) explained that sugarcane, being a crop that is grown for a long period, has a significant requirement for nutrients, particularly nitrogen (N) and potassium (K) when compared to other micronutrients. Nitrogen is more susceptible to losses due to leaching and volatilization. The application of fertilizer is more widespread in intensive sugarcane farming, which requires a significant amount of nitrogen because it produces a large quantity of biomass (Otto et al., 2022; Guo et al., 2022; Huang et al., 2022; Li et al., 2023). Therefore, a well-balanced fertilizer application, which includes a significant amount of N and K, might be a first step toward attaining these critical objectives (Kumar et al., 2023; Solanki et al., 2023). However, phosphorus is also vital for cane production (Kumar and Sinha, 2008). Split applications describe adjusting fertilizer supply to a predetermined target yield and a specific soil moisture level (Pramanick et al., 2022; Kumari et al., 2022; 2023a, 2023b).

Incorporating organic matter through green manure or crop residue retention can enhance soil fertility and microbial activity, improving nutrient cycling and soil health over time (Krishna et al., 2022; Chaudhary et al., 2023; Kumar et al., 2024; Jagadesh et al., 2024). Apart from the traditional nutrient approach, by using precision agriculture technologies, advanced nutrient management allows for the tailoring of fertilizer applications to meet the specific necessities of individual fields or even zones within a field (Verma et al., 2020a, 2020b; 2021a, 2021b, 2021c; Song et al., 2022; Guo et al., 2023). By applying nutrients in proportion to the severity of a nutrient deficiency and by restoring soil nutrient imbalance, plant nutrients applied in response to a soil test help achieve a higher response ratio and benefit: cost ratio, which in turn helps to utilize the synergistic advantages of balanced fertilization (Rao and Srivastava, 2000; Chen et al., 2023; Verma et al., 2024a, 2024b).

Soil test crop response (STCR) studies help to develop fertilizer adjustment equations and calibration charts, which are useful tools for maintaining soil fertility and prescribing fertilizers based on test results. This allows farmers to achieve their desired crop yields. According to Kumar and Santhi (2017) and Singh et al. (2012), this system adoption can also have a favorable impact on the biological, inorganic, and physical health of the soil. Soil mapping, remote sensing, and GPS (the global positioning system)-guided equipment help identify areas with varying nutrient requirements, allowing for variable rate application (VRA) of fertilizers. Advanced soil testing techniques, such as soil spectroscopy or ion-selective electrodes, may be employed for rapid and accurate nutrient analysis (Rana et al., 2017). Mathematical models simulate nutrient dynamics in the soil–plant system, considering factors such as soil type, climate, crop physiology, and management practices (Verma et al., 2020c, 2020d, 2020e; 2022a, 2022b, 2022c; 2023d). These models help optimize nutrient management strategies and predict crop responses to fertilization under different scenarios (Banerjee et al., 2015; Verma et al., 2021d, 2022d, 2022e).

Advanced irrigation systems allow for the precise delivery of water and nutrients directly to the root zone through fertigation. Foliar feeding of micronutrients can also supplement soil-applied fertilizers and address nutrient deficiencies during critical growth stages (Verma et al., 2021e, 2022f, 2022g, 2022h). The holistic approach to nutrient management involves various nutrient sources, including chemical fertilizers, organic amendments, and biofertilizers, to maximize nutrient use efficiency and minimize environmental impacts (Malviya et al., 2021, 2022; Sharma et al., 2021, 2022). Biological inoculants and microbial additives can enhance nutrient availability and soil health in the plant–ratoon system (Kumar et al., 2024a; Singh et al., 2021; Joshi et al., 2022; Zheng et al., 2023). While overall traditional nutrient management practices provide a foundation for sugarcane cultivation, adopting advanced approaches can optimize nutrient use efficiency, improve yields, and promote environmental sustainability in the plant–ratoon system. Integrating precision agriculture technologies, soil testing, and holistic nutrient management strategies can help sugarcane growers meet the demands of modern agriculture while minimizing resource inputs and environmental impacts.

6.2 CONVENTIONAL NUTRIENT APPROACHES

Conventional nutrient management practices in agriculture aim to optimize soil fertility and crop productivity through the application of synthetic fertilizers (Kumar et al., 2014; Verma et al., 2022i, 2022j, 2022k) and other amendments. Also, to optimize nutrient levels for plant growth while minimizing environmental impact. Conventional nutrient management practices include soil testing, intercropping, crop rotation, organic amendments, green manuring, etc.

6.2.1 SOIL TESTING

Soil testing is crucial for determining the nutrient status of the soil. It involves analyzing soil samples to determine nutrient levels, pH, and other properties. This information helps in formulating a targeted nutrient management plan. Soil testing is a fundamental practice in agriculture and environmental science used to assess the chemical, physical, and biological properties of soil. The primary purpose of soil testing is to provide valuable information that helps optimize soil fertility, improve crop productivity, and support sustainable land management practices. The soil test results are interpreted to provide recommendations for soil fertility management as well as best management practices tailored to specific crops, soil types, and environmental conditions (Banerjee et al., 2016). Soil testing is an ongoing process that may be conducted periodically to monitor changes in soil fertility over time and adjust management practices accordingly. Regular soil testing helps optimize nutrient management strategies, improve crop yields, and prevent environmental pollution from excessive fertilizer use. Overall, soil testing plays a critical role in informed decision-making for sustainable agriculture, enabling farmers and land managers to optimize soil fertility, protect natural resources, and enhance agricultural productivity while minimizing environmental impacts.

6.2.2 CROP ROTATION AND INTERCROPPING IN SUGARCANE AND RATOON SYSTEMS

Monoculture is generally followed in sugarcane production globally. In recent years, the emphasis has been shifted from sole cropping to inter-cropping. Since the responses in component crops of the intercropping system are greatly influenced by the fertilizer and irrigation applied to the individual crop (Kumar et al., 2015). Soil fertility may be preserved by crop rotation because it alternates nutrient demands and prevents pests and diseases from building up (Kumar et al., 2016). Soil nutrient levels can be more evenly distributed when crops are rotated since different crops have different nutrient needs (Guo et al., 2021; Song et al., 2021). Growing pulses with sugarcane crop not only increases the area under pulse crop but also reduces the intensity of weeds and provides mid-season income to households for further use of critical inputs to sugarcane, along with additional employment opportunities. The compatibility of French beans with autumn sugarcane has also been reported by Singh and Lal (2007).

Growing multiple crops together can help spread risk, as all the crops may not be susceptible to pests, diseases, or adverse weather conditions. If one crop fails or suffers from a problem, others may still thrive, providing a more stable income for the farmer. Different crops have varying root structures and nutrient needs (Chen et al., 2020; Joshi et al., 2023). Intercropping can help improve soil structure and fertility by enhancing nutrient cycling, reducing soil erosion, and promoting beneficial microbial activity. Also, it can disrupt the buildup of pests and diseases by creating a more diverse environment that is less favorable to specific pests. Intercropping can reduce worldwide requirements of synthetic N fertilizer, and therefore, it can support the development of more sustainable cropping systems (Jensen et al., 2020; Tian et al., 2023).

Intercropping can help to reduce weed growth by shading the soil (smothering effect) and outcompeting weeds for resources like sunlight, water, and nutrients. Wang et al. (2014) in their study found that sugarcane-legume intercrops effectively harness atmospheric nitrogen and soil nutrients to promote plant development while reducing reliance on fertilizers, thereby contributing to environmental protection. Solanki et al. (2018) revealed that soybean intercropping with sugarcane fields in China may increase yields, decrease carbon emissions, and decrease nitrogen inputs. According to multiple studies, intercrops have been found to improve soil fertility and the composition of microorganisms (Figure 6.1).

FIGURE 6.1 Sugarcane–legume intercropping system.

According to Li et al. (2018), agro-ecosystems must retain biodiversity and achieve high yields. However, it activates a number of biotic and abiotic processes, damages soil nutrients, decreases yields, and increases fertilizer inputs (Shoko et al., 2007). Soil bacterial diversity is altered, and intercropping systems reduce crop disease rates. It has the potential to maximize farmers' economic advantages by making better use of land and water (Tang et al., 2021a). Successful intercropping requires careful planning, crop selection, and management practices tailored to the specific conditions and goals of each farm.

6.2.3 MANURE AND COMPOST APPLICATION

Organic materials such as manure and compost are rich sources of nutrients and organic matter. They can improve soil structure, water retention, and nutrient availability. Application rates are typically based on nutrient content and soil test results. According to soil test results (Sakarvadia et al., 2021), increasing the use of inorganic fertilizers, organic manures, and biofertilizers will surely lead to higher soil health and an improvement in sugarcane yield. Balanced fertilization, their principles, suggestions, and practices have been effectively promoted among sugarcane growing farmers in India, and split nitrogen application is also extremely frequent (Bhilala et al., 2022). Simultaneously, achieve high nutrient use efficiency and high crop productivity is necessary in modern agriculture (Cassman et al., 2003).

6.2.4 INTEGRATED NUTRIENT MANAGEMENT (INM)

INM involves integrating various nutrient sources, including chemical fertilizers, organic amendments, and biological inputs, to optimize nutrient availability while minimizing environmental impact. These conventional nutrient management practices aim to sustainably optimize soil fertility and crop productivity while reducing adverse environmental effects such as nutrient runoff, soil erosion, and groundwater contamination (Figure 6.2).

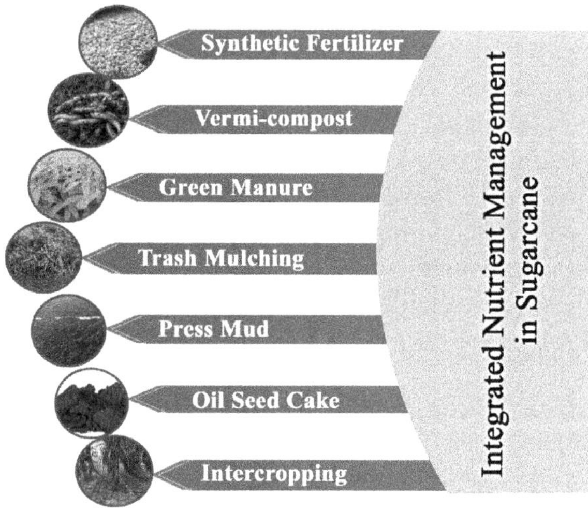

FIGURE 6.2 INM practices in sugarcane.

6.3 ADVANCED SOIL NUTRIENT MANAGEMENT APPROACHES

Approaches go beyond conventional practices to achieve even greater precision, efficiency, and sustainability. Utilizing advanced software and modeling tools, farmers can create detailed nutrient management plans tailored to specific field conditions, crop types, and yield goals. These plans integrate data from soil tests, crop sensors, historical yield data, and weather forecasts to optimize nutrient application rates and timing.

6.3.1 SOIL AMENDMENTS AND BIOFERTILIZERS

Utilizing advanced soil amendments such as biochar, microbial inoculants, and biofertilizers can enhance nutrient availability, improve soil structure, and promote beneficial microbial activity. These biologically based inputs work in synergy with conventional fertilizers to improve soil fertility and crop performance (Kumar et al., 2024).

6.3.2 PRECISION AGRICULTURE TECHNOLOGIES AND REMOTE SENSING

Due to an imbalanced application of nutrients, environmental concerns may arise, and production may become unprofitable (Mozumder et al., 2014). An alternative method, known as plant-based site-specific nutrient management (SSNM), is presently accessible. This approach emphasizes the most effective fertilizer management for a particular field or specific local circumstances (Bhilala et al., 2023), taking into account factors such as the growth duration of the crop variety, management of crop residues, previous fertilizer usage, and external nutrient inputs (Dobermann et al., 2004).

To optimize the supply and demand of a nutrient within a specific field during a particular cropping season, dynamic management of nutrients was implemented using this method (Buresh et al., 2010). According to Witt et al. (1999), the SSNM approach suggests that crop fertilizer needs should be determined by more general correlations, such as internal nutrient efficiency (defined as the quantity of cereal yield generated per unit of nutrient accumulated in above-ground plant dry matter). Precision agriculture technologies, such as GPS-guided equipment and soil sensors, enable farmers to apply nutrients more precisely, optimizing nutrient use efficiency and minimizing environmental impacts. High-tech equipment, such as variable rate fertilizer applicators and drip irrigation systems, allows for precise placement and timing of nutrient inputs directly to the root zone of crops. This reduces nutrient losses through leaching, volatilization, and runoff while ensuring optimal nutrient uptake. Satellite imagery, drones, and ground-based sensors can provide real-time data on crop health, nutrient status, and soil moisture levels. Farmers can use this information to make timely decisions about nutrient application, irrigation, and other management practices. Advanced soil health management enhances overall soil quality and resilience through reduced tillage, cover cropping, crop rotation, and organic amendments. Healthy soils with improved structure, biological activity, and organic matter content are better able to retain and supply nutrients to plants.

6.3.3 VARIABLE RATE APPLICATION

VRA involves applying fertilizers and amendments at variable rates across a field based on spatial variability in soil properties, nutrient levels, and crop needs. This precision approach maximizes nutrient use efficiency and minimizes overapplication in areas with high soil fertility while addressing deficiencies in low-fertility areas.

6.4 DIAGNOSIS AND RECOMMENDATION INTEGRATED SYSTEM (DRIS)

In the context of nutrient management, a DRIS is designed to assist farmers or agronomists in optimizing the application of nutrients to crops. The system gathers data from various sources such as soil samples, weather conditions, crop type, growth stage, and historical yield data. Using advanced algorithms and techniques, the system would analyze the collected data to identify nutrient deficiencies or excesses, as well as other factors affecting crop growth and yield. It might identify imbalances in nutrients like nitrogen, phosphorus, potassium, sulfur, or micronutrients. The system would then provide tailored recommendations for nutrient application, including the type, amount, and timing of fertilizers or soil amendments needed to address the identified deficiencies or optimize nutrient levels. These recommendations could be based on established best practices, scientific research, and localized agronomic knowledge. It could greatly benefit farmers by helping them optimize their use of fertilizers and other inputs, leading to improved crop yields, reduced environmental impact, and increased profitability.

6.4.1 SOIL TEST CROP RESPONSE

It is a soil fertility management concept utilized in agriculture. Fertilizer recommendations based on soil tests have been advocated for quite some time to improve crop productivity and nutrient utilization efficiency. Nevertheless, the process of fertilizer recommendation is not without its share of obstacles and uncertainties. These include gathering representative soil samples, determining an appropriate analytical method for the specific soil in question, and establishing a reliable correlation and calibration between the results of soil tests and the response (Mozumder et al., 2014). In addition, soil testing is a lengthy and labor-intensive procedure. Despite the availability of soil test values, determining a science-based philosophy for fertilizer recommendation remains difficult (He et al., 2015). It pertains to the correlation between the anticipated crop response to nutrients administered and the outcomes of soil tests. The basic idea is to correlate soil nutrient levels, as determined through soil testing, with the expected yield response of crops to nutrient application.

Soil samples are collected from fields and analyzed to determine nutrient levels, pH, organic matter content, and other relevant soil properties. Soil testing is crucial for effective fertilizer recommendation and crop management in agriculture. It is not just about adding nutrients but understanding the specific needs of the soil, the crop, and the environment. This holistic approach helps optimize yields, maintain soil health, and ensure sustainable agricultural practices. By integrating soil testing with crop management strategies, like the Integrated Plant Nutrition System, we can address multiple nutrient deficiencies, improve fertilizer efficiency, and ultimately enhance agricultural productivity while minimizing environmental impact. The targeted yield approach pioneered by Troug (1960) and further refined by Ramamoorthy et al. (1967) has significantly contributed to farming practices, especially in India. This approach relies on understanding soil characteristics and crop nutrient requirements to tailor fertilizer recommendations (Figure 6.3).

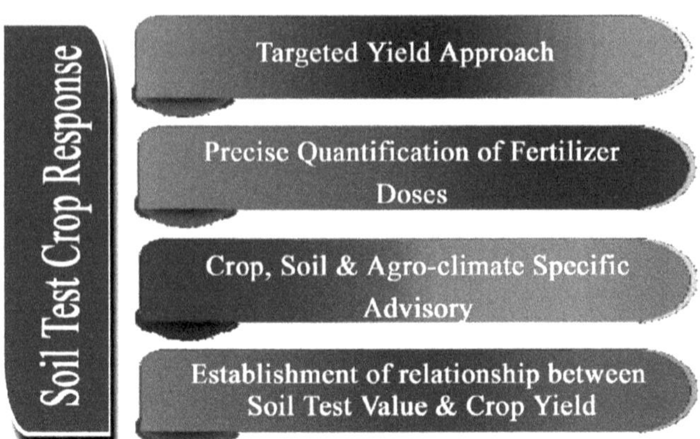

FIGURE 6.3 Concepts of STCR.

By analyzing soil test results and considering crop-specific nutrient demands, recommendations can be fine-tuned to address deficiencies and optimize nutrient levels, ensuring optimal crop growth and maximizing yields. This method aligns with the principles of precision agriculture, where inputs are applied precisely where and when needed, leading to improved resource use efficiency and sustainable production practices. This ensures that nutrient applications are tailored to the requirements of individual fields and crops, maximizing the efficiency of nutrient use and minimizing waste. STCR is a valuable tool for precision nutrient management, as it allows farmers to make informed decisions about nutrient applications based on scientific principles and field data. By understanding the relationship between soil fertility and crop response, farmers can optimize nutrient use efficiency, improve crop yields, and minimize environmental impacts associated with excessive fertilizer use.

6.4.2 COMPOSITION NUTRIENT DIAGNOSIS

Composition Nutrient Diagnosis sounds like a term related to assessing the nutrient composition of a sample and diagnosing any deficiencies or imbalances based on that composition. However, it is not a standard term in soil science or agronomy. In soil science and agriculture, nutrient diagnosis typically involves analyzing soil, plant tissue, or water samples to determine their nutrient content and assess whether it meets the needs of crops or other plants. Soil and plant tissue, such as leaves, stems, or roots, are collected from crops or other plants, and water samples are analyzed to determine their nutrient content. Plant tissue analysis provides insights into nutrient uptake, translocation, and utilization within plants, helping diagnose nutrient deficiencies, toxicities, or imbalances. Water samples from irrigation sources, drainage systems, or natural water bodies are analyzed to assess their nutrient content, including nitrogen, phosphorus, potassium, and other nutrients. Water nutrient analysis helps evaluate water quality and potential risks of nutrient leaching, runoff, or contamination. Nutrient diagnosis involves analyzing soil, plant tissue, or water samples to assess their nutrient composition and diagnose any deficiencies, excesses, or imbalances that may impact plant growth and productivity. This information guides nutrient management decisions to optimize soil fertility, crop nutrition, and environmental sustainability in agricultural and horticultural systems.

6.5 SMART FARMING SYSTEMS

Integrated smart farming systems combine data-driven decision support tools, automation, and Internet-of-Things technologies to optimize nutrient management practices in real time. These systems enable farmers to monitor soil and crop conditions remotely, automate field operations, and make data-driven decisions to maximize productivity while minimizing environmental impact (Figure 6.4).

FIGURE 6.4 Smart farming for nutrient management in sugarcane.

Crop canopy sensors and unmanned aerial vehicles (UAVs) are advanced technologies that play a crucial role in precision agriculture, including nutrient management. Crop canopy sensors measure the amount of light reflected or absorbed by crop canopies at specific wavelengths. Different nutrients affect the spectral properties of plants, allowing canopy sensors to indirectly assess crop nutrient status. Canopy sensors can detect variations in crop nutrient levels across fields by measuring differences in plant vigor, chlorophyll content, and biomass accumulation. These sensors provide real-time data on crop health and nutrient status, enabling farmers to identify nutrient deficiencies or excesses early in the growing season. Canopy sensor data can be processed, analyzed, and integrated with precision agriculture systems using decision support tools or algorithms to generate fertilizer prescription maps and implement variable rate fertilizer application. By mapping spatial variations in crop nutrient requirements, farmers can adjust fertilizer application rates within fields, optimizing nutrient use efficiency and minimizing overapplication in areas with sufficient nutrients. These maps guide site-specific fertilizer applications based on crop nutrient needs, improving accuracy and reducing input costs.

UAVs equipped with multispectral or hyperspectral cameras capture high-resolution aerial imagery of crops and fields. These images provide detailed information on crop health, vigor, and nutrient status by analyzing spectral signatures associated with different plant characteristics. UAV-based remote sensing allows for rapid and cost-effective monitoring of large agricultural areas. By flying over fields at regular intervals, UAVs can capture temporal variations in crop growth and nutrient status, enabling timely interventions and management decisions. UAVs generate high-resolution maps of crop nutrient status, including spatial variability in nutrient levels within fields. These maps help identify areas of nutrient deficiency or excess, guiding targeted fertilizer applications and corrective measures. UAV-derived imagery and data can be integrated with geographic information systems (GIS) to create detailed nutrient management plans and prescription maps. GIS-based analysis allows farmers to visualize and analyze spatial patterns in crop nutrient status, facilitating informed decision-making.

6.6 MACHINE LEARNING (ML) AND DEEP LEARNING (DL) ALGORITHMS

It offers powerful tools for real-time crop nutrition management in agriculture. ML and DL algorithms can analyze data from various sources, such as soil sensors, remote sensing imagery, and plant sensors, to detect signs of nutrient deficiencies in crops. By training algorithms on labelled datasets of healthy and nutrient-deficient crops, models can learn to identify patterns and indicators of nutrient stress in real time and can analyze historical data on soil properties, weather conditions, crop management practices, and yield outcomes to predict future crop yields. These predictions can help optimize nutrient management strategies by adjusting fertilizer application rates based on predicted yield potentials and expected nutrient requirements.

ML algorithms can analyze spatial data from satellite imagery, UAVs, or soil sensors to create prescription maps for VRA of fertilizers. By identifying areas within fields with different nutrient needs, VRA systems can optimize fertilizer application rates to match crop requirements, improving nutrient use efficiency and minimizing waste. Apart from this, techniques can be used to develop crop growth models that simulate the effects of nutrient management practices on crop growth and yield. These models can incorporate complex interactions between soil, weather, genetics, and management practices to predict the impact of different nutrient management scenarios on crop performance. ML algorithms can power decision support systems that provide real-time recommendations for nutrient management based on current soil and crop conditions and also analyze soil moisture data, crop water stress indices, and weather forecasts to optimize irrigation scheduling and nutrient delivery. These systems can integrate data from multiple sources, including soil tests, weather forecasts, and crop sensors, to generate tailored nutrient management prescriptions for farmers. By dynamically adjusting irrigation and fertigation rates based on real-time conditions, precision irrigation systems can improve water and nutrient use efficiency while minimizing environmental impacts.

DL algorithms can be deployed in automated monitoring systems that continuously analyze sensor data from fields to detect changes in crop nutrition status. These systems can trigger alerts or automatically adjust nutrient application equipment in response to nutrient deficiencies or imbalances, enabling proactive management interventions. Overall, ML and DL algorithms offer powerful tools for real-time crop nutrition management. They allow farmers to optimize nutrient applications, improve crop productivity, and enhance sustainability. Leveraging advanced analytics and automation technologies enables precision nutrient management practices that minimize environmental impacts and maximize resource use efficiency.

6.7 CONCLUSION

By leveraging traditional and advanced nutrient management, farmers can adopt precision agriculture practices that optimize fertilizer use, improve crop yields, and enhance environmental sustainability. Advanced technologies enable site-specific nutrient applications tailored to the needs of individual crops and fields, leading to more efficient use of resources and reduced environmental impacts. This targeted approach minimizes waste, reduces environmental pollution, and maximizes the efficiency of nutrient use, ultimately leading to higher crop yields and profitability for farmers. Additionally, precision agriculture promotes sustainability by optimizing resource use and minimizing the ecological footprint of agricultural activities, thereby ensuring the long-term viability of farming systems.

KEYWORDS

- **machine learning**
- **precise nutrient management**
- **soil fertility**
- **sugarcane**
- **yields**

REFERENCES

Anas M, Verma KK, Haq IU, Naeem M, Li Q, Liao F, Xiang YX, Li YR. (2021a) Characterization of exotic, native and wild-type genotypes of sugarcane (*Saccharum* spp. hybrids) for internal nitrogen use efficiency under different nitrogen levels. *Sugar Tech.* 23, 1258–1267. https://doi.org/10.1007/s12355-021-01014-1

Anas M, Verma KK, Riaz M, Qiang L, Liao F, Liu Y, Li YR. (2021b) Physio-morphological and biochemical mechanism of nitrogen use efficiency in sugarcane (*Saccharum* spp.) genotypes under different growth stages and nitrogen levels. *J. Plant Interact.* 16, 332–343.

Banerjee H, Rana L, Ray K, Sarkar S, Bhattacharyya K, Dutta S. (2016) Differential physiological response in potato (*Solanum tuberosum* L.) upon exposure to nutrient omissions. *Ind. J. Plant Physiol.* 21, 129–136.

Banerjee H, Ray K, Sarkar S, Puste AM, Mozumder M, Rana L. (2015) Impact of nitrogen nutrition on productivity and nutrient use efficiency of potato (*Solanum tuberosum* L.) in an inceptisol of West Bengal, India. SAARC. *J. Agric.* 13(2), 141–150.

Bhatt R, Verma KK, Jamwal NS. (2023a) Can paddy-compost eliminate N fertilization for climate-smart sugarcane? *Sugar Tech.* 25, 1036–1046. https://doi.org/10.1007/s12355-023-01270-3

Bhatt R, Verma KK, Kumar R, Sanghera GS. (2023b) Foliar application of potassium salt of active phosphorus (PSAP) mitigates insect pests and improves yield along with sugarcane quality in response to agroclimatic conditions of Punjab. *Sugar Tech.* 25(3), 660–669.

Bhilala S, Rana L, Kumar N, Kumar A, Jha CK, Meena SK, Singh AK. (2022) Splitting dose influence the nutrient acquisition of *Saccharum officinarum* L. under calcareous soil conditions. *Pharma. Inno. J.* 11(12), 5114–5117.

Bhilala S, Rana L, Kumar N, Kumar A, Meena SK, Singh AK. (2023) Yield and juice quality in sugarcane influenced by split application of nitrogen and potassium under subtropical climates. *Environ. Ecol.* 41(1B), 492–495.

Buresh RJ, Pampolino MF, Witt C. (2010) Field-specific potassium and phosphorus balances and fertilizer requirements for irrigated rice-based cropping systems. *Plant Soil* 335, 35–64.

Cassman KG, Dobermann A, Walters DT, Yang H. (2003) Meeting cereal demand while protecting natural resources and improving environmental quality. *Ann. Rev. Environ. Resour*. 28(1), 315–358.

Chaudhary P, Xu M, Ahamad L, Chaudhary A, Kumar G, Adeleke BS, Verma KK, Hu D-M, Siric I, Kumar P, Popescu SM, Fayssal SA. (2023) Application of synthetic consortia for improvement of soil fertility, pollution remediation, and agricultural productivity: A review. *Agronomy* 13, 643. https://doi.org/10.3390/agronomy13030643.

Chen GL, Zheng FJ, Lin B, Lao SB, He J, Huang Z, Zeng Y, Sun J, Verma KK. (2020) Phenolic and volatile compounds in the production of sugarcane vinegar. *ACS Omega* 5, 30587–30595.

Chen G-L, Zheng F-J, Lin B, Yang Y-X, Fang X-C, Verma KK, Yang L-F. (2023) Vinegar: A potential source of healthy and functional food with special reference to sugarcane vinegar. *Front. Nutr*. 10, 1145862. https://doi.org/10.3389/fnut.2023.1145862

Deng ZN, Huang HR, Huang CM, Verma KK, Luo HB, Jiang SL, Li YJ, Cao HQ, Wu KC, Xu L. (2023) Short-term limited water irrigation influences the photosynthetic pigments, enzymatic and non-enzymatic activities in *Saccharum spontaneum* L. at vegetative stage. *Sugar Tech*. 25, 110–118. https://doi.org/10.1007/s12355-022-01175-7

Dobermann A, Witt, C. et al. (2004) The evolution of site-specific nutrient management in irrigated rice systems of Asia. In: Dobermann A. (Ed.), *Increasing Productivity of Intensive Rice Systems Through Site-specific Nutrient Management*. Science Publishers, Inc., and International Rice Research Institute (IRRI): Enfield, NH, p. 410.

Dotaniya ML, Datta SC, Biswas DR, Dotaniya CK, Meena BL, Rajendiran S, et al. (2016) Use of sugarcane industrial by-products for improving sugarcane productivity and soil health. *Int. J. Recycl. Org. Waste Agric*. 5, 185–194. https://doi.org/10.1007/s40093-016-0132-8.

Guo D-J, Li D-P, Chen Z-D, Yang B, Verma KK, Singh RK, Singh P, Khan Q, Sharma A, Qin Y, Zhang B-Q, Song X-P, Li Y-R. (2023) Effect of endophytic diazotroph *Enterobacter roggenkampii* ED5 on nitrogen-metabolism-related microecology in the sugarcane rhizosphere at different nitrogen levels. *Front. Microbiol*. 14, 1132016. https://doi.org/10.3389/fmicb.2023.1132016

Guo D-J, Li D-P, Singh RK, Singh P, Sharma A, Verma KK, Qin Y, Khan Q, Lu Z, Malviya MK, Song X-P, Xing Y-X and Li Y-R. (2021) Differential protein expression analysis of two sugarcane varieties in response to diazotrophic plant growth-promoting endophyte enterobacter roggenkampii ED5. *Front. Plant Sci*. 12, 727741. https://doi.org/10.3389/fpls.2021.727741

Guo DJ, Li DP, Singh RK, Singh P, Verma KK, Sharma A, Qin Y, Khan Q, Song XP, Malviya MK, Xing YX, Li YR. (2022) Comparative transcriptome analysis of two sugarcane varieties in response to diazitrophic plant growth promoting endophyte *Enterobacter roggenkampii* ED5. J. Plant Interact. 17(1), 75–84.

He P, Yang L, Xub X, Zhao S, Chend F, Li S. (2015) Shihua Tue, Jiyun Jinb, Adrian M. Johnstonc, f. Temporal and spatial variation of soil available potassium in China (1990–2012). *Field Crops Res*. 173, 49–56.

Huang CM, Wu KC, Verma KK, Luo HB, Deng ZN, Shi GY, Cao HQ, Wu XJ, Ye LP, Wei YW, Xu L. (2024) Isolation and expression analysis of cellulose synthase 3 (*Ces3*) genes from sugarcane (*Saccharum officinarum* L.). *Chem. Biol. Technol. Agric*. 11, 71. https://doi.org/10.1186/s40538-024-00597-7

Huang HR, Gao YJ, Malviya MK, Verma KK, Solanki MK, Huang YX, Li X, Deng Y, Yan J, Tang SY, Wang LW, Xu L. (2022) Genetic diversity analysis of sugarcane (*Saccharum* spp. Hybrids) among high-sucrose clones of GT series and commonly used parents by using microsatellite markers in Guangxi, China. *Sugar Tech.* 24, 397–407. https://doi.org/10.1007/s12355-021-01033-y

Jagadesh M, Dash M, Kumari A, Singh SK, Verma KK, Kumar P, Bhatt R, Sharma SK. (2024) Revealing the hidden world of soil microbes: Metagenomic insights into plant, bacteria, and fungi interactions for sustainable agriculture and ecosystem restoration. *Microbiol. Res.* 285, 127764. https://doi.org/10.1016/j.micres.2024.127764

Jensen ES, Carlsson G, Nielsen HH. (2020) Intercropping of grain legumes and cereals improves the use of soil N resources and reduces the requirement for synthetic fertilizer N: A global-scale analysis. *Agron. Sustain. Dev.* 40, 5. https://doi.org/10.1007/s13593-020-0607-x

Joshi A, Verma KK, Rajput VD, Arora J. (2023) Metabolic engineering of microorganisms in advancing biofuel production. In: *Advances in Lignocellulosic Biofuel Production Systems*. pp. 181–202. https://doi.org/10.1016/B978-0-323-91192-4.00016-X

Joshi A, Verma KK, Rajput VD, Minkina T, Arora J. (2022) Recent advances in metabolic engineering of microorganisms for advancing lignocellulose-derived biofuels. *Bioengineered* 13(4), 8135–8163. https://doi.org/10.1080/21655979.2022.2051856.

Krishna D, Sachan HK, Verma KK. (2022) Abiotic stresses in plants: An overview. In: *Understanding Abiotic Stresses*. Nova Science Publishers, Inc.: New York.

Kumar N, Kumar V, Kishor K, Singh AK. (2024a) Optimizing nutrient application and nitrogen transformation to maximize the growth and productivity of bud chip transplanted sugarcane. *J. Plant Nutri.* 47(16), 2583–2596. https://doi.org/10.1080/01904167.2024.2354201

Kumar N, Kumar V. (2020) Production potential and nitrogen fractionation of sugarcane-based cropping system as influenced by planting materials and nitrogen nutrition. *Sugar Tech.* 22(4), 622–629.

Kumar N, Kumari G, Kumar A, Paswan S. (2016) Influence of manuring and irrigation scheduling on system productivity, resource use efficiency, nutrient uptake and incidence of early shoot borer (*Chilo infuscatellus*) in sugarcane (*Saccharum* spp. hybrid complex) and garlic (*Allium sativum*) intercropping system. *Ind. J. Agron.* 61(3), 297–306.

Kumar N, Kumari G, Kumar V. (2017) Enhancing crop productivity and soil health from intercropping coriander with autumn sugarcane through scheduling of fertilizer and irrigation. *J. Ind. Soc. Soil Sci.* 65(3), 300–307.

Kumar N, Kumari G, Paswan S. (2015) Optimizing nutrient and irrigation requirement of sugarcane (*Saccharum* species hybrid complex) and French bean (*Phaseolus vulgaris*) intercropping system. *Ind. J. Agron.* 60(4), 524–533.

Kumar N, Rana L, Singh AK, Pramanick B, Gaber A, Alsuhaibani AM, Skalicky M and Hossain A. (2023) Precise macronutrient application can improve cane yield and nutrient uptake in widely spaced plant-ratoon cycles in the Indo-Gangetic plains of India. *Front. Sustain. Food Syst.* 7, 1223881. https://doi.org/10.3389/fsufs.2023.1223881

Kumar N, Singh H, Kumar R, Kumari G. (2014) Nutrient uptake, sugarcane yield and economics of high sugar early genotypes of sugarcane (*Saccharum* spp. hybrid complex) under various planting seasons and fertility levels in Bihar. *Ind. J. Agric. Sci.* 84(4), 444–451.

Kumar N, Sinha UP. (2008) Response of spring-planted sugarcane (*Saccharum officinarum*) to phosphorus and sulphur application. *Ind. J. Agron.* 53(2), 145–148.

Kumar N, Sow S, Rana L, Kumar V, Kumar J, Pramanick B, Singh AK, Alkeridis LA, Sayed S, Gaber A, Hossain A. (2024) Productivity, water use efficiency and soil properties of sugarcane as influenced by trash mulching and irrigation regimes under different planting systems in sandy loam soils. *Front. Sustain. Food Syst.* 8, 1340551. https://doi.org/10.3389/fsufs.2024.1340551

Kumar N. (2012) Productivity, quality and nutrient balance in spring sugarcane (*Saccharum* spp. hybrid complex) under organic and inorganic nutrition. *Ind. J. Agron.* 57(1), 68–73.

Kumar S, Santhi R. (2017) Soil test based integrated plant nutrition system for pearl millet on an Inceptisol. *Res. Crops.* 18, 21–28.

Kumari A, Dash M, Singh SK, Jagadesh M, Mathpal B, Mishra PK, Pandey SK, Verma KK. (2023a) Soil microbes: A natural solution for mitigating the impact of climate change. *Environ. Monit. Assess.* 195, 1436. https://doi.org/10.1007/s10661-023-11988-y

Kumari A, Lakshmi GA, Krishna GK, Patni B, Prakash S, Bhattacharyya M, Singh SK, Verma KK. (2022) Climate change and its impact on crops: A comprehensive investigation for sustainable agriculture. *Agronomy* 12, 3008. https://doi.org/10.3390/agronomy12123008

Kumari A, Singh SK, Mathpal B, Verma KK, Garg VK, Bhattacharyya M, Bhatt R. (2023b) The biosynthesis, mechanism of action, and physiological functions of melatonin in horticultural plants: A review. *Horticulturae* 9, 913. https://doi.org/10.3390/horticulturae9080913

Li A-M, Liao F, Wang M, Chen Z-L, Qin C-X, Huang R-Q, Verma KK, Li Y-R, Que Y-X, Pan Y-Q, et al. (2023) Transcriptomic and proteomic landscape of sugarcane response to biotic and abiotic stressors. *Int. J. Mol. Sci.* 24, 8913. https://doi.org/10.3390/ijms24108913

Li Q, Chen J, Wu L, Luo X, Li N, Arafat Y, et al. (2018) Belowground interactions impact the soil bacterial community, soil fertility, and crop yield in maize/peanut intercropping systems. *Int. J. Mol. Sci.* 19, 622. https://doi.org/10.3390/ijms19020622

Malviya MK, Li C-N, Lakshmanan P, Solanki MK, Wang Z, Solanki AC, Nong Q, Verma KK, Singh RK, Singh P, Sharma A, Guo D-J, Dessoky ES, Song X-P, Li Y-R. (2022) High-throughput sequencing-based analysis of rhizosphere and diazotrophic bacterial diversity among wild progenitor and closely related species of sugarcane (*Saccharum* spp. inter-specific hybrids). *Front. Plant Sci.* 13, 829337. https://doi.org/10.3389/fpls.2022.829337.

Malviya MK, Solanki MK, Li CN, Wang Z, Zeng Y, Verma KK, Singh RK, Singh P, Huang HR, Yang LT, Song XP, Li YR. (2021) Sugarcane-legume intercropping can enrich the soil microbiome and plant growth. *Front. Sustain. Food Syst.* 5, 606595. https://doi.org/10.3389/fsufs.2021.606595.

Mozumder M, Banerjee H, Ray K, Paul T. (2014) Evaluation of potato (*Solanum tuberosum*) cultivars for productivity, N requirement and eco-friendly indices under different nitrogen levels. *Ind. J. Agron.* 59, 327–335.

Oliveira MA, Verma KK, Bhatt R, Oliveira TBA. (2022) Impact of green and organic fertilizers on soil fertility and sugarcane productivity. In: Verma KK. et al. (Eds.), *Agro-industrial Perspectives on Sugarcane Production Under Environmental Stress*. https://doi.org/10.1007/978-981-19-3955-6_11.

Otto R, Ferraz-Almeida R, Sanches GM, Lisboa IP, Cherubin MR. (2022) Nitrogen fertilizer consumption and nitrous oxide emissions associated with ethanol production – a national-scale comparison between Brazilian sugarcane and corn in the United States. *J. Clean. Prod.* 350, 131482. https://doi.org/10.1016/j.jclepro.2022.131482

Pramanick B, Kumar M, Naik BM, Kumar M, Singh SK, Maitra S, et al. (2022) Long-term conservation tillage and precision nutrient management in maize–wheat cropping system: Effect on soil properties, crop production, and economics. *Agronomy* 12, 2766. https://doi.org/10.3390/agronomy12112766

Ramamoorthy B, Narasimhan RL, Dinesh RS. (1967) Fertilizer application for specific yield targets of Sonara-64. *Ind. Farming* 5, 43–45.

Rana L, Banerjee H, Dutta SK, Ray K, Majumdar K, Sarkar S. (2017) Management practices of macronutrients for potato for smallholder farming system at alluvial soil (Entisols) of India. *Arch. Agron. Soil Sci.* 63, 1963–1976.

Rao S, Srivastava S. (2000) Soil test-based fertilizer urea must for sustainable agriculture. *Fertilizer News* 45, 25–38.

Reddy TP, Srijaya T, Madhavi A, Firdoz Sahana D, Lakshmi V, Ravindhar K. (2022) Validation of soil test and yield target based fertilizer prescription equation for sugarcane on vertisols of northern Telangana zone. *Biol. Forum–An Int. J.* 14(4a), 385–389.

Sakarvadia HL, Vekaria LC, Ponkia HP, Jadeja AS, Parakhia DV. (2021) Soil test-based fertilizers application for targeted yield of soybean (*Glycine max* L.) in Saurashtra region of Gujarat. *Agric. Sci. Digest* 41(3), 464–467.

Sharma A, Singh RK, Singh P, Vaishnav A, Guo DJ, Verma KK, Li DP, Song XP, Malviya MK, Khan N, Lakshmanan P, Li YR. (2021) Insights into the bacterial and nitric oxide-induced salt tolerance in sugarcane and their growth-promoting abilities. *Microorganisms* 9, 2203. https://doi.org/10.3390/microorganisms9112203.

Sharma A, Song X-P, Singh RK, Vaishnav A, Gupta S, Singh P, Guo D-J, Verma KK, Li Y-R. (2022) Impact of carbendazim on cellular growth, defence system and plant growth promoting traits of *Priestia megaterium* ANCB-12 isolated from sugarcane rhizosphere. *Front. Microbiol.* 13, 1005942. doi:10.3389/fmicb.2022.1005942

Shoko MD, Tagwira F, Zhou M. (2007) The potential of reducing nitrogen fertilizers in a soybean-sugarcane production system in Zimbabwe. *Afr. J. Agric. Res.* 2, 475–480.

Singh AK, Lal M. (2007) Assessment of system productivity, cane physiology and economic viability of sugarcane (*Saccharum* complex hybrid)–based crop diversification optionsthrough on–station trials. *Ind. J. Agri. Sci.* 77(12), 866–889.

Singh M, Dwivedi BS, Datta SP. (2012) Integrated nutrient management for enhancing productivity, nutrient use efficiency and environmental quality. In: Goswami NN. (Ed.), *Soil Science in the Service of Nation*, New Delhi, pp. 55–67.

Singh P, Singh RK, Li HB, Guo DJ, Sharma A, Verma KK, Solanki MK, Upadhyay SK, Lakshmanan P, Yang LT, Li YR. (2023) Nitrogen fixation and phytohormone stimulation of sugarcane plant through plant growth promoting diazotrophic *Pseudomonas*. Biotechnol. Genet. Eng. Rev. 40(1), 15–35. https://doi.org/10.1080/02648725.2023.2177814

Singh P, Singh RK, Song QQ, Li HB, Guo DJ, Malviya MK, Verma KK, Song XP, Lakshmanan P, Yang LT, Li YR. (2021) Comparative analysis of protein and differential responses of defense-related gene and enzyme activity reveals the long-term molecular responses of sugarcane inoculated with *Sporisorium scitamineum*. *J. Plant Interact.* 16(1), 12–29.

Singh RK, Singh P, Li HB, Song QQ, Guo DJ, Solanki MK, Verma KK, Malviya MK, Song XP, Lakshmanan P, Yang LT, Li YR. (2020) Diversity of nitrogen-fixing rhizobacteria associated with sugarcane: A comprehensive study of plant-microbe interactions for growth enhancement in *Saccharum* spp. *BMC Plant Biol.* 20, 220. https://doi.org/10.1186/s12870-020-02400-9.

Solanki MK, Verma KK, Dastogeer KMG, Mora-Poblete F, Mundra S. (2023) Microbial resilience in plant nutrient management towards sustainable farming. *Front. Microbiol.* 14, 1280811. https://doi.org/10.3389/fmicb.2023.1280811

Solanki MK, Wang FY, Wang Z, Li CN, Lan TJ, Singh RK, et al. (2018) Rhizospheric and endospheric diazotrophs mediated soil fertility intensification in sugarcane-legume intercropping systems. *J. Soil Sediments* 19, 1911–1927. https://doi.org/10.1007/s11368-018-2156-3

Song X-P, Mo F, Verma KK, Wei J, Zhang X, Yang L, Li YR. (2019) Effect of sugarcane smut (*Ustilago scitaminea* Syd.) on ultrastructure and biochemical indices of sugarcane. *Biomed J. Sci. Tech. Res.* 002950. https://doi.org/10.26717/BJSTR.2019.17.002950.

Song XP, Mo F, Yan M, Zhang X, Zhang B, Huang X, Huang D, Pan Y, Verma KK, Li YR. (2022) Effect of smut infection on the photosynthetic physiological characteristics and related defense enzymes of sugarcane. *Life* 12, 1201. https://doi.org/10.3390/life12081201

Song XP, Verma KK, Tian DD, Zhang XQ, Liang YJ, Huang X, Li CN, Li YR. (2021) Exploration of silicon functions to integrate with biotic stress tolerance and crop improvement. *Biol. Res.* 54, 19. https://doi.org/10.1186/s40659-021-00344-4

Tang X, Jiang J, Huang Z, Wu H, Wang J, He L, et al. (2021a) Sugarcane/peanut intercropping system improves the soil quality and increases the abundance of beneficial microbes. *J. Basic Microbiol.* 61, 165–176. https://doi.org/10.1002/jobm.202000750

Tian D, Qin L, Verma KK, Wei L, Li J, Li B, Zhou W, He Z, Wei D, Huang S, Long S, Huang Q, Li C, Wei S. (2023) Transcriptomic and metabolomic differences between banana varieties which are resistant or susceptible to *Fusarium wilt. PeerJ* 11, e16549. https://doi.org/10.7717/peerj.16549

Troug E. (1960) Fifty years of soil testing, transactions of 7th International Congress of Soil Science. Vol. III, Commission IV. Paper No. 746–53.

Verma KK, Joshi A, Song X-P, Liang Q, Xu L, Huang H-r, Wu K-C, Seth CS, Arora J and Li Y-R (2024a) Regulatory mechanisms of plant rhizobacteria on plants to the adaptation of adverse agroclimatic variables. *Front. Plant Sci.* 15, 1377793. https://doi.org/10.3389/fpls.2024.1377793

Verma KK, Joshi A, Song X-P, Singh S, Kumari A, Arora J, Singh SK, Solanki MK, Seth CS, Li Y-R. (2024b) Synergistic interactions of nanoparticles and plant growth promoting rhizobacteria enhancing soil-plant systems: A multigenerational perspective. *Front. Plant Sci.* 15, 1376214. https://doi.org/10.3389/fpls.2024.1376214

Verma KK, Li DM, Singh M, Rajput VD, Malviya MK, Minkina T, Singh RK, Singh P, Song XP, Li YR. (2020a) Interactive role of silicon and plant-rhizobacteria mitigating abiotic stresses: A new approach for sustainable agriculture and climate change. *Plants* 9, 1055. https://doi.org/10.3390/plants9091055.

Verma KK, Rajput VD, Singh M, Verma CL, Minkina T. (2022a) Influence of waterlogging on plant physiology and molecular biology. In: *Understanding Abiotic Stresses*. Nova Science Publishers, Inc.: New York.

Verma KK, Siddiqui N, Singh M. (2022b) Understanding the plant physiological and molecular adaptation strategies to soil salinity. In: *Understanding Abiotic Stresses*. Nova Science Publishers, Inc.: New York. https://doi.org/10.52305/UBUJ4024

Verma KK, Singh P, Song X-P, Malviya MK, Singh RK, Chen G-L, Solomon S, Li YR. (2020b) Mitigating climate change for sugarcane improvement: Role of silicon in alleviating abiotic stresses. *Sugar Tech.* 22(5), 741–749. https://doi.org/10.1007/s12355-020-00831-0.

Verma KK, Singh RK, Song QQ, Singh P, Zhang B-Q, Song X-P, Chen G-L, Li YR. (2019a) Silicon alleviates drought stress of sugarcane plants by improving antioxidant responses. *Biomed. J. Sci. Tech. Res.* 002957. https://doi.org/10.26717/BJSTR.2019.17.002957

Verma KK, Song X, Degu HD, Guo D, Joshi A, Huang H, Xu L, Singh M, Huang D, Rajput VD, Li YR. (2023a) Recent advances in nitrogen and nano-nitrogen fertilizers for sustainable crop production: A mini-review. *Chem. Biol. Technol. Agric.* 10, 1–14. https://doi.org/10.1186/s40538-023-00488-3

Verma KK, Song X, Verma CL, Huang H, Singh M, Xu L, Wu J, Khan ZH, Zhang B, Li Y. (2023b) Mathematical modeling of climate and fluoride effects on sugarcane photosynthesis with silicon nanoparticles. *Plant Physiol. Biochem.* 204, 108089. https://doi.org/10.1016/j.plaphy.2023.108089

Verma KK, Song XP, Budeguer F, Nikpay A, Enrique R, Singh M, Zhang B, Wu, JM, Li YR. (2022c) Genetic engineering: An efficient approach to mitigating biotic and abiotic stresses in sugarcane cultivation. *Plant Signal. Behav.* 17(1), e2108253.

Verma KK, Song XP, Joshi A, Rajput VD, Singh M, Sharma A, Singh RK, Li DM, Arora J, Minkina T, Li YR. (2022d) Nanofertilizer possibilities for healthy soil, water and food in future: An overview. *Front. Plant Sci.* 13, 865048. https://doi.org/10.3389/fpls.2022.865048

Verma KK, Song XP, Joshi A, Tian DD, Rajput VD, Singh M, Arora J, Minkina TM, Li YR. (2022e) Recent trends in nano-fertilizer for advancing sustainable agriculture under the era of climate change to ensure food security for future generations. *Nanomaterials* 12, 173. https://doi.org/10.3390/nano12010173.

Verma KK, Song XP, Li DM, Singh M, Wu JM, Singh RK, Sharma A, Zhang BQ, LI YR. (2022f) Silicon and soil microorganisms improves rhizospheric soil health with bacterial community, plant growth, performance and yield. *Plant Signal. Behav.* 17(1), e2104004.

Verma KK, Song XP, Lin B, Guo DJ, Singh M, Rajput VD, Singh RK, Singh P, Sharma A, Malviya MK, Chen GL, Li YR. (2022g) Silicon induced drought tolerance in crop plants: Physiological adaptation strategies. *Silicon* 14(6), 2473–2487. https://doi.org/10.1007/s12633-021-01071-x

Verma KK, Song XP, Rajput VD, Boldyreva V, Zhang BQ, Minkina T, Li YR. (2022h) Morpho-physiological, biochemical and ultrastructural modifications on sugarcane to prolonged water deficit. In: Verma KK. et al. (Eds.), *Agro-industrial Perspectives on Sugarcane Production Under Environmental Stress*. https://doi.org/10.1007/978-981-19-3955-6_11.

Verma KK, Song XP, Rajput VD, Solomon S, Li YR, Rao GP. (2022i) *Agro-industrial Perspectives on Sugarcane Production Under Environmental Stress*. Springer Nature: Singapore, https://doi.org/10.1007/978-981-19-3955-6

Verma KK, Song XP, Singh M, Tian DD, Rajput VD, Minkina T, Li YR. (2023c) Association of silicon and soil microorganisms induces stress mitigation, increasing plant productivity. In: de Mello Prado R. (Ed.), *Benefits of Silicon in the Nutrition of Plants*, pp. 299–328. https://doi.org/10.1007/978-3-031-26673-7_17

Verma KK, Song XP, Tian DD, Guo DJ, Chen ZL, Zhong CS, Nikpay A, Singh M, Rajput VD, Singh RK, Minkina T, Li YR. (2021a) Influence of silicon on biocontrol strategies to manage biotic stress for crop protection, performance and improvement. *Plants* 10, 2163. https://doi.org/10.3390/plants10102163

Verma KK, Song XP, Tian DD, Singh M, Verma CL, Rajput VD, Singh RK, Sharma A, Singh P, Malviya MK, Li YR. (2021b) Investigation of defensive role of silicon during drought stress induced by irrigation capacity in sugarcane: Physiological and biochemical characteristics. *ACS Omega* 6, 19811–19821.

Verma KK, Song XP, Verma CL, Chen ZL, Rajput VD, Wu KC, Liao F, Chen GL, Li YR. (2021c) Functional relationship between photosynthetic leaf gas exchange in response to silicon application and water stress mitigation in sugarcane. *Biol. Res.* 54, 15. https://doi.org/10.1186/s40659-021-00338-2

Verma KK, Song XP, Verma CL, Malviya MK, Guo DJ, Rajput VD, Sharma A, Wei KJ, Chen GL, Solomon S, Li YR. (2021d) Predication of photosynthetic leaf gas exchange of sugarcane (*Saccharum* spp.) leaves in response to leaf positions to foliar spray of potassium salt of active phosphorus under limited water irrigation. *ACS Omega* 6, 2396–2409.

Verma KK, Song X-P, Xu L, Huang H-R, Liang Q, Seth CS, Li Y-R. (2023d) Nano-microplastic and agro-ecosystems: A mini-review. *Front. Plant Sci.* 14, 1283852. https://doi.org/10.3389/fpls.2023.1283852

Verma KK, Song XP, Yadav G, Degu HD, Parvaiz A, Singh M, Huang HR, Mustafa G, Xu L, Li YR. (2022j) Impact of agroclimatic variables on proteo-genomics in sugarcane (*Saccharum* spp.) plant productivity. *ACS Omega* 7, 22997–23008.

Verma KK, Song XP, Zeng Y, Guo DJ, Sing M, Rajput VD, Malviya MK, Wei KJ, Sharma A, Li DP, Chen GL, Li YR. (2021e) Foliar application of silicon boosts growth, photosynthetic leaf gas exchange, antioxidative response and resistance to limited water irrigation in sugarcane (*Saccharum officinarum* L.). *Plant Physiol. Biochem.* 166, 582–592.

Verma KK, Song XP, Zeng Y, Li DM, Guo DJ, Rajput VD, Chen GL, Barakhov A, Minkina TM, Li YR. (2020c) Characteristics and correlation of leaf stomata and its relationship with photosynthesis on *Saccharum* spp. under different irrigation and silicon application. *ACS Omega* 5, 24145–24153.

Verma KK, Vatsal S, Singh M, Rajput VD, Minkina T. (2022k) Coping with water stress: Effects and adaptive mechanisms to plants. In: *Understanding Abiotic Stresses*. Nova Science Publishers, Inc.: New York.

Verma KK, Wu KC, Verma CL, Li DM, Malviya MK, Singh RK, Singh P, Chen GL, Song XP, Li YR. (2020d) Developing mathematical model for diurnal dynamics on photosynthetic responses in sugarcane responsive to different irrigation and silicon application. *PeerJ* 8, e10154. https://doi.org/10.7717/peerj.10154.

Verma KK, Liu X-H, Wu K-C, Singh RK, Song QQ, Malviya MK, Song X-P, Singh P, Verma CL, Li YR. (2020e) The impact of silicon on photosynthetic and biochemical responses of sugarcane under different soil moisture levels. *Silicon* 12, 1355–1367.

Verma KK, Wu K-C, Singh P, Malviya MK, Singh RK, Song X-P, Li YR. (2019b) The protective role of silicon in sugarcane under water stress: Photosynthesis and antioxidant enzymes. *Biomed. J. Sci. Tech. Res.* 002685. https://doi.org/0.26717/BJSTR.2019.15.002685

Wang, ZG, Jin X, Bao XG, Li XF, Zhao JH, Sun JH, et al. (2014) Intercropping enhances productivity and maintains the most soil fertility properties relative to sole cropping. *PLoS One* 9, e113984. https://doi.org/10.1371/journal.pone.0113984

Witt C, Dobermann A, Abdul rachman S, Gines HC, Guanghuo W, Nagarajan R, Satawatananont S, Son TT, Tan PS, Tiem L, et al. (1999) Internal nutrient efficiencies of irrigated lowland rice in tropical and subtropical Asia. *Field Crops Res.* 63, 113–138.

Wu K-C, Huang C-M, Verma KK, Deng Z-N, Huang H-R, Pang T, Cao H-Q, Luo H-B, Jiang S-L, Xu L. (2022) Transcriptomic responses of *Saccharum spontaneum* roots in response to polyethylene glycol–6000 stimulated drought stress. *Front. Plant Sci.* 13, 992755. https://doi.org/10.3389/fpls.2022.992755

Xu L, Deng Z-N, Wu K-C, Malviya MK, Solanki MK, Verma KK, Pang T, Li Y-J, Liu X-Y, Kashyap BK, Dessoky ES, Wang W-Z, Huang H-R. (2022) Transcriptome analysis reveals a gene expression pattern that contributes to sugarcane bud propagation induced by Indole-3-Butyric acid. *Front. Plant Sci.* 13, 852886. https://doi.org/10.3389/fpls.2022.852886

Zheng F-J, Lin B, Yang Y-X, Fang X-C, Verma KK, Chen G-L. (2023) Efficacy and functionality of sugarcane original vinegar on mice. *Front. Microbiol.* 14, 1224666. https://doi.org/10.3389/fmicb.2023.1224666

CHAPTER 7

Mechanization in Sugarcane Cultivation: An Overview

JYOTIRMAY MAHAPATRA[1*], JAGJEET SINGH[1*], RAMESH K. SAHNI[1,2], and VIKAS PAGARE[1]

[1]*ICAR-Central Institute of Agricultural Engineering, Bhopal, Madhya Pradesh, India*

[2]*Center for Precision and Automated Agricultural Systems, Washington State University, Pullman, WA, USA*

*Corresponding author

ABSTRACT

Sugarcane is an important cash crop grown in tropical and subtropical regions of the globe. The demand for the crop is increasing with population and diversified application areas. Nowadays, it is not only an input for sugar production but also for molasses, ethanol, rum, bagasse, bioplastics, pulp, paper, livestock feed, and organic fertilizer. The bioenergy sector is boosting to solve issues like international problems of declining crude oil and global warming. More sugarcane needs to be produced from limited resources like land, water, seed cane, agrochemicals, labor, and so on. Mechanization can support it by providing a substitute for human labor, reducing input cost and environmental impact by precise placements of inputs as per requirements. Sugarcane crop requires several operations, including tillage and harvesting. Residue and ratoon management after harvesting is also mechanized. Many types of tools and equipment have been developed based on power sources like humans, animals, engines, power tillers, and tractors. Semiautomatic machinery is available in the market, while

Sugarcane Production: Environmental Impacts and Socioeconomic Issues.
Krishan K. Verma, Xiu-Peng Song, Lin Xu, Rajan Bhatt, and Yang-Rui Li (Eds.)

DOI: 10.1201/9781779641137-7

reliable automatic systems are still under research and development. This chapter discusses different farm operations and sugarcane cropping equipment.

7.1 INTRODUCTION

Sugarcane is native to tropical, subtropical, and warm temperate regions worldwide (Chen et al., 2020). It is the world's largest crop by production quantity, with around 1.9 billion tons in 2020, with Brazil and India accounting for 40% and 20%, respectively. The sugarcane production is not sufficient to meet the demand of the current world population due to its various applications, such as industrial processing of the stalks for alcohol, food industries, paper industries, feedstock, biofuel, and so on (Verma et al., 2021a, 2021b, 2022a, 2023a, 2023b; Huang et al., 2024). Sugarcane production and its productivity are highly affected by various factors such as climate, geographical factors, technological involvement, and so on (Verma et al., 2019a, 2019b; Singh et al., 2020, 2021; Kumari et al., 2022, 2023a, 2023b; Sharma et al., 2021, 2022; Guo et al., 2022; Bhatt et al., 2023a, 2023b). Climate and geography are not much in humans' control, but the adaptation of appropriate technologies could play a significant role in increasing sugarcane production (Verma et al., 2020a, 2020b, 2021c, 2021d, 2023c, 2023d; Oliveira et al., 2022; Zheng et al., 2023).

The practices involved in sugarcane production are different from those of other crops (Verma et al., 2020c, 2020d, 2020e; Solanki et al., 2023; Singh et al., 2023). Different methods are adapted as per suitability to specific conditions (Krishna et al., 2022). The seedbed may be prepared following the ridge and furrow or ring pit methods. The setts may be prepared with one to three nodes. The planting may be done using any one of bud, sett, whole cane planting, and transplanting (Song et al., 2019; Verma et al., 2021e, 2022b, 2022c, 2022d; Joshi et al., 2022, 2023; Wu et al., 2022; Xu et al., 2022). Farmers can do detrashing before or after harvesting. The residues left may be managed for *in situ* or *ex situ* utilization. Any one method is to be selected at each stage. A general sequence of operation flowcharts in sugarcane cultivation is given in Figure 7.1.

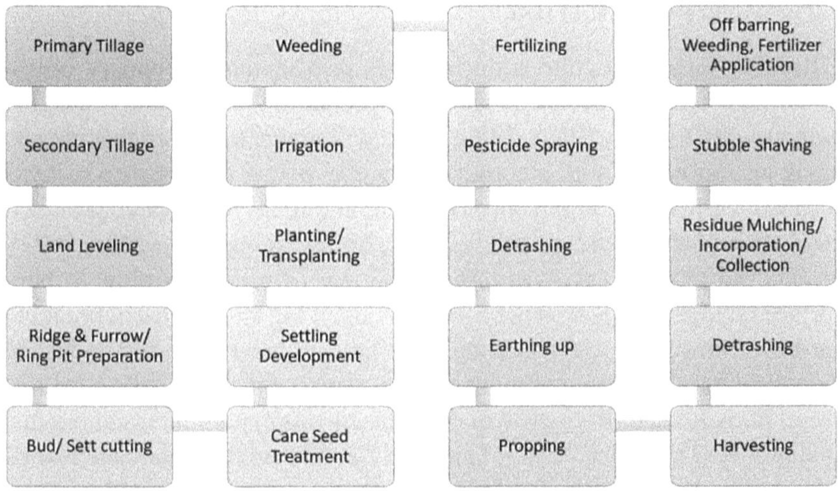

FIGURE 7.1 Sequence of farm operations in sugarcane cultivation.

Over the period, considerable research and development have been conducted to design and develop agricultural machinery on various operational aspects of sugarcane (Verma et al., 2022e, 2022f, 2022g). Modern machinery uses in sugarcane cultivation can increase sugarcane production and productivity, save labor, carry out timely operations, improve the quality of work, reduce operation costs, and ensure effective utilization of resources. A manual or automatic sugarcane bud cutter is used before the plantation by a semiautomatic or automatic sugarcane sett planter. Planters with integrated bud cutters and whole cane planters are also available. Ridge planters develop ridges on tilled land and plants. Settling transplanters can be used as an alternative to sugarcane sett planters or whole planters. Interculture operations are mechanized with a weeder, an earthing hoe, high-clearance sprayers, a wide-swath boom sprayer, a leaf cutter machine, and a detrasher. Detrashier cum detopper, manual cane harvesting knife, sugarcane harvester, multipurpose stubble shaver, ratoon management device, and plant residue shredder are equipment used for harvesting and residue management.

This chapter discusses in detail the machines involved in sugarcane production, such as sugarcane planting, interculture, plant protection, harvesting, and ratoon management.

7.2 LAND PREPARATION

Tillage in sugarcane cultivation involves primary and secondary operations, each serving specific purposes in preparing the soil and managing crop growth. Primary tillage is the initial soil preparation step that aims to break up compacted soil, incorporate organic matter, and create a suitable seedbed for planting sugarcane (Huang et al., 2022; Malviya et al., 2021, 2022; Verma et al., 2022h, 2022i, 2022j, 2022k; Song et al., 2021, 2022). Traditional plowing involves turning the soil over using a plow to bury surface residues and loosen the soil. Looking at the importance of breaking hardpan formed due to the movement of heavier machinery and deposited agrochemicals, deep penetrating tillage equipment like a subsoiler and chisel plow is used. Root growth depends on water, nutrients, and oxygen availability (Deng et al., 2023; Guo et al., 2023). As the sugarcane plant's root moves deeper, maximum root growth occurs in an average range of 0.6–1.8 m. Deep tillage enhances better root development.

Secondary tillage is practiced for weed control, soil aeration, and soil preparation before planting. The specific tillage practices employed can vary depending on factors such as soil conditions, weed pressure, and the overall management approach of the farm (Anas et al., 2021a, 2021b). Generally, a cultivator, disc harrow, clod crusher, and rotavator are used. However, on the other side, conservation tillage practices can also be employed with multiple advantages, like reduced soil erosion, maintaining soil structure and organic matter, reducing input costs, and reducing environmental impact (Guo et al., 2021; Li et al., 2023; Chen et al., 2023). Sugarcane is planted in a ridge and furrow system. Hence, a double-mold board or double-disc ridge forms ridges in well-tilted soil. Ridges and furrows can also be formed while planting using a ridge planter. This equipment has a ridge operating ahead of planting; hence, both jobs are carried out simultaneously. Combining two or more processes has several advantages, including saving time and money and reducing soil compaction by avoiding multiple passes.

7.3 SEEDING AND PLANTING

Sugarcane is planted in different ways, and the appropriate method must be selected based on factors like land area, sugarcane variety, available machinery, soil conditions, and so on. Sugarcane buds, setts or billets of 1/2/3 buds, whole sugarcane, or sugarcane seedlings may be used for planting.

To prepare these, different machines are used for cutting sugarcane, treating to protect the seed cane, and nursery growing to produce seedlings.

7.3.1 SEED CANE PREPARATION

A sugarcane bud cutter is a tool specifically designed to cut sugarcane buds or setts for planting. Sugarcane propagation primarily involves planting these setts, essentially segments of the sugarcane stalk containing one or more buds or nodes. These buds serve as the starting point for new sugarcane growth. The bud cutter is used to precisely cut the sugarcane stalk into sections, ensuring each section contains one or more viable buds. This tool helps maintain uniformity in the setts' size and quality, which is essential for successful planting and subsequent crop growth.

The first step is to select healthy sugarcane stalks free from disease and damage. These stalks are used to obtain the setts. The sugarcane bud cutter cuts the selected sugarcane stalks into sections. Human-powered and electric motor-based bud and sett cutters are generally used. The four-way bud-cutting machine shown in Figure 7.2(b) can produce about 10,000 buds per hour, and the bud size can be adjusted as per requirements. Currently, research is focused on developing machine vision-based precise buds or sett cutters to produce healthy cane seeds and reduce unnecessary losses. A mechatronic-operated system was developed by Nare et al. (2019). Another such equipment was developed and analyzed by Zhou et al. (2020), where a bud recognition rate of 93% was achieved. A single cutting unit can deliver 2400 buds/h.

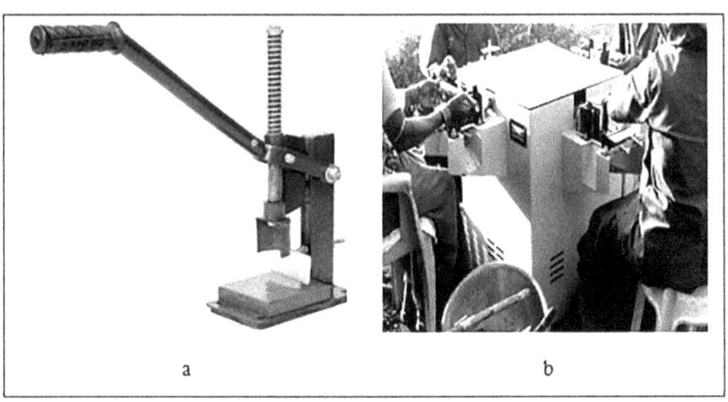

FIGURE 7.2 Hand-operated sugarcane bud chipper (a) and four-way bud cutting machine (b).

Once the setts are cut, they are usually allowed to dry for a short period to promote calluses, which helps to reduce the risk of rot and disease. However, the prepared seed canes should be treated with moist, hot air or hot water. In this method, seed canes are treated at 54°C for about 150 min at 95–99% humidity. This reduces primary infections from the parent cane. After this, the setts are ready for planting in the prepared soil.

7.3.2 BUD SETTS, WHOLE CANE PLANTERS

Planters can be categorized as bud-cutter planters, sett planters, and whole-cane planters. These can be further classified as semiautomatic or automatic based on human labor requirements for their operation. Semi-automatic planters are more popular due to their lower price and precise placement. These planters use a rotating disk where human workers manually place seedlings or setts. Belt-type metering systems are dominantly employed for automatic planting. A sugarcane cutter planter cuts sugarcane to produce setts and simultaneously plants them. Different types of planters are also available based on the type of soil-manipulating tools and the depth of placement of the seed cane. Ridger type, trench planter, paired row planter, and deep furrow planter are some examples.

Singh et al. (2017) developed a sugarcane cutter planter with a deep furrow planting arrangement and evaluated it. This planter produces 30 cm long setts by cutting whole sugarcane and then plants these into 25 cm long furrows. The effective field capacity was 0.16 ha/h with a 72-mm mean overlapping between two successive setts. About 63% of input costs can be reduced using this planter instead of conventional planting. The power take-off (PTO) of a tractor, ground wheel, or hydraulic motor powers planters. Considering several advantages of film mulching, a bud planter with soil covering and a film mulching arrangement was evaluated by Li et al. (2021).

The sugarcane cultivation ring pit method can be mechanized using a tractor-operated double-bottom pit digger. Singh et al. (2017) developed a tractor-mounted double-bottom pit digger. It can make digs with a diameter of 75 cm and a depth of 30 cm. The capacity of the equipment was evaluated to be 150 pits/h, covering 0.13 ha/day. The application of this equipment alone can save about 426 man-days/year.

7.3.3 SETTLING TRANSPLANTER

Seedlings are grown in portrays from sugarcane bud chips, and these are fed manually to a disk-type metering mechanism. Many researchers have modified the semiautomatic vegetable transplanters to use them for sugarcane settling. A single-row machine with a ground wheel for mechanical sugarcane transplantation has been developed by Abd El Mawla et al. (2014). A 35-hp tractor-mounted mechanical transplanter for settlings grown from sugarcane bud chips was designed by Naik et al. (2013). Presently, automatic-type sugarcane planters are under development. Technologies like programmable logic control-based automatic seedling extractor developed by Khadatkar et al. (2024) may also be used for sugarcane. Several studies have been conducted on the precise metering of buds or setts. Elwakeel et al. (2023) developed a variable-rate sugarcane planter to plant sugarcane seedlings. Further development of precision metering systems for sugarcane planting requires attention. Figure 7.3 shows nursery seedlings grown in a portray as well as a semiautomatic seedling transplanter.

a b

FIGURE 7.3 Sugarcane settlings grown in portray (a) and sugarcane settling transplanter with disk-type metering (b).

7.4 INTERCULTURAL OPERATION IN SUGARCANE

Intercultural operation is not a single operation process; it involves several operations, such as weeding, earthing, detrashing, mulching, and propping, to control weeds, save moisture, encourage microbial action, and improve the overall conditions for plant development.

7.4.1 NEED OF INTERCULTURAL OPERATION

- In sugarcane, weed infestation happens quickly, and the amount of crop loss owing to weeds can range from 10% to 100% crop failure.
- To control weeds, save moisture, encourage microbial action, and improve the overall conditions for plant development.
- Reduce competition between weeds and tender seedlings for light, water, and soil nutrients.
- To create the ideal environment for vigorous growth of sugarcane crops, cane yield must ultimately be increased.

7.4.2 METHODS ADAPTED FOR INTERCULTURAL OPERATION

There are many methods, namely, biological, chemical, and mechanical, that can be used to control weed growth and create favorable conditions for vigorous growth or sprouting of sugarcane seedlings.

Biological methods are generally adopted in small fields to control the sturdy attack of insect pests because they are time-consuming. Therefore, it is not frequently used on a wide scale in sugarcane. On the other hand, chemical approaches are very effective for controlling diseases and insect pests using different spraying systems, such as knapsack and boom sprayers. Still, weedicides can be costly and harmful to both humans and crops. Hence, it is preferable to employ a mechanical method instead of a chemical one.

7.4.3 MECHANIZATION IN SUGARCANE FOR INTERCULTURAL OPERATIONS

7.4.3.1 MACHINERIES FOR WEEDING OPERATION

Mechanical techniques also help preserve the soil's physical properties, enhancing plant growth and effectively reducing sugarcane yield loss. Common practices include hand-weeding with a khurpa or shovel, intercultivation with bullocks, and cultivators pulled by tractors are followed.

The most efficient method is manual weeding, but it is also the most expensive. Manual or mechanical hoeing is helpful at 30, 60, and 90 days following planting to reduce yield loss. On the other hand, a tractor-operated three-row rotary weeder for widely separated crops was recently introduced to the sugarcane interculture equipment market. The market is filled with high-capacity, effective tools for intercultural operations and weeding, including power tillers, self-propelled power weeders, and mini-tractors of all shapes and sizes. At the Indian Institute of Sugarcane Research, Lucknow, versatile intercultural equipment and a sugarcane manager have been built for local needs. A detailed description of this mechanical equipment is provided below.

7.4.3.2 TRACTOR DRAWN CULTIVATOR

A cultivator is an implement used for intercultivation operations with laterally adjustable tines to work in sugarcane crop, as shown in Figure 7.4(a). The main objective is to destroy the weeds and reduce competition for light, nutrients, and water among sugarcane plants and weeds. This implement is commercially available with spring-loaded and rigid tines with 7, 9, 11, 13, or more tynes, depending on the requirements. The animal-drawn cultivators have three and five tynes.

FIGURE 7.4 Weeding machinery for sugarcane (a–d).

7.4.3.3 TRACTOR-MOUNTED MULTIPURPOSE IMPLEMENT FOR SUGARCANE

The multipurpose implement for sugarcane is developed in two versions, as shown in Figure 7.4(b). First, it was developed as a three-row sugarcane planter. With little modification, this apparatus can be utilized as a (1) nine-tyne cultivator, (2) for intercultural operations, and (3) for earthing-up operations. With a few small accessories, it can also be used as a paddy puddler and seed drill. Second, it is produced as a sugarcane planter with two rows and utilized as a tractor-driven nine-tyne cultivator for preparing land as well as cross-cultural operations in sugarcane fields. In interculture mode, the field capacity is 0.72 ha/h.

7.4.3.4 TRACTOR-MOUNTED THREE-ROW ROTARY WEEDER

A tractor-mounted rotating weeder (Figure 7.4c) consisted of a primary frame, gearbox, three assemblies of rotating weeding blades, a shaft that transfers power from the gearbox to the rotating assemblies, and a series of chains and sprockets. The main square shaft of the tractor receives power from the PTO through a gearbox positioned on the main frame, a series of sprockets, and a chain. Row-to-row spacing can be easily adjusted from 675 to 1165 mm. The machine has a field capacity of 0.24 ha/h and an 83–87% weeding efficiency. Comparing the machine to the old approach, the savings are 54% labor and 74% operating costs.

7.4.3.5 SELF-PROPELLED ROTARY WEEDER/POWER TILLERS

It is used for interculturing in between crop rows, like sugarcane. The rotary weeder is attached to the entire power unit and receives power from the engine through a chain and sprocket mechanism. This machine saves a lot of labor and drudgery compared to the traditional system of manual weeding. A power tiller is also a multipurpose machine used for roto-tilling between sugarcane crop rows, as shown in Figure 7.4(d). Most available power tillers on the market have 10–15 hp. The working width is normally 60 cm, and the field capacity is 0.06 ha/h with fuel consumption of 1.0 l/h.

7.4.4 DETRASHING

All these leaves on the sugarcane stalk are not productive in nature; only a few leaves are required at the top of the cane stalk (around 8–10 leaves) to carry out optimum photosynthesis for cane stalk development.

Therefore, regular detrashing is essential in the sugarcane crop to attain the following advantages:

- The bottom green leaves are parasitic on the upper productive leaves and deplete the food reserves (photosynthesis), often used to grow sugarcane stalks.
- It improves air circulation and increases CO_2 within the crop canopy, creating the perfect microclimate for unconstrained cane growth.
- It helps sustain clean fields, which reduces the number of mice, rats, and squirrels in the field that may harm the crop.
- It decreases the threat of infestation by various insect pests, such as mealybugs, scales, and whiteflies.
- It lessens the bud sprouting caused by water buildup inside the sheath.
- Detrashed leaves are suitable for mulching the field to conserve moisture.

In sugarcane, it is important to remove dry and green leaves from the entire sugarcane stalk at regular intervals, except for a few leaves at the top of the stalk. The detrashing of sugarcane is generally carried out using a manual detrashing tool, as shown in Figure 7.5(a). A sugarcane detrashier is a basic hand tool that consists of two strip blades connected at the lower end through

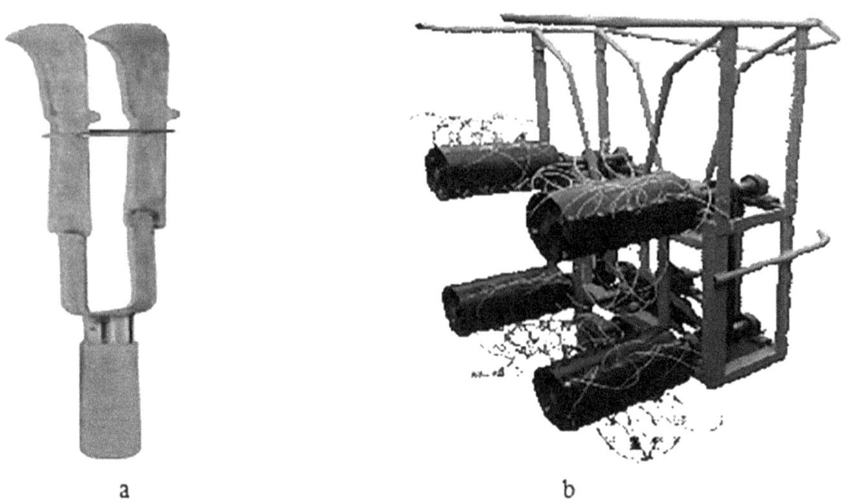

FIGURE 7.5 Sugarcane detrashier. (a) Detrashier manual tool and (b) tractor-operated.

a wooden handle. The intended person has to hold the tool from the handle, insert the cane stalk between strips of the tool in the standing sugarcane crop, and push the tool downward to tear off the unwanted leaves from the cane stalk. The sugarcane detrashier makes it easy and danger-free for individuals to remove the sugarcane leaves without risking injury or hand itching. Tractor-operated nylon-wire-based impact strippers are also available. Rotary power is supplied from the tractor PTO to all four drums, and it is attached to the three-point linkage of the tractor for forward movement in the sugarcane field, as shown in Figure 7.5(b). Small-powered tools like brush cutter-based sugarcane strippers have been discussed by Mahapatra (2024).

7.4.4.1 DETRASHING BEFORE HARVESTING

It is done manually in standing crops using manual detrashier tools to remove unwanted dry and green leaves from the lower portion of almost every sugarcane plant, as shown in Figure 7.6.

FIGURE 7.6 Sugarcane stalk at field without detrashing (a), detrashing at initial stage of sugarcane stalk development (b), and detrashed sugarcane in the final stage (c).

- This process is often done at regular intervals to protect sugarcane from insects, pests, or diseases that attack sugarcane due to the presence of excessive straw leaves around the stalk, as shown in Figure 7.6(a–c). Under proper management practices, the number of leaves on a sugarcane stalk equals the number of internodes.
- It should be accomplished only after cane stalk formation, about 150 days after sugarcane planting. Afterward, it may be completed every two months up to the final growth stage of the cane stalk.

7.4.4.2 DETRASHING AFTER HARVESTING

When the detrashing process is not carried out within the entire life cycle of sugarcane, it becomes essential to detrash all the leaves from the cane stalk after harvesting. After this operation, it is brought to the market for sale. Complete detrashing can be performed at once using a sugarcane detrasher mechanical machine. A tractor, engine, or electric motor-powered machine removes dry waste and green tops from the harvested sugarcane stalks. A sugarcane detrasher was developed by Singh and Solomon (2015), which separates the top by breaking it from the weakest point. When feeding two to three sugarcane stalks at a time, the detrasher's output is 2.4 t/h, and its trash removal efficiency ranges from 77.5 to 94.5%.

7.4.5 EARTHING OPERATION IN SUGARCANE DEVELOPMENT

In the earthing-up operation, the soil between two adjacent lines is pulverized, then the soil is moved away to form a furrow, and that soil is piled up at the root zone of sugarcane. This is done to increase the cane sprouting for improved productivity and prevent the sugarcane crop lodging. The earthing-up operation is also known as "hilling up." This operation is carried out in two or three stages. The first earthing-up operation is known as "partial earthing up," and the second or third operation is known as "full earthing up." This operation converts the furrows into ridges and ridges into furrows. This operation could be done either manually or by using mechanical machinery.

A number of tools and machines have been developed to mechanize the whole earthing process, such as spades, animal-drawn sugarcane earthing hoes, earthing machines powered by power tillers, single-bottom ridgers equipped with mini-tractors, tractor-drawn ridgers (three-bottom), and so on.

7.4.5.1 MANUAL TOOL FOR EARTHING UP

The manual earthing-up operation is generally done using a spade, which consists of a cutting blade and a long wooden handle, as shown in Figure 7.7(a). However, it is highly time-consuming and laborious work and is mostly adopted for small-scale earthing-up operations.

a b c

FIGURE 7.7 Earthing operation for sugarcane spade (a), earthing-up machine operated by power tiller (b), and earthing ridger with mini-tractor (c).

7.4.5.2 ANIMAL-DRAWN SUGARCANE EARTHING-UP HOE

It works perfectly to push soil toward the upright plant and loosen it between the rows. It can be managed by keeping the row-to-row spacing between 900 and 1200 mm. It comprises three iron bars, a scraper, and adjustable earthing-up wings to loosen soil. It saves about 78% of the running costs and works well for earthing-up sugarcane crops with row spacing between 900 and 1200 mm.

7.4.5.3 EARTHING-UP MACHINE OPERATED BY POWER TILLER

The earthing-up machine consisted of a cutting unit (tine blades mounted on a shaft powered by a power tiller through a belt or chain mechanism) and a soil compactor (Figure 7.7(b)). The cutting unit is fitted in the forward position of the soil compactor unit. The cutting unit pulverizes soil in between two rows of sugarcane plants. The rear-mounted soil compactor displaced the pulverized soil near each sugarcane stalks root and provided a well-defined shape to the furrow. A power tiller-based, low-cost earthing machine was developed and evaluated by Ferdows et al. (2023). The effective field capacity was reported as 0.16 ha/h with a field efficiency of 77.41%.

7.4.5.4 RIDGER EQUIPPED WITH MINI-TRACTOR

The single-bottom ridge is attached to the back of the mini-tractor, as shown in Figure 7.7(c). It displaces the loose soil in two opposite directions for earthing up the soil along the roots of the sugarcane plants.

7.4.5.5 FERTILIZER BAND PLACEMENT CUM EARTHING-UP MACHINE

The tractor-operated (26 kW and higher) earthing-up/fertilizer band placement equipment was designed and developed by Rana et al. (2021). When it comes to crops like maize, sugarcane, potatoes, and others with row spacing greater than 0.50 m, this device is perfect for applying fertilizer, tilling the soil, and pulling weeds all at once. The amount of urea fertilizer applied per hectare ranges from 60 to 250 kg. It facilitates applying fertilizer at 50 to 100 mm from the plant. The machine has a field capacity of 0.56 ha/h and field efficiency of 82.4%. Compared to the old method, there are significant labor, time, and fertilizer savings.

7.4.6 PROPPING OPERATION

It is an operation of tying the leaves together using the bottom dry and green leaves, known as propping. It is primarily done to avoid the lodging of cane. Propping can be done for each row, or two rows can be brought together and tied. It is done at the age of 210 days of the crop. For this operation, no specific tool or machine is available. It is done manually by using local materials for tying purposes, such as the leaves of sugarcane itself, rope, and so on.

7.5 SUGARCANE HARVESTING

Sugarcane harvesting is done using a variety of harvesting knives that vary in size, shape, and weight according to location. Manual sugarcane harvesting with traditional harvesting tools is the most expensive and labor-intensive of all the sugarcane production techniques. Harvesting sugarcane using traditional methods requires between 850 and 1000 man-hours/ha as reported by Pandey (2019). The process of harvesting sugarcane entails base cutting, stripping and retracing the stalks, detopping, forming bundles, and lastly, transporting the stalks to the sugar mills. The timely and reasonably priced harvesting of sugarcane is progressively becoming problematic, especially in tropical regions. It is now difficult to oversee a large fleet of workers to provide the mill with a timely supply of cane. However, sugarcane quality, production, juice quality, and sugar recovery are all impacted by delayed harvesting. Therefore, the sugar business is searching for alternative mechanical methods to harvest sugarcane at a fair wage rate under these conditions. The goal of developing effective harvesting equipment for sugarcane is to increase the crop's cutting quality and output per man per day with minimum energy. For now, partial mechanization appears to be a workable solution, especially in the subtropical region, as reported by Singh et al. (2010). The many methods for harvesting sugarcane can be divided into five groups:

- Whole stalk linear windrowing
- Whole stalk transverse windrowing
- Self-propelled harvester
- Whole stalk bundling machine
- Chopper-type cane harvesters

7.5.1 MANUAL CUTTING

Proper manual cutting frequently achieves the best product quality and the least amount of cane loss. By providing appropriate training, an appropriate cane knife, a balanced diet, and other resources, manual cutters' efficiency and effectiveness can be increased.

7.5.2 MECHANICAL CUTTING

There are now cane-cutting attachments available for agricultural tractors. However, the cane stalks need to be detopped, wrapped, and then loaded into the carts, which means that this harvesting method still requires a significant amount of labor. These aids are not very helpful for heavy reclining cane and only partially remove the dry leaves.

7.5.3 SELF-PROPELLED HARVESTER

Modern equipment like the sugarcane harvester is now available in the market, which can chop the cane stalk from the base and the top portion, as shown in Figure 7.8. The front divider facilitates the separation of the lodging or standing sugarcane crop. The top cutter cut the top portion of the sugarcane, and the remaining portion was fed to the existing chopper in the sugarcane combine harvester. The chopper cuts most of the trash on the sugarcane and delivers it from the outlet using an elevator. Mechanical loaders or manual loaders can be used for loading, depending on labor availability. The cane that has been harvested still contains 8–10% of the cane trash.

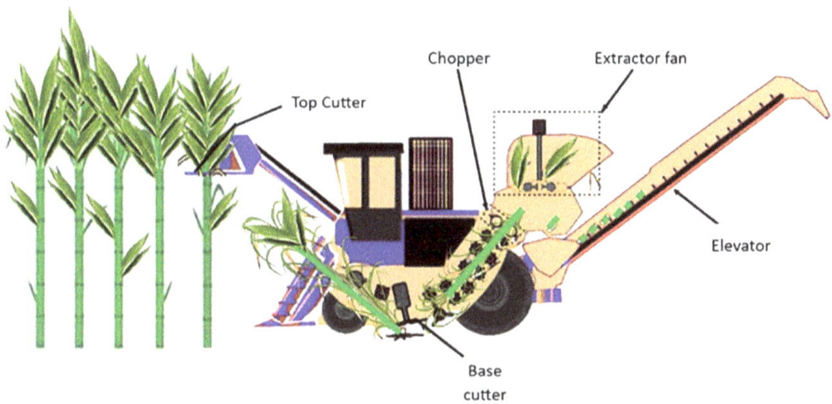

FIGURE 7.8 Self-propelled harvester.

7.5.4 CHOPPER-TYPE HARVESTER

With the growing labor shortage, the rate of mechanical harvesting may be stepped up, and chopper harvester models from developed countries may be tested and introduced after being modified to meet local requirements in developing countries. Limited trials have been conducted in certain regions with imported chopper harvester types. Without a doubt, it is possible to easily operate even a heavy-tonnage lying-down crop and produce 25–30 t/h. Some studies have been carried out to improve the performance of chopper-type harvesters. Xie et al. (2019) assessed the optimized inputs of rotational speed of chopping drums, overlapping length of upper and lower chopping blades, and bevel angle based on chopping quality and power requirement. Yadav et al. (2002) developed a sugarcane chopper harvester and reported satisfactory field capacity and net cane output performance.

The main obstacles to successfully introducing these imported machines in many countries are the small fields, the high starting price of the machine, the wider row spacing requirement of 1.5 m, and the sophisticated systems. The situation is made worse by delays in crushing after harvest. An indigenous form of sugarcane harvester must be developed, looking at labor availability, land conditions, field size, economic conditions, skill requirements, and adaptability.

7.6 MECHANIZATION OF SUGARCANE RATOON MANAGEMENT OPERATIONS

Ratoon management in sugarcane refers to the practices employed to manage the regrowth of sugarcane after the initial harvest. Unlike many crops replanted each season, sugarcane can be harvested multiple times from the same planting, producing secondary or ratoon crops. Effective ratoon management is crucial for maintaining high yields and overall crop health. Proper management of some of the affecting key parameters is essential, which includes soil nutrients, weeds, irrigation, stubble, and excess tillers (desuckering).

After the initial harvest, the soil's nutrient levels may be depleted. It is important to conduct soil tests and apply fertilizers accordingly to ensure proper nutrient levels for ratoon growth. Weeds can compete with sugarcane for nutrients, water, and sunlight, reducing yields. Properly managing the stubble left after the initial harvest is important for ratoon success.

7.6.1 MULTIPURPOSE STUBBLE SHAVER

To address the above points, a tractor-operated, two-row multipurpose stubble shaver is used. In a single pass, the machine simultaneously accomplishes stubble shaving, off-barring, interculturing, and fertilizer delivery. The stubble shaver comprises an iron frame with a three-point linkage system, a gear drive unit, universal joints, a rotating disc with blade holders and shaving blades, and an auxiliary frame with fertilizer metering. Following the crop's harvest, any waste in the field is appropriately removed or used as mulch, depending on the situation.

Trash management is crucial for getting the most out of the machine, and the field's moisture content needs to be at the ideal level for effective shaving without uprooting the clump. Planting in straight rows enables better equipment application, as uprooting can be avoided. After examining the field conditions, the equipment should be inspected for wear on the off-barring and intercultural shovels and the sharpness of the shaving blades.

When the multipurpose device was applied, the performance was assessed by Choudhary et al. (2017). It improved cane yield by 16% over the conventional practice of trash burning. Similarly, 11% more cane yield was observed when the equipment was used in comparison to trash chopping, followed by three stages of nitrogen application: basal (0.45), earthing up (0.45), and a set of monsoon rains (0.10). The initial two-split nitrogen application was modified and placed at once by the equipment, where cane yield was improved by 22% and nitrogen use efficiency by 11% over the conventional practice of trash burning.

Recently, Singh et al. (2024) developed a tractor-operated trash mulcher cum stubble shaver device that cuts and mulches the trash, followed by the spraying of liquid decomposer and stubble shaver. The stubble shaver saves two rows of stubbles in one pass without splitting sugarcane and clogging. The trash shredding uniformity was 95.8%, with a spreading uniformity of 92.6%. It can be operated at a forward speed of 2.5 km/h and cover 0.3 ha/h. A power tiller-operated stubble shaver has also been developed by the National Innovation Foundation, India (Anonymous, 2024).

7.7 RESIDUE MANAGEMENT

Residue management in sugarcane cultivation is a crucial aspect of sustainable farming practices. Sugarcane residue, also known as sugarcane trash or mulch, refers to the leaves, tops, and other plant materials left behind after harvesting the sugarcane stalks. Proper management of sugarcane residue is essential for maintaining soil health, preventing erosion, conserving moisture, and enhancing overall crop productivity. Some common practices for managing sugarcane residues are mulching, incorporation, and collection for *ex situ* utilization.

Tractor and engine-operated trash shredders cut the residues into smaller pieces for faster decomposition. In addition to the reduction in size, mulchers spread it around the planting area, which helps to conserve soil moisture, suppress weed growth, and improve soil structure. Rotary slashers are commonly used to chop sugarcane residue into smaller pieces. They can be adjusted to different cutting heights to manage residue levels effectively. Plows can be used to incorporate residues into the soil for faster decomposition, improving soil fertility. Rakes are used to gather sugarcane residue into windrows, making subsequent operations like baling or chopping easier. Balers can be used to bale sugarcane residue into compact bales for easy handling and transport. These bales can also be used for bioenergy production or as livestock feed.

7.8 TROUBLE SHOOTING IN MECHANIZED OPERATIONS FOR SUGARCANE

Some of the troubles faced in the field and needing immediate remedial measures are listed below.

- **Improper ripper deep penetration:** This problem can be resolved by adjusting the top link of the three-point linkage of the tractor.
- **Stubbles being uprooted:** The PTO rotational speed needs to be corrected through the engine throttle or through the drive system. The tilting angle of the spinning blades must also be ensured. The serrated blades must also be examined for choking and cutting-edge sharpness.
- **Manure or fertilizer being disbursed improperly:** Ensure that the metering unit is not clogged and that the material is not moist or has stones or other foreign materials.
- **Liquid chemicals are not being dropped desirably:** Ensure that the liquid solution is free from debris and that the orifice, tube, or nozzle is not choking.
- **Soil cover is not in order:** Ensure that all the earthing-up units are at the same level and are in order.

7.9 FARM MECHANIZATION'S CONSTRAINTS IN SUGARCANE

Sugarcane is a long-season crop with various processes from field preparation to its rotoon management. However, partial mechanization has been achieved for certain operations involved in the sugarcane crop. Certain constraints, as discussed below, stop progress in the field of farm mechanization in sugarcane.

- Small holding capacity of the farmers due to continued land fragmentation as a result of increasing world population.
- Poor implementation of the latest sugarcane-related technology is due to a lack of ground-level demonstration and the expensiveness of these sugarcane machines.
- Insufficient investment in research and development from private and government.
- Inadequate facilities for the repair and maintenance of the existing sugarcane machinery, especially in rural areas.
- Poor penetration of custom-hiring facilities.
- The absence of national policies for the implementation of complete mechanization.

7.10 FUTURE STRATEGIES FOR MECHANIZATION IN SUGARCANE

Mechanization could significantly improve sugarcane production and productivity. It might help meet the current demand for sugar and encourage further sugarcane-based industries to create employment. However, the partial mechanization of the entire life cycle of sugarcane has only been accomplished across the world, which leads to insufficient sugarcane production, poor quality of sugarcane, labor drudgery in the various sugarcane operations, excessive labor requirements, wastage of input resources, and inappropriate handling of sugarcane and its residue. Therefore, there is an urgent need for a future strategy to improve the level of mechanization in sugarcane. Singh et al. (2016) described the effective policy for sugarcane mechanization as given below.

- The government should promote eco-friendly, user-friendly sugarcane technology for sustainable development by providing subsidies and easy credit facilities.
- The government has to increase investment in research and development relevant to sugarcane-related machinery by allocating specific funds under the National Research Foundation.
- We must encourage packages of improved sugarcane machinery for different categories of farmers, especially small and marginal farmers, and location-specific machinery.

- The government and private industries should financially support cane growers in purchasing sugarcane machinery. Banks may relax their loan guidelines to allow farmers and young people from rural areas to improve farm mechanization in sugarcane.
- Sugar mills and agroindustry service centers should actively participate in developing the infrastructure necessary to provide growers with expensive machinery and equipment through specialized recruiting services.
- Cane farmers should regularly attend training sessions and short courses on the use, repair, and maintenance of sugarcane machines.
- In rural regions along the sugarcane growing belt, self-help organizations should be established as programs for entrepreneurial growth. These groups should focus on manufacturing, maintenance, marketing, and hiring custom machinery.
- Critical information on improved sugarcane machines must be disseminated through training, kisan melas, television programs, and radio talks.
- Manufacturers, extension organizations, R&D centers, and other stakeholders must collaborate to properly execute the mechanization strategy.

7.11 CONCLUSION

Sugarcane mechanization has the potential to significantly boost sugarcane productivity while also enhancing the standard of labor in rural areas. The goal of sugarcane productivity can be attained using integrated mechanization technology in conjunction with an efficient mechanization strategy. The sugarcane cultivation method is significantly different and requires specific equipment for each operation. Conversely, the expense of purchasing specialized machinery for every step of a given crop's production raises the cost of machinery investment. Prioritizing the development of versatile equipment is necessary to address these kinds of problems. To address labor dependency, shortages, and rising wages, new technologies are needed for automation in cane seed production, planting, and transplanting. Attachments for small powered tools need to be developed for small and marginal farmers.

KEYWORDS

- **sugarcane**
- **mechanization**
- **bud planter**
- **ratoon management**

REFERENCES

Abd-El Mawla HA, Hemida B, Mahmoud WA. (2014) Study on the mechanization of sugarcane transplanting. *Int. J. Eng. Technol. Res.* 2(8), 237–241.

Anas M, Verma KK, Haq IU, Naeem M, Li Q, Liao F, Xiang YX, Li YR. (2021a) Characterization of exotic, native and wild-type genotypes of sugarcane (*Saccharum* spp. Hybrids) for internal nitrogen use efficiency under different nitrogen levels. *Sugar Tech.* 23, 1258–1267. https://doi.org/10.1007/s12355-021-01014-1

Anas M, Verma KK, Riaz M, Qiang L, Liao F, Liu Y, Li YR. (2021b) Physio-morphological and biochemical mechanism of nitrogen use efficiency in sugarcane (*Saccharum* spp.) genotypes under different growth stages and nitrogen levels. *J. Plant Interact.* 16, 332–343.

Anonymous. (2024) from: https://nif.org.in/innovation/Power_Tiller_Operated_Stubble_Shaver_Sugarcane-Trash_-Grinder_Tractor_Operated_Stubble_Shave/325. Access 25 March, 2024.

Bhatt R, Verma KK, Jamwal NS. (2023a) Can paddy-compost eliminate N fertilization for climate-smart sugarcane? *Sugar Tech.* 25, 1036–1046. https://doi.org/10.1007/s12355-023-01270-3

Bhatt R, Verma KK, Kumar R, Sanghera GS. (2023b) Foliar application of potassium salt of active phosphorus (PSAP) mitigates insect pests and improves yield along with sugarcane quality in response to agroclimatic conditions of Punjab. *Sugar Tech.* 25(3), 660–669.

Chen GL, Zheng FJ, Lin B, Lao SB, He J, Huang Z, Zeng Y, Sun J, Verma KK. (2020) Phenolic and volatile compounds in the production of sugarcane vinegar. *ACS Omega* 5, 30587–30595.

Chen G-L, Zheng F-J, Lin B, Yang Y-X, Fang X-C, Verma KK, Yang L-F. (2023) Vinegar: A potential source of healthy and functional food with special reference to sugarcane vinegar. *Front. Nutr.* 10, 1145862. https://doi.org/10.3389/fnut.2023.1145862

Choudhary RL, Wakchaure GC, Minhas PS, Singh AK. (2017) Response of ratoon sugarcane to stubble shaving, off-barring, root pruning and band placement of basal fertilisers with a multipurpose drill machine. *Sugar Tech.* 19, 33–40. https://doi.org/10.1007/s12355-016-0438-x

Deng ZN, Huang HR, Huang CM, Verma KK, Luo HB, Jiang SL, Li YJ, Cao HQ, Wu KC, Xu L. (2023) Short-term limited water irrigation influences the photosynthetic pigments, enzymatic and non-enzymatic activities in *Saccharum spontaneum* L. at vegetative stage. *Sugar Tech.* 25, 110–118. https://doi.org/10.1007/s12355-022-01175-7

Elwakeel AE, Mazrou YS, Eissa AS, Okasha AM, Elmetwalli AH, Makhlouf AH, Elsayed S. (2023) Design and validation of a variable-rate control metering mechanism and smart monitoring system for a high-precision sugarcane transplanter. *Agriculture* 13, 2218. https://doi.org/10.3390/agriculture13122218

Ferdows MN, Rahman MA, Ali MR, Hanif MA, Tabriz SS, Hossen MS, Rokonuzzaman M. (2023) Development and performance evaluation of low-cost 2WT-operated earthing-up machine for sugarcane cultivation in Bangladesh. *Agri. Eng.* 5(3), 1327–1346. https://doi.org/10.3390/agriengineering5030084

Guo D-J, Li D-P, Chen Z-D, Yang B, Verma KK, Singh RK, Singh P, Khan Q, Sharma A, Qin Y, Zhang B-Q, Song X-P, Li Y-R. (2023) Effect of endophytic diazotroph *Enterobacter roggenkampii* ED5 on nitrogen-metabolism-related microecology in the sugarcane rhizosphere at different nitrogen levels. *Front. Microbiol.* 14, 1132016. https://doi.org/10.3389/fmicb.2023.1132016

Guo D-J, Li D-P, Singh RK, Singh P, Sharma A, Verma KK, Qin Y, Khan Q, Lu Z, Malviya MK, Song X-P, Xing Y-X and Li Y-R. (2021) Differential protein expression analysis of two sugarcane varieties in response to diazotrophic plant growth-promoting endophyte *Enterobacter roggenkampii* ED5. *Front. Plant Sci.* 12, 727741. https://doi.org/10.3389/fpls.2021.727741

Guo DJ, Li DP, Singh RK, Singh P, Verma KK, Sharma A, Qin Y, Khan Q, Song XP, Malviya MK, Xing YX, Li YR. (2022) Comparative transcriptome analysis of two sugarcane varieties in response to diazitrophic plant growth promoting endophyte *Enterobacter roggenkampii* ED5. *J. Plant Interact.* 17(1), 75–84.

Huang CM, Wu KC, Verma KK, Luo HB, Deng ZN, Shi GY, Cao HQ, Wu XJ, Ye LP, Wei YW, Xu L. (2024) Isolation and expression analysis of cellulose synthase 3 (*Ces3*) genes from sugarcane (*Saccharum officinarum* L.). *Chem. Biol. Technol. Agric.* 11, 71. https://doi.org/10.1186/s40538-024-00597-7

Huang HR, Gao YJ, Malviya MK, Verma KK, Solanki MK, Huang YX, Li X, Deng Y, Yan J, Tang SY, Wang LW, Xu L. (2022) Genetic diversity analysis of sugarcane (*Saccharum* spp. Hybrids) among high-sucrose clones of GT series and commonly used parents by using microsatellite markers in Guangxi, China. *Sugar Tech.* 24, 397–407. https://doi.org/10.1007/s12355-021-01033-y

Joshi A, Verma KK, Rajput VD, Arora J. (2023) Metabolic engineering of microorganisms in advancing biofuel production. In: *Advances in Lignocellulosic Biofuel Production Systems*. pp. 181–202. https://doi.org/10.1016/B978-0-323-91192-4.00016-X

Joshi A, Verma KK, Rajput VD, Minkina T, Arora J. (2022) Recent advances in metabolic engineering of microorganisms for advancing lignocellulose-derived biofuels. *Bioengineered* 13(4), 8135–8163. https://doi.org/10.1080/21655979.2022.2051856.

Khadatkar A, Magar AP, Sawant CP, Modi RU. (2024) Development and testing of automatic seedling extractor in robotic transplanter using mechatronics for nursery seedlings. *Discov. Appl. Sci.* 6(2), 1–13. https://doi.org/10.1007/s42452-024-05670-2

Krishna D, Sachan HK, Verma KK. (2022) Abiotic stresses in plants: An overview. In: *Understanding Abiotic Stresses*. Nova Science Publishers, Inc.: New York.

Kumari A, Dash M, Singh SK, Jagadesh M, Mathpal B, Mishra PK, Pandey SK, Verma KK. (2023a) Soil microbes: A natural solution for mitigating the impact of climate change. *Environ. Monit. Assess.* 195, 1436. https://doi.org/10.1007/s10661-023-11988-y

Kumari A, Lakshmi GA, Krishna GK, Patni B, Prakash S, Bhattacharyya M, Singh SK, Verma KK. (2022) Climate change and its impact on crops: A comprehensive investigation for sustainable agriculture. *Agronomy* 12, 3008. https://doi.org/10.3390/agronomy12123008

Kumari A, Singh SK, Mathpal B, Verma KK, Garg VK, Bhattacharyya M, Bhatt R. (2023b) The biosynthesis, mechanism of action, and physiological functions of melatonin in horticultural plants: A review. *Horticulturae* 9, 913. https://doi.org/10.3390/horticulturae9080913

Li A-M, Liao F, Wang M, Chen Z-L, Qin C-X, Huang R-Q, Verma KK, Li Y-R, Que Y-X, Pan Y-Q, et al. (2023) Transcriptomic and proteomic landscape of sugarcane response to biotic and abiotic stressors. *Int. J. Mol. Sci.* 24, 8913. https://doi.org/10.3390/ijms24108913

Li K, Li S, Teng X, Deng Z, Huang W, Gan F, Ma F. (2021) Integrated design and evaluation of a soil-covering and film-mulching device for sugarcane transverse planters. *Agronomy* 11(7), 1382. https://doi.org/10.3390/agronomy11071382

Mahapatra J. (2024) Brush cutter: A multipurpose solution for mechanised agriculture. *Biotica Res. Today* 6(1), 15–17.

Malviya MK, Li C-N, Lakshmanan P, Solanki MK, Wang Z, Solanki AC, Nong Q, Verma KK, Singh RK, Singh P, Sharma A, Guo D-J, Dessoky ES, Song X-P, Li Y-R. (2022) High-throughput sequencing-based analysis of rhizosphere and diazotrophic bacterial diversity among wild progenitor and closely related species of sugarcane (*Saccharum* spp. inter-specific hybrids). *Front. Plant Sci.* 13, 829337. https://doi.org/10.3389/fpls.2022.829337.

Malviya MK, Solanki MK, Li CN, Wang Z, Zeng Y, Verma KK, Singh RK, Singh P, Huang HR, Yang LT, Song XP, Li YR. (2021) Sugarcane-legume intercropping can enrich the soil microbiome and plant growth. *Front. Sustain. Food Syst.* 5, 606595. https://doi.org/10.3389/fsufs.2021.606595.

Naik R, Annamalai SJK, Nair NV, Prasad NR. (2013) Studies on mechanisation of planting of sugarcane bud chip settlings raised in portrays. *Sugar Tech.* 15, 27–35. https://doi.org/10.1007/s12355-012-0187-4

Nare B, Tewari VK, Chandel AK, Kumar SP, Chethan CR. (2019) A mechatronically integrated autonomous seed material generation system for sugarcane: A crop of industrial significance. *Ind. Crops Prod.* 128, 1–12. https://doi.org/10.1016/j.indcrop.2018.10.001

Oliveira MA, Verma KK, Bhatt R, Oliveira TBA. (2022) Impact of green and organic fertilizers on soil fertility and sugarcane productivity. In: Verma KK. et al. (Eds.), *Agro-industrial Perspectives on Sugarcane Production Under Environmental Stress*. https://doi.org/10.1007/978-981-19-3955-6_11.

Pandey A, Bista DR, Bhandari T, Panta HK, Devkota S. (2020) Profitability and resource-use efficiency of sugarcane production in Nawalparasi west district, Nepal. *Cogent Food Agric.* 6(1), 1857592. https://doi:10.1080/23311932.2020.1857592

Rana L, Navnit KN, Kumar A, Meena SK, Kumar M. (2021) Mechanization in sugarcane: Need of the hour. *Just Agric. Multidiscip. e-Newslett.* 1(8).

Sharma A, Singh RK, Singh P, Vaishnav A, Guo DJ, Verma KK, Li DP, Song XP, Malviya MK, Khan N, Lakshmanan P, Li YR. (2021) Insights into the bacterial and nitric oxide-induced salt tolerance in sugarcane and their growth-promoting abilities. *Microorganisms* 9, 2203. https://doi.org/10.3390/microorganisms9112203.

Sharma A, Song X-P, Singh RK, Vaishnav A, Gupta S, Singh P, Guo D-J, Verma KK, Li Y-R. (2022) Impact of carbendazim on cellular growth, defence system and plant growth promoting traits of *Priestia megaterium* ANCB-12 isolated from sugarcane rhizosphere. *Front. Microbiol.* 13, 1005942. doi:10.3389/fmicb.2022.1005942

Singh AK, Sharma MP, Gupta R. (2017) Development of tractor operated double bottom pit digger for mechanizing ring pit method of sugarcane planting. *Sugar Tech.* 19(5), 510–516. https://doi.org/10.1007/s12355-016-0488-0

Singh AK, Sharma MP, Singh J. (2010) *Sugarcane Harvesters and their Utility in Indian Conditions.* Souvenir: National Seminar on Mechanization of Sugarcane Cultivation organized at IISR: Lucknow, pp. 13–20.

Singh AK, Solomon S. (2015) Development of a sugarcane de-trasher. *Sugar Tech.* 17, 189–194. https://doi.org/10.1007/s12355-014-0306-5

Singh MK, Modi RU, Singh AK, Singh RD. (2024) Design and development of tractor-operated trash mulcher-cum-stubble shaver device for sugarcane ratoon crop. *Sugar Tech.* 26, 573–584. https://doi.org/10.1007/s12355-024-01361-9

Singh P, Singh RK, Li HB, Guo DJ, Sharma A, Verma KK, Solanki MK, Upadhyay SK, Lakshmanan P, Yang LT, Li YR. (2023) Nitrogen fixation and phytohormone stimulation of sugarcane plant through plant growth promoting diazotrophic *Pseudomonas. Biotechnol. Genet. Eng. Rev.* 40(1), 15–35. https://doi.org/10.1080/02648725.2023.2177814

Singh P, Singh RK, Song QQ, Li HB, Guo DJ, Malviya MK, Verma KK, Song XP, Lakshmanan P, Yang LT, Li YR. (2021) Comparative analysis of protein and differential responses of defense-related gene and enzyme activity reveals the long-term molecular responses of sugarcane inoculated with *Sporisorium scitamineum. J. Plant Interact.* 16(1), 12–29.

Singh RK, Singh P, Li HB, Song QQ, Guo DJ, Solanki MK, Verma KK, Malviya MK, Song XP, Lakshmanan P, Yang LT, Li YR. (2020) Diversity of nitrogen-fixing rhizobacteria associated with sugarcane: A comprehensive study of plant-microbe interactions for growth enhancement in *Saccharum* spp. *BMC Plant Biol.* 20, 220. https://doi.org/10.1186/s12870-020-02400-9.

Singh S, Singh PR, Singh AK, Gupta R. (2016) Present status and future need of mechanizing sugarcane cultivation in India. *Agric. Mech. Asia, Afri. Latin Amer.* 47(1), 75–81.

Solanki MK, Verma KK, Dastogeer KMG, Mora-Poblete F, Mundra S. (2023) Microbial resilience in plant nutrient management towards sustainable farming. *Front. Microbiol.* 14, 1280811. https://doi.org/10.3389/fmicb.2023.1280811

Song X-P, Mo F, Verma KK, Wei J, Zhang X, Yang L, Li YR. (2019) Effect of sugarcane smut (*Ustilago scitaminea* Syd.) on ultrastructure and biochemical indices of sugarcane. *Biomed J. Sci. Tech. Res.* 002950. https://doi.org/10.26717/BJSTR.2019.17.002950.

Song XP, Mo F, Yan M, Zhang X, Zhang B, Huang X, Huang D, Pan Y, Verma KK, Li YR. (2022) Effect of smut infection on the photosynthetic physiological characteristics and related defense enzymes of sugarcane. *Life* 12, 1201. https://doi.org/10.3390/life12081201

Song XP, Verma KK, Tian DD, Zhang XQ, Liang YJ, Huang X, Li CN, Li YR. (2021) Exploration of silicon functions to integrate with biotic stress tolerance and crop improvement. *Biol. Res.* 54, 19. https://doi.org/10.1186/s40659-021-00344-4

Verma KK, Li DM, Singh M, Rajput VD, Malviya MK, Minkina T, Singh RK, Singh P, Song XP, Li YR. (2020a) Interactive role of silicon and plant-rhizobacteria mitigating abiotic stresses: A new approach for sustainable agriculture and climate change. *Plants* 9, 1055. https://doi.org/10.3390/plants9091055.

Verma KK, Rajput VD, Singh M, Verma CL, Minkina T. (2022a) Influence of waterlogging on plant physiology and molecular biology. In: *Understanding Abiotic Stresses*. Nova Science Publishers, Inc.: New York.

Verma KK, Siddiqui N, Singh M. (2022b) Understanding the plant physiological and molecular adaptation strategies to soil salinity. In: *Understanding Abiotic Stresses*. Nova Science Publishers, Inc.: New York. https://doi.org/10.52305/UBUJ4024

Verma KK, Singh P, Song X-P, Malviya MK, Singh RK, Chen G-L, Solomon S, Li YR. (2020b) Mitigating climate change for sugarcane improvement: Role of silicon in alleviating abiotic stresses. *Sugar Tech*. 22(5), 741–749. https://doi.org/10.1007/s12355-020-00831-0.

Verma KK, Singh RK, Song QQ, Singh P, Zhang B-Q, Song X-P, Chen G-L, Li YR. (2019a) Silicon alleviates drought stress of sugarcane plants by improving antioxidant responses. *Biomed J. Sci. Tech. Res*. 002957. https://doi.org/10.26717/BJSTR.2019.17.002957

Verma KK, Song X, Degu HD, Guo D, Joshi A, Huang H, Xu L, Singh M, Huang D, Rajput VD, Li YR. (2023a) Recent advances in nitrogen and nano-nitrogen fertilizers for sustainable crop production: A mini-review. *Chem. Biol. Technol. Agric*. 10, 1–14. https://doi.org/10.1186/s40538-023-00488-3

Verma KK, Song X, Verma CL, Huang H, Singh M, Xu L, Wu J, Khan ZH, Zhang B, Li Y. (2023b) Mathematical modeling of climate and fluoride effects on sugarcane photosynthesis with silicon nanoparticles. *Plant Physiol. Biochem*. 204, 108089. https://doi.org/10.1016/j.plaphy.2023.108089

Verma KK, Song XP, Budeguer F, Nikpay A, Enrique R, Singh M, Zhang B, Wu, JM, Li YR. (2022c) Genetic engineering: An efficient approach to mitigating biotic and abiotic stresses in sugarcane cultivation. *Plant Signal. Behav*. 17(1), e2108253.

Verma KK, Song XP, Joshi A, Rajput VD, Singh M, Sharma A, Singh RK, Li DM, Arora J, Minkina T, Li YR. (2022d) Nanofertilizer possibilities for healthy soil, water and food in future: An overview. *Front. Plant Sci*. 13, 865048. https://doi.org/10.3389/fpls.2022.865048

Verma KK, Song XP, Joshi A, Tian DD, Rajput VD, Singh M, Arora J, Minkina TM, Li YR. (2022e) Recent trends in nano-fertilizer for advancing sustainable agriculture under the era of climate change to ensure food security for future generations. *Nanomaterials* 12, 173. https://doi.org/10.3390/nano12010173.

Verma KK, Song XP, Li DM, Singh M, Wu JM, Singh RK, Sharma A, Zhang BQ, Li YR. (2022f) Silicon and soil microorganisms improves rhizospheric soil health with bacterial community, plant growth, performance and yield. *Plant Signal. Behav*. 17(1), e2104004.

Verma KK, Song XP, Lin B, Guo DJ, Singh M, Rajput VD, Singh RK, Singh P, Sharma A, Malviya MK, Chen GL, Li YR. (2022g) Silicon induced drought tolerance in crop plants: Physiological adaptation strategies. *Silicon* 14(6), 2473–2487. https://doi.org/10.1007/s12633-021-01071-x

Verma KK, Song XP, Rajput VD, Boldyreva V, Zhang BQ, Minkina T, Li YR. (2022h) Morpho-physiological, biochemical and ultrastructural modifications on sugarcane to prolonged water deficit. Verma KK. et al. (Eds.), *Agro-industrial*

Perspectives on Sugarcane Production Under Environmental Stress. https://doi.org/10.1007/978-981-19-3955-6_11.
Verma KK, Song XP, Rajput VD, Solomon S, Li YR, Rao GP. (2022i) *Agro-industrial Perspectives on Sugarcane Production Under Environmental Stress*. Springer Nature: Singapore, https://doi.org/10.1007/978-981-19-3955-6
Verma KK, Song XP, Singh M, Tian DD, Rajput VD, Minkina T, Li YR. (2023c) Association of silicon and soil microorganisms induces stress mitigation, increasing plant productivity. In: de Mello Prado R. (Ed.), *Benefits of Silicon in the Nutrition of Plants*. pp. 299–328. https://doi.org/10.1007/978-3-031-26673-7_17
Verma KK, Song XP, Tian DD, Guo DJ, Chen ZL, Zhong CS, Nikpay A, Singh M, Rajput VD, Singh RK, Minkina T, Li YR. (2021a) Influence of silicon on biocontrol strategies to manage biotic stress for crop protection, performance and improvement. *Plants* 10, 2163. https://doi.org/10.3390/plants10102163
Verma KK, Song XP, Tian DD, Singh M, Verma CL, Rajput VD, Singh RK, Sharma A, Singh P, Malviya MK, Li YR. (2021b) Investigation of defensive role of silicon during drought stress induced by irrigation capacity in sugarcane: Physiological and biochemical characteristics. *ACS Omega* 6, 19811–19821.
Verma KK, Song XP, Verma CL, Chen ZL, Rajput VD, Wu KC, Liao F, Chen GL, Li YR. (2021c) Functional relationship between photosynthetic leaf gas exchange in response to silicon application and water stress mitigation in sugarcane. *Biol. Res*. 54, 15. https://doi.org/10.1186/s40659-021-00338-2
Verma KK, Song XP, Verma CL, Malviya MK, Guo DJ, Rajput VD, Sharma A, Wei KJ, Chen GL, Solomon S, Li YR. (2021d) Predication of photosynthetic leaf gas exchange of sugarcane (*Saccharum* spp.) leaves in response to leaf positions to foliar spray of potassium salt of active phosphorus under limited water irrigation. *ACS Omega* 6, 2396–2409.
Verma KK, Song X-P, Xu L, Huang H-R, Liang Q, Seth CS, Li Y-R. (2023d) Nano-microplastic and agro-ecosystems: A mini-review. *Front. Plant Sci*. 14, 1283852. https://doi.org/10.3389/fpls.2023.1283852
Verma KK, Song XP, Yadav G, Degu HD, Parvaiz A, Singh M, Huang HR, Mustafa G, Xu L, Li YR. (2022j) Impact of agroclimatic variables on proteo-genomics in sugarcane (*Saccharum* spp.) plant productivity. *ACS Omega* 7, 22997–23008.
Verma KK, Song XP, Zeng Y, Guo DJ, Sing M, Rajput VD, Malviya MK, Wei KJ, Sharma A, Li DP, Chen GL, Li YR. (2021e) Foliar application of silicon boosts growth, photosynthetic leaf gas exchange, antioxidative response and resistance to limited water irrigation in sugarcane (*Saccharum officinarum* L.). *Plant Physiol. Biochem*. 166, 582–592.
Verma KK, Song XP,Zeng Y, Li DM, Guo DJ, Rajput VD, Chen GL, Barakhov A, Minkina TM, Li YR. (2020c) Characteristics and correlation of leaf stomata and its relationship with photosynthesis on *Saccharum* spp. under different irrigation and silicon application. *ACS Omega* 5, 24145–24153.
Verma KK, Vatsal S, Singh M, Rajput VD, Minkina T. (2022k) Coping with water stress: Effects and adaptive mechanisms to plants. In book: In: *Understanding Abiotic Stresses*. Nova Science Publishers, Inc.: New York.
Verma KK, Wu KC, Verma CL, Li DM, Malviya MK, Singh RK, Singh P, Chen GL, Song XP, Li YR. (2020d) Developing mathematical model for diurnal dynamics on

photosynthetic responses in sugarcane responsive to different irrigation and silicon application. *PeerJ* 8, e10154. https://doi.org/10.7717/peerj.10154.

Verma KK, Liu X-H, Wu K-C, Singh RK, Song QQ, Malviya MK, Song X-P, Singh P, Verma CL, Li YR. (2020e) The impact of silicon on photosynthetic and biochemical responses of sugarcane under different soil moisture levels. *Silicon* 12, 1355–1367.

Verma KK, Wu K-C, Singh P, Malviya MK, Singh RK, Song X-P, Li YR. (2019b) The protective role of silicon in sugarcane under water stress: Photosynthesis and antioxidant enzymes. *Biomed J. Sci. Tech. Res.* 002685. https://doi.org/0.26717/BJSTR.2019.15.002685

Wu K-C, Huang C-M, Verma KK, Deng Z-N, Huang H-R, Pang T, Cao H-Q, Luo H-B, Jiang S-L, Xu L. (2022) Transcriptomic responses of *Saccharum spontaneum* roots in response to polyethylene glycol–6000 stimulated drought stress. *Front. Plant Sci.* 13, 992755. https://doi.org/10.3389/fpls.2022.992755

Xie L, Wang J, Cheng S, Zeng B, Yang Z. (2019) Performance evaluation of a chopper system for sugarcane harvester. *Sugar Tech.* 21, 825–837. https://doi.org/10.1007/s12355-019-00714-z

Xu L, Deng Z-N, Wu K-C, Malviya MK, Solanki MK, Verma KK, Pang T, Li Y-J, Liu X-Y, Kashyap BK, Dessoky ES, Wang W-Z and Huang H-R. (2022) Transcriptome analysis reveals a gene expression pattern that contributes to sugarcane bud propagation induced by Indole-3-Butyric acid. *Front. Plant Sci.* 13, 852886. https://doi.org/10.3389/fpls.2022.852886

Yadav RNS, Sharma MP, Kamthe SD, Tajudin A, Yadav S, Tejra RK. (2002) Performance evaluation of sugarcane chopper harvester. *Sugar Tech.* 4, 117–122. https://doi.org/10.1007/BF02942692

Zheng F-J, Lin B, Yang Y-X, Fang X-C, Verma KK, Chen G-L. (2023) Efficacy and functionality of sugarcane original vinegar on mice. *Front. Microbiol.* 14, 1224666. https://doi.org/10.3389/fmicb.2023.1224666

Zhou D, Fan Y, Deng G, He F., Wang M. (2020) A new design of sugarcane seed cutting systems based on machine vision. *Comput. Electron. Agric.* 175, 105611. https://doi: 10.1016/j.compag.2020.105611

CHAPTER 8

Major Biotic Stress and Their Management in Sugarcane Crop

SUJEET PRATAP SINGH*, S. P. SINGH, and S. K. VISHWAKARMA

Division of Plant Pathology, Sugarcane Research Institute, U.P. Council of Sugarcane Research, Shahjahanpur, Uttar Pradesh, India

*Corresponding author

ABSTRACT

Sugarcane has been cultivated globally. Plant pathogenic diseases are associated for more socioeconomic damage in the cane agroindustries around the globe. Management and observation of pathogens are important to minimize disease spreading and facilitate significant approaches. In India, about sixty diseases were recorded; among them, red rot, smut, wilt, and pokkah boeng diseases are the major diseases seriously affecting sugarcane production in India. After the 1850s, the sugar industry expanded rapidly, leading to new threats in red rot spread by the dreaded fungal pathogen, such as *Colletotrichum falcatum*. The red rot disease in the nation over the decades resembled to 'boom and bust cycle' monitored in cereal rust diseases. Domination of alone genotype associated with rainy season favored red rot disease. Red rot disease and resistant cultivars played a significant role in managing the disease and sustaining cane cultivation in India over a century. Efficient management of plant diseases mainly depends on the availability of disease-resistant genotypes, accurate and timely detection of the pathogens, the severity of their infection, and cost-effective control of disease. Upgradation of cultivar resistance efficiency is the more significant approaches to minimize cane diseases.

Sugarcane Production: Environmental Impacts and Socioeconomic Issues.
Krishan K. Verma, Xiu-Peng Song, Lin Xu, Rajan Bhatt, and Yang-Rui Li (Eds.)

DOI: 10.1201/9781779641137-8

8.1 INTRODUCTION

The founder of plant pathology was built and strengthened by Anton de Bary (1831–1888). The founder of plant pathology in India was Sir Edwin John Butler (1874–1943). About 240 diseases caused by various pathogens like fungi, bacteria, virus, phytoplasma, and nematodes have been reported from various parts of the world, causing low to severe losses yearly (Rott et al., 2000; Song et al., 2019, 2021, 2022; Tian et al., 2023). About 180 diseases have been reported from India, and 60 diseases have been reported (Alexander and Viswanathan, 2002). The causes of disease are classified into three parts such as: (1) Disease caused by parasitic organisms or biotic causes (Fungi, Algae, Bacteria, Phanerogams, Phytoplasma (MLO), Protozoa, Rickettsia-like organisms, Nematodes); (2) Mesobiotic causes (Viruses and viroids); and (3) Abiotic causes or nonparasitic or physiological diseases are discussed in Table 8.1.

TABLE 8.1 Factors Affecting Disease Development

Host Factor	Pathogen Factors	Environmental Factors	Biotic Factors
Host reaction	Aggressiveness	Temperature	Rhizosphere
Plant nature	Virulence	Moisture	Phyllosphere
Plant type	Parasite fitness	Relative humidity	Vectors
Plant vigor	Variability	Shade	–
Plant density	Latent infection	Light	–
Plant age	Inoculum potential	Soil fertility	–
Host exudates	–	Soil reaction	–
	–	Deficiencies or excess of nutrients	–

The disease is widely distributed throughout the country. The vital disease such as red rot was not recorded for several decades in Maharashtra and Karnataka. Still, in Maharashtra, traces of red rot are noticed (Rao et al., 2002). Transportation of pathogen-infected seed material resulted in the introduction of red rot to South Gujarat, Maharashtra, and Northern Karnataka. Similarly, leaf scald disease was introduced to Tamil Nadu state, and constricted sugarcane cultivation was practised in coastal Tamil Nadu (Viswanathan et al., 1997). Since many pathogens do not exhibit clear visual symptoms on the canes, it is impossible to diagnose the pathogen infection in the seed pieces. Highly sensitive diagnostic techniques need to be developed to precisely detect pathogens. Alternative of an present known cultivars with a significant solution is a big task under field conditions in the event of its failure due to diseases. Unlike in seed propagated crops, vegetatively propagated crop as sugarcane required a long duration to multiply a cultivar and more expensive by the agroindustries and farmers. Although the multiplication of sugarcane varieties through meristem culture has become popular, it is also cost-prohibitive for farmers. In these seedlings, systemically infected viruses are not eliminated. Seed materials are exchanged within and across the countries for research and development activities. Most of the countries stipulate certain quarantine norms to prevent the introduction of new diseases into their country. However, quarantine regulations require high-precision diagnostic techniques utilizing sophisticated equipment. In many of the developing countries, such facilities are lacking, which results in delays in getting the germplasm accessions for economic breeding.

8.2 RED ROT

Dr. FAFC Went in Java (Indonesia) first reported red rot in sugarcane in 1893 (Went, 1893). He proposed "Het rood snot (Red smut)" as a common name for the disease. Dr. C.A. Barber (Economic Botanist of Madras Presidency) early studies during 1901–1906 indicated that it was more important sugarcane disease in the Madras Presidency in India. He first observed this disease in the cultivar Red Mauritius in the Godawari Delta of Andhra Pradesh (Namalu, Kheli). The name "Red rot" was first coined by Dr EJ Butler in 1906 (Butler, 1906) (Table 8.2). The disease occurs in all the sugarcane-growing continents. The disease is widespread in 77 countries (Singh and Lal, 2000). The red rot pathogen *Colletotrichum falcatum* Went is a facultative parasite that keeps mutating in nature, and as a result, new races of pathogens are frequently emerging. The pathogen has both anamorphic and teleomorphic stages. The anamorphic stage (Asexual) of *C. falcatum* caused the standing cane. The teleomorphic stage (sexual) of *C. falcatum* was first reported by Chona and Srivastava (1952). This teleomorphic stage of the fungus *Glomerella tucumanensis* occurred on the leaf lamina, midribs, and leaf sheath of dry foliage in nature (Chona and Bajaj, 1953).

TABLE 8.2 Breakdown of the Red Rot Disease

Report by	Country	Year	Remark
Dr. FAFC Went	Java (Indonesia)	1893	*Colletotrichum falcatum* Went. Het rood snot (Red smut)
Dr. Cobb	New South Wales	1893	–
Dr. Massee	West Indies	1893	–
Dr. Spegazzini	Argentina	1896	The perfect stage was named as *Physalospora tucumenensis* (Speg)
Dr C.A Barber (Economic Botanist of Madras Presidency)	India	1901	Cultivar Red Mauritius in the Godawari Delta of Andhra Pradesh (Namalu, Kheli)
Dr A. Haward	West Indies	1904	–
Dr EJ Butler	India	1906	Coined "Red rot", critically studied this devastating disease
Dr. L. Lewton-Brain	*Hawaii*	1908	–
Dr. C.W. Enderton	USA	1910	–
Dr. Kirtikar	India	1959	Physiological specialization

Red rot disease is severely infected by *C. falcatum* pathogen in arid and semiarid growing areas of sugarcane. The low sugar yield and other agronomic aspects are the main concerns for sugar industries and cane growers. Red rot disease is one of the severe problems for the production of sugarcane. *C. falcatum* is more variable. Cane scientists monitored the major loss in yield production in India last few decades. In the beginning, red rot caused devastating damage to *Saccharum officinarum* genotypes and subsequently, many hybrid varieties were severely affected, leading to their removal from cultivation. In the history of Indian sugarcane, the country witnessed a series of red rot epidemics and the loss of many elite varieties such as Co 210, Co 213, Co 312, Co 313, Co 419, Co 421, Co 453, Co 658, Co 785, Co 997, Co 1148, Co 62175, Co 6304, Co 7717, CoC 671, CoC 92061, CoC 24, CoJ 64, CoS 8436, CoLk 8102, BO 11, BO 17, BO 54, and Co 0238.

Each of such red rot epidemic caused huge losses to the sugar industry and farmers; however, continued research efforts to develop red rot-resistant varieties successfully managed sugarcane cultivation through the epidemics. This disease infects any part of the standing crop and effect loss of yield and 8–15% losses in germination in susceptible cultivars (Agnihotri et al., 1979). Till now, there is no well-known chemical control for this disease. Using disease-resistant cultivars is the easiest and more convenient to control pathogenic diseases. The sugar industry in the subtropics faces serious problems due to the sudden outbreak of red rot. The known variety Co 0238 was released for commercial cultivation in 2009, and due to its optimum productivity and sucrose content, both growers and millers patronized this variety. The average cane yield of 80 t/ha as against 68 t/ha of the verified variety CoJ 64 has motivating influence on the farmers, whereas its high sucrose content, i.e., 18% in juice by 300 days and its steady increase to 20% during March–April (against 17.9% in the check variety CoJ 64) fascinated the sugar mills to indiscriminately expand Co 0238 in their command areas. In the state of Uttar Pradesh, this variety was grown in only 3.1% of the area in 2013–2014, due to its all-round socioeconomic effect on cane growers and agroindustries, the cultivating area rapidly expanding to more than 10 times in 2016–2017 (Ram and Hemaprabha, 2020) and further enhanced to about monopoly of a single genotype/variety now.

In the 2019–2020 season, Co 0238 cultivated 2.2 mha in Uttar Pradesh, India approximately 82% of cane area in the state totally. In subtropical states, this wonder variety was cultivation in the area of about 3 mha in Punjab, Haryana, Uttar Pradesh, Uttarakhand, and Bihar, which is 79% of the sugarcane area in these five states during the 2019–2020 season. At all India levels, the area covered by Co 0238 is around 53.2%, the highest percentage ever covered by a single variety. Besides five subtropical states, Co 0238 has also occupied in the states of Gujarat, Madhya Pradesh, and Odisha (Ram, 2020). The increase in sugar recovery and cane yield has led to tremendous sugar production by many sugar mills in north India. The variety has beneficial for the cane industries, with the sugar recovery exceeding 12%. In the 2019–2020 season, a few sugar mills in Uttar Pradesh recorded more than 13% sugar recovery, the highest ever in subtropical India (Ram, 2020).

The red rot disease infecting very fast and enhancing incidences are mentioned from the different sugar industries of Central and Eastern Uttar Pradesh and Bihar, especially in the predominant cv Co 0238 during the recent years (Figure 8.1). According to ICAR IISR, Lucknow, UPCSR, Shahjahnapur, and ICAR-SBI have revealed that breakdown of varietal resistance against red rot is due to the origin of a new virulent pathotype in cv Co 0238. The highly virulent pathotype specifically causes susceptible reactions on cultivar Co 0238 and is recently developed as a new pathotype CF13 (Viswanathan et al., 2022).

FIGURE 8.1 Red rot affected the sugarcane crop (cultivar Co 0238) and caused severe foliage drying at Hariyawan, a sugar mill area.

A large incidence of red rot disease was observed on Co 0238 cultivar up to 1–100% in Hariyawan, Ajbapur, Kumbhi, Khambarkheda, Loni, Faridpur, Nigohi, Aira, Gola, Maksoodapur, Rosa, Pilibhit, and Hargaon sugar milling areas (plot-wise data) (Table 8.3). Trace incidence of smut was monitored in Co 0238, Co 0118, CoS 13231, CoS 09232, CoS 13235, UP 05125, and CoLk 14201 up to 5% in various sugar mill areas, including research farms. The incidence of wilt was observed in Gola and Shahjahanpur, ranging from 0.5% to 20% on varieties Co 0238, Co 15023, Co 0118, CoS 13235, and CoS 08272. The occurrence of grassy shoot disease was recorded to be up to 5% in the varieties Co 0238, CoS 13231, CoS 09232, and UP 05125 in many sugar mill areas. The incidence of pokkah boeng was recorded from 0.5 to 20% on Co 0238, Co 0118, Co 98014, Co 15023, CoLk 14201, CoLk 94184, CoS 13231, CoS 8436, CoS 13235, and CoS 08272 at various sugar factory areas. The sharp cut stage of pokkah boeng disease was monitored on different varieties (Figures 8.1 and 8.2).

TABLE 8.3 Status of Red Rot in Various Sugar Mill Areas of Uttar Pradesh, India

Sl. No.	Name of Sugar Mill Area Surveyed	District	% Incidence	Varieties Affected	Crop Stage (months)
1	Rosa	Shahjahanpur	5–30	Co 0238	7
2	Nigohi	Shahjahanpur	3–80	Co 0238	8
3	Maksoodapur	Shahjahanpur	2–20	Co 0238	7
4	Ajbapur	Lakhmpur Kheri	2–60	Co 0238	8
5	Kumbhi	Lakhmpur Kheri	1–80	Co 0238	8
			0.5–20	Co 0118	8
6	Khambarkheda	Lakhmpur Kheri	2–8	Co 0238	6
7	Gola	Lakhmpur Kheri	3–50	Co 0238	7
			1–3	CoJ 85	5
8	Aira	Lakhmpur Kheri	1–30	Co 0238	8
			0.5–2	CoPk 05191	7
9	Loni	Hardoi	1–70	Co 0238	8
10	Hariyawan	Hardoi	1–100	Co 0238	9
11	Rupapur	Hardoi	0	Co 0238	8
12	Pilibhit	Pilibhit	0.5–15	Co 0238	9
			0.5–10	CoS 8436	9
13	Hargaon	Sitapur	2–45	Co 0238	8
			0.5–2	CoLk 94184	7
14	Faridpur	Bareilly	0–5	Co 0238	6

FIGURE 8.2 Infection on midrib and eye-shaped spot on leaf. Reddening symptoms on leaf sheath.

The severity of yellow leaf disease was assessed from traces to 20% on almost all the varieties planted at Shahjahanpur farm and in different districts. Leaf fleck (*Sugarcane bacilliform virus*) was observed on the

CoLk 14201, CoS 13235, Co 05011, and CoPant 97222 at Shahjahanpur and Muzaffarnagar farms. The other fungal, bacterial, and virus diseases such as red strip/top rot, mosaic, leaf scald, banded sclerotial, eye spot, and leaf binding were also observed up to various extend in numerous varieties. Banded chlorosis was also noticed at Shahjahanpur farm (Figures 8.1 and 8.2).

8.2.1 SYMPTOMS

While red rot can affect various parts of the sugarcane plants, it is usually considered to be a disease that affects the stalk and seed portions. Red rot symptoms can differ significantly based on the environment and the sugarcane varieties susceptibility. Primarily in the initial stages of the disease, symptoms might not always be apparent in the field. Standing cane may deteriorate due to red rot in the later stages of the disease. Some of the visible symptoms associated with red rot disease include yellowing and drying of the leaves from the margin to the midrib, drying of the entire top, including the crown, loss of natural color, significant shrinkage of the stalk, and the emergence of reddish lesions on the rind. The most effective way to identify diagnostic symptoms is to break the stalk lengthwise. The diseased tissues are a dull-red shades with sporadic white patches spaced throughout the stalk. The presence of lesions or reddish discolored areas on the interior tissue scattered with white horizontal patches is the most distinctive and diagnostic sign of the infection. These white spots or patches can differ in size and volume, and occasionally they might be so many that the tissue appears mottled. Longitudinal pith cavities are caused by the disease's progressive darkening and drying of the interior tissues. Sugarcane red rot disease emits a vinegar-like odor in the infected stalk. The pathogen may cause reddish patches on the leaf sheaths, small dark spots occasionally, and elongated red lesions on the midribs (Figures 8.2–8.4) (Viswanathan, 2010, 2011).

After sufficient stalk production, the first outwardly apparent symptom develops on the stalk during the gradual grand growth stage. The first indications of this are one or two leaves becoming yellow at the top, followed by more leaves beginning to yellow and wither at the margins. It turns orange-yellow and starts to yellow the third or fourth open leaves.

FIGURE 8.3 Entire top becomes light orange yellowish and dies. Infected and healthy eye bud.

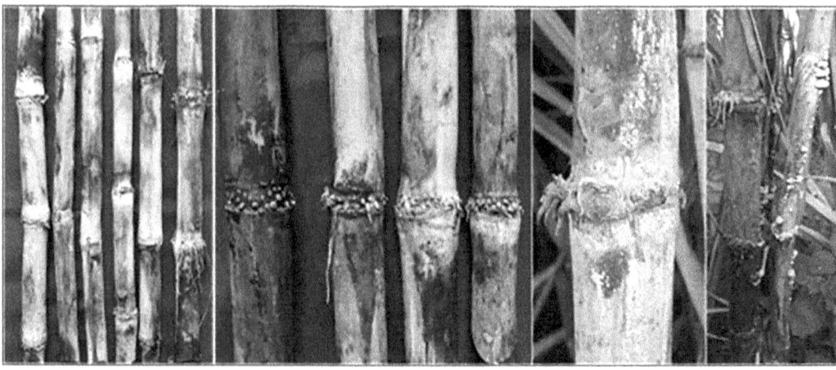

FIGURE 8.4 Discoloration of cane epidermis (upper stem surface) and sporulation at the nodal region.

Eventually, the entire top dies and turns pale orange-yellow (Figure 8.2). The rind discoloration on the stalk may sometimes range from reddish to purplish (Figures 8.3 and 8.4). The distinctive red rot symptoms appear in the internodal tissues and are only detectable by slicing the stalk longitudinally. The interior tissues become reddened and speckled with white, usually opposite to the stalk long axis (Figure 8.5). In sensitive cultivars, the entire stalk may be red, possibly highlighted by just a little of gray. In addition to the stalk symptoms, the disease appears on the backside of the leaf mid rib as red elongated lesions with ashy gray center. With the progress of the disease, small cavities are formed in the intermodal tissue (mostly the white spot area from the cavity), in which the fungus often sporulates (Figure 8.4). The other characteristic symptom of red rot is nodal rotting. The node is invariably badly damaged in red rot, and the affected cane can break easily. *C. falcatum* proliferates within the cane tissue and emerges to sporulate when the cane starts drying. Usually, it sporulates on the nodal region, especially on the root primordial (Figures 8.2–8.5).

FIGURE 8.5 Reddening of the internal tissues with white spots and cavity formation.

8.2.2 DISEASE SPREADING

Stalks that are diseased produce a lot of inoculum. The inoculum is dispersed by wind, rain, strong dews, and irrigation water. Plant material that is infected can spread rapidly and cause secondary diseases. Stubble or crop waste can also serve as an inoculum for a subsequent crop. Spores that wash into the soil from the fungus may infect planted seed pieces, despite the fact that the fungus is not a real soil-borne organism. Other than sugarcane, hosts are not thought to be significant sources of inoculum. Climatic factors affect red rot's spread and severity (Viswanathan, 2010).

Red rot may affect mature cane stem and leaf midribs and cause planting material rot, resulting in substantial reduces production and cane quality (Rao, 2004). The fungus is sett-borne. Red rot is perpetuated through infected setts/canes, diseased stubble/debris, and resting propagules in the soil. Red rot in sugarcane recurs every year mainly because of infected seed cane and stubble, which spread the disease to new crops (Butler and Khan, 1913; Agnihotri et al., 1979). The role of soil-borne inoculum in disease recurrence is perhaps negligible as the fungus is not a true soil-borne organism and does not survive in the soil for more than 5–6 (Chona and Nariani, 1954; Singh et al., 1977). The pathogen attacks the cane plant from the germination and germination failure or causes the death of germinates (Agnihotri, 1996).

8.2.3 PROPAGATION

Expression of red rot depends on the habit of infection, time/duration, and the prevailing environment. Symptoms are observed during the monsoon or post-monsoon period (Agnihotri and Singh, 1977; Singh and Singh, 1989). The disease begins to appear as pre-monsoon showers come in, and when the monsoon comes at an appropriate time due to optimal weather, the condition fully appears (Duttamajumder, 2008). The favorable conditions for the pathogenesis of red rot are following as under.

- The average temperature ranges from about 29–31°C is optimal for developing the disease.
- Optimum ambient air humidity (90%).
- Limited water supply during crop development stage.
- Soil flooded condition.
- Absence of management applications.
- Continuous cultivation of the same genotypes.
- Recommended susceptible cultivars are available in the environment.

8.2.4 SECONDARY INFECTION

The fungus infects the mid-rib, lamina, leaf sheath, and stalk during the monsoon season by secondary transmission that is facilitated by irrigation, precipitation, and splashing rain; in the winter, air currents contribute to the pathogen's spread (Agnihotri, 1990; Singh and Lal, 1996). Conidia that develop on the rind become infected through nodes as they wash down with water. Because the spore mass is mucilaginous, wind-mediated inoculum dissemination seems to be more challenging. On the other hand, the fact that the disease is appearing in the upper portion of the canes suggests that the inoculum is dispersing aerially. The fungus is still able to infect the crop despite the wintertime environmental conditions. They may not present a significant risk of infecting the cane crop, but they can cause early infections in the nodal region (buds, bud scales, root origins, etc.) (Satyavir, 2003). When cane is utilized as seed, these illnesses are the main cause (Agnihotri, 1996). Additionally, borer assist in the pathogen's secondary transmission (Edgerton, 1959). According to Steib and Chilton (1951), Srinivasan and Alexander (1966), and Singh et al. (1983), the pathogen mostly infects canes through nodes, with leaf scars, growth rings, root primordia, and buds serving as the main portals of entry. In addition, the virus may enter the stalk through cut sett ends, growing cracks, and rootlets (Singh and Singh, 1989). In genotypes that are somewhat resistant, the fungus may cause gum production after invading the stalk tissues and spreading its mycelium from cell to cell. According to Sharma and Tamta (2015), red rot spreads rapidly across the vascular bundles (Figure 8.6).

FIGURE 8.6 Red rot affected the crop.

8.2.5 DISTRIBUTION OF RED ROT

The disease is widely distributed throughout the country. Economic losses are caused by reduced stalk mass and the sucrose level of affected mature stalks and may be more severe. Severe losses have been reported in many countries. The enormous damage potential of red rot disease is reflected from even the first documented epidemic in 1895–1901 in the Godavari delta of India, pushing sugarcane cultivation to the brink of extinction. The red rot disease was responsible for the elimination of different commercial cultivars, that is, Co 312, Co 658, Co 997, Co 1148, Co 6304, CoC 671, CoC 8001, CoC 85061, CoC 86062, CoC 90063, CoC 92061, CoJ 64, etc in the past in India. About 100 % yield loss has been reported in the recent red rot epiphytotics in Tamil Nadu (Viswanathan and Samiyappan, 1999, 2000). The epidemic of 1938–1940 in India was disastrous when the more broadly cultivated varieties in Uttar Pradesh and Bihar states. Red rot is still regarded as the severe disease of sugarcane in India (Singh et al., 2016). Red rot was not recorded for several decades in Maharashtra and Karnataka. The alarming situation of red rot in Karnataka from 1998 to 2000 was noticed through the introduction of Co 92061. Still, in Maharashtra, traces of red rot are noticed in CoC 671 and Co 8011 (Rao et al., 2002). Transportation of pathogen-infected seed material resulted in the introduction of red rot to South Gujarat, Maharashtra and Northern Karnataka in the last decade. Detail of red rot epidemics occurred in various states of India and different districts of UP (Table 8.4, Figures 8.7 and 8.8).

TABLE 8.4 Red Rot Epidemics of Sugarcane in Various States of India

Year	Area/States	Variety Affected
1895–1901	Godawari Delta (Andhra Pradesh)	Namalu, Kheli, Red Mauritius, Striped Mauritius
1902–1913	Champaran, Muzaffarpur (Bihar)	Bourbon, Striped Mauritius
1922	Jammu (Kashmir)	Bourbon, Striped Mauritius
1925	Panjab	Bourbon, Striped Mauritius
1932	Pusa (Bihar)	Co 210
1936	Inthiathoke (Gonda, UP)	Co 210
1937	Gola, Nagina (UP)	Co 210
1938–1940	Uttar Pradesh and Bihar	Co 213
1946–1947	Uttar Pradesh and Bihar	Co 312, Co 313, Co 421, Co 453, BO 11, BO 17, BO 54, Mamchuah
1947–1950	Uttar Pradesh and Northern Bihar	Co 313, Co 421, Co 453, BO 11
1950–1951	Punjab	Co 312, Co 313
1960–1970	Andhra Pradesh	Co 421, Co 419, Co 997, Co 62175
1971–1975	Tamil Nadu	Co 419, Co 658, Co 6304
1978–1983	Andhra Pradesh	CoC 671, CoA 7701
1981–1985	Kerala	Co 997, Co 419, Co 785
1975–1985	Uttar Pradesh, Haryana, Punjab, Bihar	Co 1148, CoJ 64, BO 54, Co 7717
1992–1997	Gujarat	CoC 671
1991–2005	Uttar Pradesh	CoS 767, CoJ 64, CoS 8439
1992–2000	Andhra Pradesh, Karnataka, Orissa, Pondicherry, Tamil Nadu	Co 6304, CoC 671, CoC 90063, CoC 91061, CoC 92061, CoSi 86071, CoSi 96071
2006–2008	Orissa	Co 7805, Co 87044, CoA 89082
2008–2009	Uttar Pradesh, Haryana	CoS 8436

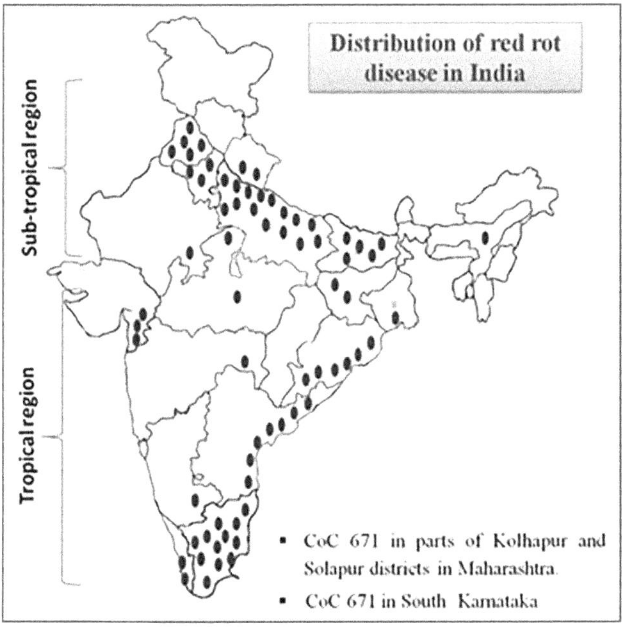

FIGURE 8.7 Distribution of red rot in various states of India.

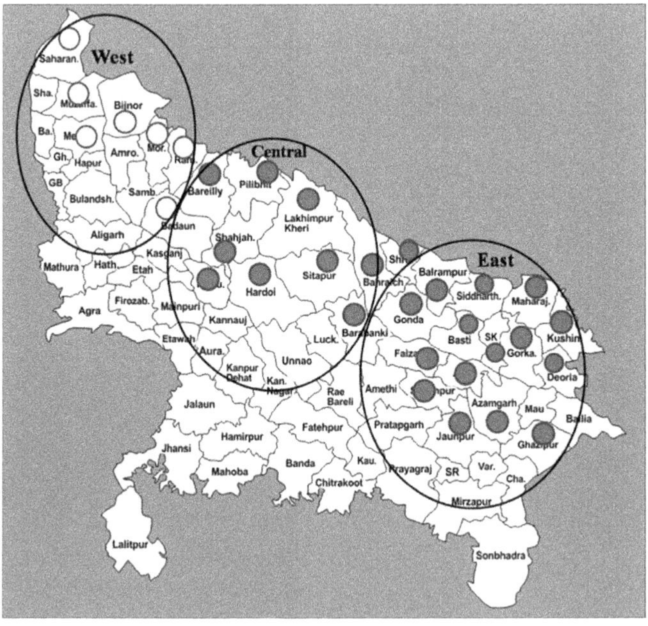

FIGURE 8.8 Distribution of red rot in various districts of Uttar Pradesh of India.

8.3 SMUT

The fungal disease known as *Sporisorium scitamineum* is the source of sugarcane smut, a global health concern that results in significant reductions in sugarcane yield and sucrose content (Hoy et al., 1986; Padmanaban et al., 1998; Que et al., 2012). The first disease incidence reported in Natal, South Africa, in 1877 (Luthra et al., 1940). Before it was discovered in Argentina, it was thought to be limited to the eastern hemisphere. The disease rapidly spread around the world, affecting most of the regions that produced sugarcane, including Indonesia, Mauritius, Rhodesia, Java, Sulawesi, and Sumbawa. Smut only affected Asia up until the 1950s, with a small population in Argentina. It has since expanded across South, Central, East, and West Africa, many of which need an exclusive breeding program for smut resistance. It expanded to Hawaii, the Caribbean, the United States mainland, Central America, and Southern Brazil in the 1970s and 1980s (Table 8.5). These outbreaks prompted much demonstrations on smut of sugarcane plants.

TABLE 8.5 Breakdown of the Smut Disease

Year	Country	Remarks
1877	Natal Around (South Africa)	Luthra et al., Dr Sydow
1906	India	The first report by Sir EJ Butler
1930	India, Sri Lanka, Bangladesh, Taiwan, Philippines, Indonesia, Japan	Serious constraints in the eastern hemisphere.
1940	Argentina	Western hemisphere.
1957	Brazil, Paraguay, Bolivia	First report
1971	Hawaii (USA)	First report
1974	Guyana	First report
1978	Louisiana, Florida (USA)	First report
1981	Caribbean, S. America, Central America, Puerto Rico, Texas	First report
1998 (July)	Western Australia (Ord River Irrigation Area-ORIA)	Report by Dr. I.T. Riley

After that, it was determined that sugarcane smut occurred in Iran (Banihashemi, 1995) and Morocco (Akalach, 1994). Antoine initially brought attention to the existence, frequency, and significance of the sugarcane smut pathogen in 1961. The disease was common, affecting several nations in East Africa, the Pacific, and the Caribbean islands. In the latter region, a serious epidemic of the disease caused enormous losses to sugarcane plantations. A comprehensive examination of the symptoms, yield reduction, causative organism, physiological races of the smut fungus, epidemiology, host resistance, and treatment of sugarcane smut was conducted by Lovick (1978). According to reports, the sugar industry and cane growers in the Caribbean have been affected by significant smut epidemics. To manage smut infection, a number of strategies have been suggested, including chemical control and efficient agronomic practices. However, according to Xu et al. (2004), only resistant cultivars have shown to be useful and economic. Crop protection through genetically engineering disease resistance is emerging as a promising technique to overcome the drawbacks of conventional strategies, mainly to the advancements in biotechnological approaches (Singh et al., 2007).

8.3.1 SPREADING OF SMUT

Smut is considered a major sugarcane disease in India, next to red rot. The smut disease is widely prevalent in the country. Severe crop losses were recorded during the last three decades due to the disease in many parts of North Karnataka and Maharashtra. This disease caused significant losses in yield and quality in popular varieties like Co 740, Co 1158, Co 7527, CoC 671, Co 97009, and CoSi 95071 in different regions (Viswanathan, 2000a). This disease was noticed to varying level in CoS 94270, CoS 95255, CoS 94257, CoS 96250 and CoSe 95436, CoS 92269, UP 9862, CoSe 94636, CoSe 92423, CoS 91269, BO 91, CoS 91269, CoS 95255, CoS 96268, CoS 94257, CoS 96436, CoS 96275, CoS 96269, CoS 97264, CoSe 92423, CoSe 98231, CoSe 01434, CoSe 01434, CoSe 03234, UP 0097, CoPant 84212, CoS 767, CoSe 95255, CoSe 98231, CoSe 01424, UP 9530, and BO 91 in UP. Similarly, it was noticed in Haryana on CoH 110 and BO 128, Co 1158, Co 87263, and Co 8270 in Bihar. In tropical regions like Tamil Nadu, CoSi 95071 affected very severe smut in plants and ratoon about 100% (Table 8.6 and Figure 8.9).

TABLE 8.6 Epidemic of Smut

Year	Area	Genotype/Variety Affected
1942–1943	Bihar (66% cane area)	Co 453, Co740, Co 975, BO 11
1947–1948	Bellary (Northern Karnataka), Maharashtra regions	Co 419, Co 740 (severe yield loss to sugarcane for a long time)
1948	Gujarat	Severe smut is noticed on Co 86002 and CoSi 95071
1950–1952	Bihar	Co 213, Co 243, Co 419, Co 513, BO 11, CoS 510

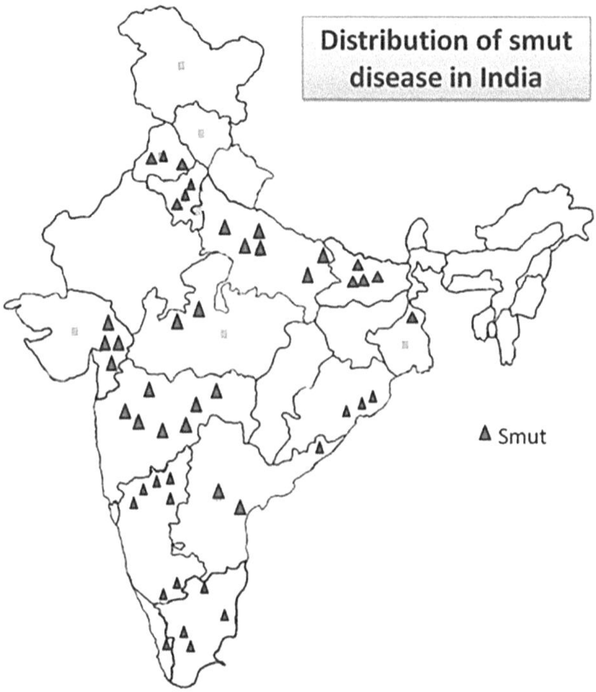

FIGURE 8.9 Distribution of smut in various states of India.

8.3.2 SYMPTOMS

Smut usually refers to the sugarcane disease that is easiest to diagnose due to its unique symptoms. The distinctive sign is a brown, whip-shaped fungal sorus that can grow to be more than 1 m long and forms at the tip of an infected stem. When a lengthy, extended whip appears on sugarcane infected with smut, it is the most apparent diagnostic sign. The form of the whip varies, ranging from short to long, twisted, many whips, and black under a thin layer of silvery membrane. When this silvery membrane ruptures, millions of smut fungal reproductive spores—which are present in a powdery mass—are released. On maturity, it ruptures, and millions of tiny black smut spores (teliospores) are liberated and disseminated by the wind. Infected sugarcane plants may tiller widely, producing thin, narrow leaves on spindly, more erect shoots that provide the cane a "grass-like" appearance and poor cane production (Figure 8.10). Bud growth and galls on the leaves and stems are further signs. The disease's growth and severity are influenced by the environmental factors and resistance of the different sugarcane types, and it can result in considerable losses in cane tonnage and juice quality (Chona, 1943).

FIGURE 8.10 Various whip symptoms of smut.

8.4 WILT

Wilt is also an important fungal disease of sugarcane. The term "wilt" is a misnomer in sugarcane as affected plants dry up gradually without displaying conventional wilt syndrome. The causal organism of this disease has been shrouded in controversy for long. Still, now it is believed that it is a "disease complex" produced by different pathogens rather than the act of a single pathogen species. *Fusarium moniliforme* var. subglutinans, *Fusarium sacchari* (= *Cephalosporium sacchari*), and *Acremonium* spp are the fungi implicated in wilt disease. (*Acremonium implicatum* in particular). One of the most serious diseases in India is sugarcane wilt disease. A significant stalk disease that impact sugarcane yield in India and other South Asian nations is sugarcane wilt.

Butler (1906) reported first report from Bihar, extensive findings were described by Butler and Khan (1913), and they identified the pathogen as *C. sacchari*. A similar disease named "*Fusarium* sett" or "Stem rot" was observed in Barbados by Bourne (1922). Wilt of sugarcane caused by *F. sacchari* has been recorded a century ago in India. Like red rot caused by *C. falcatum*, this disease also affects the stalks in sugarcane. Stalk infections by these fungal pathogens cause significant losses to the growers and make the canes unfit for juice extraction and milling. Unlike red rot, wilt infection shows desiccation of stalk tissues and leads to pith formation and drying of canes.

Bihar suffered the disease for a very long period. From that time, it expanded to other states including Punjab, Haryana, and Uttar Pradesh as a result of the ongoing movement of contaminated seed. Many commercial cultivars, including Co 245, Co 321, Co 527, Co 951, Co 1107, and Co 1223, were eliminated from cultivation as a result of wilt outbreaks that occurred in India over the last century (Kirtikar et al., 1972). The cultivars of Co 419, Co 449, Co 453, Co 527, Co 951, Co 89003, and Co 1122 that were cultivated in some parts of Tamil Nadu between 1955 and 1956 suffered significant damage due to an infection in tropical India. Wilt caused the cultivation of numerous types, including Co 449, Co 658, Co 997, and Co 1007, to be discontinued in subtropical areas (Singh and Singh 1974, Viswanathan, 2013). In many instances, complete loss to cane production occurred, such as in "CoC 671" in Gujarat, Co89003 in Punjab, and other varieties in Bihar states (Agnihotri and Rao 2002).

In India, 60% wilt on Co7717, 5–10% on CoJ 64, CoJ 79, and CoS 767 in the state of Uttar Pradesh, serious wilt affect in the combination with red rot disease in Bihar state, and on Co 89003 and moderate wilt on Co 7717, CoS 8436, and CoS 88230 in the state of Punjab, different levels of wilt in a lot of care cultivars, such as Co 85002, Co 85004, Co 85019, Co 86002, Co 86010, Co 86032, Co 97009, CoC 671, CoC 91061, CoJn 86141, CoSi 95071, and CoSi 98071 in the state of South Gujarat, mild wilt on Co740, Co 7257, Co 8011, CoC 671, and Co 86032 in the state of Maharashtra, and 5–19% wilt disease observed on Co 1148, CoLk 8001, CoC 671, Co 1305, and Co 1307 in the state of Madhya Pradesh, India. Different commercial cultivars were drawn from cultivation in North and southern India due to wilt disease according to Agnihotri and Rao (2002) and Viswanathan and Rao (2011) as shown in Figure 8.11.

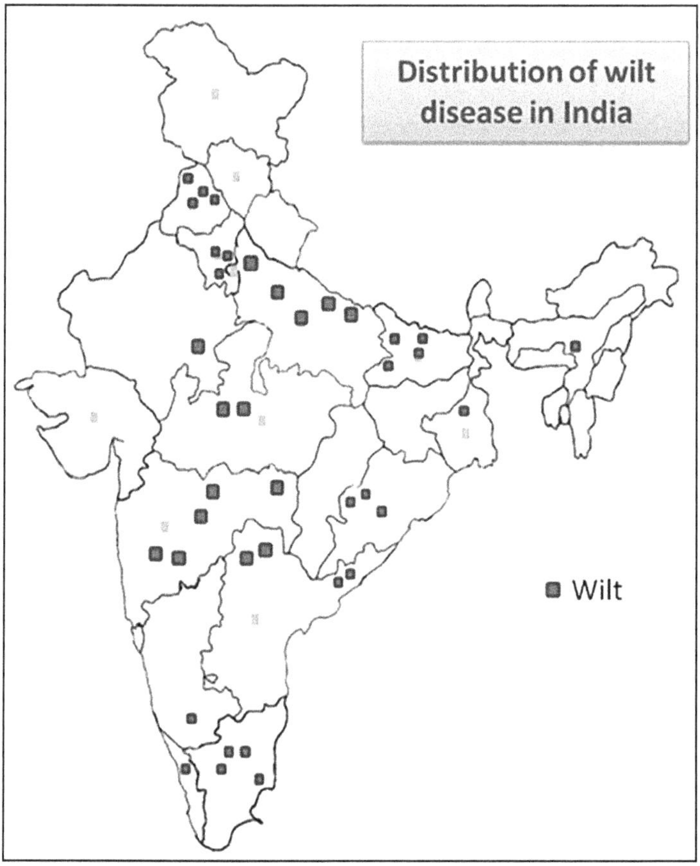

FIGURE 8.11 Distribution of wilt in various states of India.

The disease adversely affects germination when wilt-infected setts are used for planting. Besides yield reduction, wilt also causes a loss of 15–30% in the extraction of juice and about 20% in sugar recovery (Agnihotri and Rao, 2002). Also, partial infection drastically decreases extraction and quality of juice, as in red rot infections. Although the disease occurs in all parts of the country, its epidemic in subarid areas, South Gujarat, and east coastal regions (Viswanathan et al., 2006). In addition, combined infection of red rot and wilt causes more loss to the crop than their infection alone (Viswanathan, 2010). Similarly, infestations of sugarcane by borer pests aggravate wilt and cause more damage to the crop (Viswanathan, 2012; Viswanathan and Rao, 2011)

8.4.1 SYMPTOMS

The disease manifests itself mostly in the mature cane. The apparent wilt syndrome in sugarcane appears after the end of the grand growth phase, that is, from October onward (after the recession of monsoon rains). Affected plants appear pale yellowish-green with a marked drying of lower leaves. Symptoms first appear in sugarcane at 4–5 old. Eventually, plants become completely dry and die with completely hollow internodes, leading to the loss of millable canes. The infected cane shows shrinking and stalk breaking at any point, depending on this verity. In wilt, the affected canes simply dry out, and canes do not break easily, whereas in red rot, the nodal rotting cane is broken with little pressure. Disease found to be severe in field-shaving stress situations, which can be environmental stresses.

Distinctive internal signs may be monitored after splitting the infected stalks longitudinally. Whitish to dark reddish or brownish-red coloration can be seen in internal tissues, depending on the stage of the disease. The intense discoloration of tissues at the nodes might be due to the profuse effect of fungal hyphae in the tissues immediately above the ring of growth. The vascular tissues show as dark reddish-brown streaks that pass through nodal tissues from one to another internodes. The wilt-infected canes do not emit any smelling vinegar-like odor in the infected stalk, and no white spots in intermodal tissues typical of red rot infection caused by *C. falcatum*. However, shrinkage of intermodal tissues occurs due to the depletion of moisture, resulting in the formation of longitudinal, spindle-shaped cavities in the midst of the internodes. The disease's characteristic diagnostic symptoms, tissue discoloration and the formation of boat-shaped cavities, are comparatively more common in the afflicted canes. Plants that suffer from the disease first show no symptoms at all, but later on the affected roots die and the vascular bundles of the shoot turn brownish as the stalks die. Additionally, wilted canes may exhibit root borer tunnels, primarily at lower internodes, that are habitat to borer larvae. We are unable to differentiate the symptoms from red rot because the infected stalk had signs of both conditions (Viswanathan et al., 2012) (Figure 8.12).

8.4.2 IMPACT

Wilt losses are usually determined using the volume of dead or dried canes that are discovered in the field following harvest; this amount can range from 2 to 10 tonnes/ha (Parthasarathy, 1972). Yield losses might reach

FIGURE 8.12 Symptom of sugarcane wilt.

65%, and ratoons were more likely than plant crops to experience the disease's occurrence (Sarma, 1976). Diseased plants produce fewer tillers than healthy ones (Agnihotri and Singh, 1989). Juice quality deteriorates because of a reduction in sucrose concentration (Khanna and Chacravarti, 1949) and an increase in reducing sugars, gums, titrable acidity, flavonoids, and soluble salts (Singh and Waraitch, 1981). These factors also negatively impact the mills' ability to manage white sugar. In addition to decreasing cane production, wilt reduces juice extraction by 14.6–25.8% and sugar recovery by 3–20% (Gupta and Gupta, 1976; Chen et al., 2020; Zheng et al., 2023). When the disease incidence in cultivars Co 658 and Co 449 was just 6%, Subba Raja and Natarajan (1972) reported a 9.97% drop in recovery.

8.5 POKKAH BOENG

The Javanese word for a deformity or twisted top is pokkah boeng. It was originally first recorded in Java by Went in 1896 and later reported by Edgerton (1955) and Martin et al. (1961). It became a major sugarcane disease in Java with the spread of commercial cultivars, that is, POJ 2878, which was extensively studied by Bolle (1937). Pokkah boeng of sugarcane is caused by *Fusarium moniliformae (F. verticillioides)* according to Sheldon (1904) and the suitable phage of the pathogen is *Gibberella fujikuroi*. Several workers now confirm *Fusarium* as a pathogenic carrier of pokkah boeng in cane crop in Asia as a well-established pathogen by many workers (Govender et al., 2010; Mohammadi et al., 2012).

8.5.1 DISTRIBUTION

Pokkah boeng of sugarcane is an economically important disease worldwide where sugarcane is grown commercially (Martin et al., 1961). Pokkah boeng is an air-borne fungal disease in all cane-growing states of India. This disease was first reported during 1983-84 on varieties Co 7219 and CoC 671 in Maharashtra (Patil and Hapase, 1987). This disease is becoming a serious concern in the different stated, such as Uttar Pradesh, Bihar, Haryana, Punjab, and other states of India (Vishwakarma et al., 2013). Pokkah boeng disease is spreading and affecting nearly all the sugarcane genotypes/cultivars suggested for normal cultivation of various environmental regions. It was spread and affected so many commercial cultivars such as Co 0238, CoS 8436, CoS 88230, CoS 94257, CoS 767, CoS 94270, CoSe 95422, CoSe 98231, and CoJ 64 in various districts of Uttar Pradesh. Verma et al. (1984) first reported the knife-cut stage of this disease in Uttar Pradesh.

8.5.2 SYMPTOMS

The most obvious effect of pokkah boeng is to distort the cane tops. Pokkah boeng is present in most sugarcane-producing areas of the world and symptoms of the disease tend to develop during periods of rapid crop growth when rainfall is high (Martin et al., 1961). Favorable conditions for pokkah boeng becomes 20–30°C temperature and the mean ambient air humidity over 70–80% with a cloudy condition, drizzling rains favors the growth of pathogen. The result is damaged top and stalk. Malformation or distortion of the young leaves is usually accompanied by a distinct wrinkling, twisting and narrow and short relative to normal cane leaves. Irregular reddish stripes may develop within the chlorotic areas. The general symptoms of pokkah boeng were mainly caused by two types, that is, the chlorotic phase and the acute phase of the top rot phase (Patil and Hapase, 1987). The most severe and modern phases of pokkah boeng is top rot, where the growing point is killed and the entirety top of the plant dies (Martin et al., 1961). Infection may progress into the stalks, where internal and external ladder-like (knife-cut) lesions can grow up. The plants usually recover with little damage if the infection is limited to the leaves. This disease adversely affects growth because of improper photosynthesis without crown leaves.

8.5.3 CHLOROTIC STAGE

A chlorotic condition near the base of the young leaves, and occasionally on other areas of the leaf blade, is the first indication of pokkah boeng. A noticeable wrinkling, twisting, and shortening of the leaves as well as a distortion of the stem are often seen in conjunction with the malformation or distortion of the young leaves. The severity of symptoms varies with the susceptibility of the variety and with existing environmental conditions governing the development of the organism. The base of affected leaves is often narrower than that of normal leaves. As the affected leaves mature, irregular reddish stripes and specks develop within the chlorotic parts, and these may also be present to some extent in green portions of the lower sections of the leaf blade. The reddish areas sometimes develop into lens or rhomboid-shaped holes which have no definite arrangement or form ladder-like lesions, often with dark edges in longitudinal rows. The tissue along the edges and tips of the leaves may also be affected and develop irregular areas that turn dark reddish-brown to black, often producing a burned appearance. In some instances, when the young leaves do not unroll normally, a crinkling of the leaves is observed. Plants generally recover when only the leaves are infected. Leaf-sheaths affected by the disease become chlorotic and develop irregular necrotic areas of a reddish color, similar to those in the leaf blades. Ladder-like lesions (Figures 8.13 and 8.14) may develop on both sheaths and midribs (Patil, 1995).

8.5.4 ACUTE PHASE

The more advance and severe phage of pookah boeng is top rot; the young spindle is killed, and the entire top of the plant dies. Leaf infection sometimes moves downward and penetrates the stalk through the growing point, and a dark reddish streak may extend through several internodes.

FIGURE 8.13 Chlorotic phase of pokkah boeng.

FIGURE 8.14 Chlorotic stage of pokkah boeng.

In the node, these streaks appear as fine lines, while in the internodes, they sometimes form a ladder-like appearance. In the advanced infection stage, the whole base of the spindle and even the growing point showed malformation of leaves, pronounced wrinkling, twisted, and rotting of spindle leaves. Red specks and stripes also appeared, and the entire phage of the spindle rotted and dried up very fast. Finally, it formed a top rot of tender tissues of the apical cane part (Figure 8.15). Sprouting of the buds was also observed in severe infection. Occasionally, symptoms of knife cuts were observed in the acute phase of the disease, characterized by one or more narrow transverse cuts in the rind of the stalk in such a uniform manner as if the tissue was removed with a sharp knife. This is an exaggerated phage of the typical ladder lesion of pokkah boeng disease (Patil, 1995).

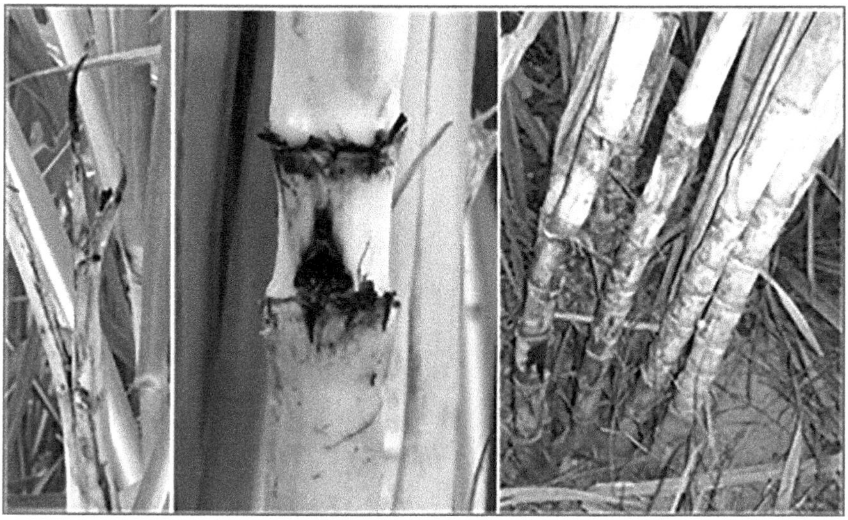

FIGURE 8.15 Acute and knife cut stage of pokkah boeng.

8.5.5 KNIFE-CUT PHASE

Knife-cut disease is an advanced stage of pokkah boeng disease of sugarcane. Knife cut disease of sugarcane was observed in association with cultivars Co 7219, CoC 671, Co 8014, Co 1158, CoS 767, CoS 8315, and CoS 8436 (Patil, 1995). Previously, the acute phase was considered the most advanced and serious stage of pokkah boeng disease. However, critical observations on symptomology revealed that the acute phase (top rot) of pokkah boeng is always associated with knife-cut infection. Such cane exhibited staggered appearance. Large horizontal conspicuous angular cuts were noticed on stripping off the leaves, which were smaller than those of knife-cut disease (Figure 8.15). In general, the infection occurs uniformly on the cane tops, which is responsible for the spread of disease. Based on symptomology, causal organism, and their pathogenic behavior, it is confirmed that the knife cut disease is not a separate infection; however, it may be regarded as an advanced stage that is always associated with pokkah boeng disease.

8.6 LEAF SCALD OF SUGARCANE

Leaf scalds are caused by bacteria called *Xanthomonas albilineans*. It was discovered as a bacterial vascular disease of sugarcane in the 1920s in Java, Australia, and Fiji. It was rapidly discovered in Philippines (1923), Mauritius (1928), and Hawaii (1930). According to the majority of early accounts, the disease was likely existing long before it was identified, which is likely because infected sugarcane varieties were widely used for breeding and ultimately led to its spread to other nations. The noble cane (*S. officinarum*) kinds that were commonly grown in the early 1900s suffered severe losses. By substituting resistant cultivars—hybrids of Saccharum spontaneum—for the susceptible noble canes, the disease was managed (Rott and Davis, 2000).

Sugarcane leaf scald is a disease that can be fatal. As a result, it may significantly impact sugarcane produce and severely restrict the cultivation of sensitive varieties. The disease might have a latent (asymptomatic) phase that lasts for or even years, which makes it deceptive. Leaf scald is further complex because it can manifest in a chronic stage (progressive rise of disease degree) or an acute phase (unexpected appearance of the infection and plant damage).

8.6.1 SYMPTOMS

The leaf scald pathogen does not usually cause symptoms on the outside of affected plants. These plants have been reported to be latently infected; the exact process underlying latent infection remains unknown. There are also cases of apparent recovery in which signs subside and do not become apparent until ratoon plant regrowth after infected cane seeds. Still, throughout this apparent recovery, the disease is in persistent infection in the damaged cane. The chronic phase has been defined by a number of outward signs. The most common symptom is a streak of white pencil line that is 1–2 mm wide and runs parallel to the veins on the leaf, from the midrib to the leaf border (Figure 8.1). A variably wide, diffuse yellow border runs parallel to the straight-line streak. There could be spots of reddish discoloration along a portion of the pencil line length. Necrosis begins at the leaf margin or tip and spreads throughout the leaf as the disease intensifies (Figures 8.16 and 8.17). The disease receives its name from the way leaves appear scorched and curl inward, giving the foliage a scalded appearance. The plant ultimately perishes as its entirety.

Additionally, leaf scald can result in partial or total chlorosis, or bleaching, of the leaf blade (Figure 8.17). Upon further inspection, the characteristic of its reddish necrotic parts may become visible (Davis et al., 1997).

FIGURE 8.16 Symptoms of leaf scald in sugarcane plants.

FIGURE 8.17 Symptom of leaf scald in sugarcane plant.

Additionally, the disease may result in stunted and wilted shoots. Affected leaves typically get a dull blue-green shade prior to extensive browning, especially is a late disease sign. The entire stool could perish in stressful circumstances. The spindle leaves on mature stalks generate moderate to profuse side shoots and turn necrotic from the tips. Side shoots begin to emerge at the base of the stalk and develop higher. Usually, these side shoots have white pencil lines or streaks (Figures 8.16 and 8.17). Necrosis of the vascular bundles inside the affected stalks might result in bright to dark red streaks. These stripes are typically seen at the intersection of the stalk and side shoots, where they are most noticeable at the nodes. Mature stalks suddenly wilt and die during the acute phase, frequently without any prior symptom appearance. This condition appears to be limited to very vulnerable types, but it usually follows a period of stress, particularly after extended drought (Rott and Davis, 2000).

8.7 DISEASE MANAGEMENT IN SUGARCANE

Red rot management has proven difficult. It has been noted that managing the disease becomes difficult once it appears in the field. As a result, the majority of advised management techniques focus on proactive measures to reduce pathogen accumulation in the soil. IDM should be employed since different diseases have different effects, making it unfeasible to reduce red rot losses with a single technique.

8.7.1 AVOID MONO-CROPPING

In the all-sugar mill control area, mono-cropping with a single variety has played a major role in the development of particularly virulent pathotypes of any fungus. The state of Uttar Pradesh has had historical setbacks with dominant cultivars including Co 213, Co 312, Co 313, Co 421, Co 453, Co 513, Co 1148, BO 11, and so on throughout the last few decades. The emergence of pathogenic pathotypes and varietal breakdown was attributed for each of these failures. The crop was drastically damaged due to monocropping of these types, a phenomenon known as the "boom" and "bust" cycle. Therefore, for sustainable sugarcane production in the future, no variety should be grown in more than 50% of a mill area. Even if the industry and farmers have benefited greatly from using the cv Co 0238 in recent seasons (Ram, 2020), the harm that cane production is currently causing to thousands of hectares of cane fields in the state exceeds the advantages. Furthermore, the pathogen's greater potency poses an important barrier to the successful cultivation of cane because it will reduce the new varieties' resistance to red rot due to the virulent pathotype (Viswanathan et al., 2022). It might take some time to replace Co 0238 with other suggested kinds because its acreage still varies from 40 to 60% in several state sugar mills (Viswanathan et al., 2024).

8.7.2 CLEAN CULTIVATION

Clean cultivation is the primary prerequisite for growing a healthy sugarcane crop. This starts with selecting and planting healthy setts, preferably in a healthy field. This practice reduces the load of initial inoculum and helps in checking sett-borne diseases. This practice of sett selection should be religiously followed if one really desires to check the damage from the diseases. Removal of diseased plants (roguing), as and when detected, also helps reduce the inoculum available for the secondary spread. Interculture operations like earthing up, wrapping, and propping, and sanitation practices like removing dead plant residues, old leaves, and weeds should be implemented for good crop and clean cultivation. This practice also prevents the crop from being attacked by pests and diseases by rendering the microclimate uncongenial for pathogens (Viswanathan, 2021b).

8.7.3 VARIETY RESISTANCE

The use of resistant cultivars is the main approach for controlling disease in sugarcane. It is the goal of screening trials against some of the most significant diseases and frequent inspection of new varieties at different selection phases in the breeding program to eliminate sensitive varieties. It ensures that the cultivars eventually developed for growers have a high measure of general disease resistance. The sensitivity of newly cultivars cannot be permanent. New problems or the reappearance of previously important diseases may occur, and cultivars, once adequately resistant, cannot remain during adverse environmental conditions. Breeding and selecting new varieties to meet the industry's requirements is a lengthy process. It also takes a long duration for growers to replace existing varieties with ones more resistant to disease, especially if the susceptible cultivars are broadly grown. Therefore, various products cannot always be rapidly withdrawn from production. Sugar mills must only take varieties from other areas with proper phytosanitory certificates from competent authorities. The various measures will alleviate the disease, such as selecting resistant varieties, a healthy nursery program, selection of healthy setts, fungicide treatment of setts, irrigation management, and rouging and drenching the soil with fungicides.

8.7.4 SEED NURSERY PROGRAM

Strictly enforcing a clean nursery program is essential to remove pathogens from primary seed materials. Seed material must be free from diseases and pests, and combination of other cultivars should be ensured. Even with a red rot incidence of as low as 1%, the nurseries must be deleted from the seed program. Infected seeds should be replaced with healthy-quality seeds. Robust seed replacement program has to be initiated in the affected area. The farmers must be adequately educated on how to use quality seed from any research center or authorized center. Farmers can also be suggested to have seed cane nurseries. Quality seed should be raised using a single bud of healthy cane by treating better canes using moist hot air of about 54°C for 2.5 hours and ambient air humidity of 95–99% (Satyavir, 2003) or hot water treatment (52°C for 2 hours) and subsequent generations are raised with the inspection. Hence, to improve efficacy, systemic fungicides (Thiophanate Methyl (0.1%)) may be mixed with warm water and may kill deep-seated fungal infections, if any. The single bud sett should be soaked for half an hour in water suspension of systemic fungicide (Thiophanate Methyl, 1g/l) before planting (Viswanathan, 2021b).

8.7.5 DISEASE MANAGEMENT THROUGH SETT TREATMENT DEVICE (STD)

Most fungal diseases are transmitted through the planting materials in sugarcane, and they serve as the primary source for the pathogen inoculum in the field. This facilitates the buildup of diseases in the early crop phase, leading to disease development at an epidemic level depending on the variety and prevailing climatic conditions. A clean seed program is essential for effective fungal disease management. Although we had effective fungicides for sugarcane, an efficient delivery system was unavailable. The new method of sett treatment, "mechanized sett treatment for fungicides," ensures protection of the crop from the primary source of *C. falcatum* and sett-borne fungal inoculum in early phases under field conditions. The new method of fungicide treatment would greatly reduce red rot and other fungal diseases in the disease-endemic regions in the country (Malathi et al., 2017).

ICAR-SBI has developed a new STD, Coimbatore (Figure 8.18), which is more potential. This approach has been demonstrated to be efficient in eliminating red rot sett-borne diseases. The setts are treated with thiophanate methyl (0.1%) for 20 minutes in a STD. In addition to inhibiting soil-borne inoculum entry, this treatment produces red rot in immature crops and planted setts that persists for ninety days. The STD controls red rot in sugarcane by facilitating the efficient diffusion of fungicides inside the setts (Malathi et al., 2017, 2022; Viswanathan et al., 2016). Disease accumulation in the field is prevented by providing young crops with effective protection through sett treatment (Figure 8.19).

FIGURE 8.18 Sett treatment device.

FIGURE 8.19 The effect of STD on red rot-infected soil and the production of healthy sugarcane plants (left treated with STD) and (infected cane).

8.7.6 PRODUCTION OF DISEASE-FREE HEALTHY SUGARCANE THROUGH STD

The best way to ensure the production of disease-free sugarcane nurseries is through the primary treatment of sugarcane buds at planting time. A clean nursery program is essential to remove sett-born pathogens from primary seed materials. Effectively soaking sugarcane setts or single buds with carbendazim + imidacloprid or micronutrient components or microbial combinations is essential to producing disease-free, healthy sugarcane through the STD. Plant vigor and shoot length improve with STD. The STD could produce disease-free, healthy seed nurseries with pesticides and micronutrients.

This experiment of UP Council of Sugarcane Research, Shahjahanpur, revealed that the treated single bud setts in STD with different various microorganisms such as *Trichoderma, Stenotrophomonas maltophilia* (B2132), *Pseudomonas stutzeri* (B2133), *Pseudomonas rubrilineans* (PSB), *Azotobacter croccocum,* and *Tricoderma harziannum* gave better results to produce healthy settlings with improved vigor for nursery program. Hence, microbes with water at a vacuum level of 200 Hg/mm were observed early, with good germination and settling vigor. At the same time, the late germination in untreated control with poor settling growth (Figure 8.20). The other yield contributing traits were also found to be good with STD in compression to untreated control (Viswanathan et al., 2023).

FIGURE 8.20 Healthy settling treated with STD on sugarcane plant seedlings.

8.7.7 ROGUING OF DISEASED PLANTS

The traditional practice of roguing (uprooting) of the diseased plants has become the main stay in managing sugarcane diseases. Systematic roguing of the diseased sugarcane plants is important in maintaining a healthy ratoon and plant crop.

8.8 AGRONOMIC PRACTICES

Effective drainage and the removal of excess trash from fields are two cultural practices that can lower the risk of red rot. Accelerated germination agronomic techniques are critical for minimizing seed rotting and producing high-quality stands. The effective tips to manage red rot as outlined are:

- Affected fields should not be permitted to ratoon even if negligible red rot incidence is noticed in the plant crop.
- Crop rotation of sugarcane followed by rice for one season or with other crops for two seasons will reduce the severity of red rot. Farmers should start crop rotation after disease incidence by taking a crop of rice or wheat.

- Green manuring with *Dhaincha* is also mandatory for infected field sanitation. This agronomic practice not only restores the lost vigor of the soil but also reduces the number of propagules of pathogens like *Rhizoctonia*, *Pythium*, *Fusarium,* and *Colletotrichum* and thus ensures a good sugarcane crop.
- The affected field and adjoining field should be harvested immediately. The affected field should be plowed up without any delay, and the field should be made from the stubbles. The stubble and other sugarcane crop residues may be preferably destroyed by burning.

8.8.1 BIOLOGICAL CONTROL

Using antagonistic biological control agents, like *Pseudomonas* species and *Trichoderma*, red rot losses have been effectively minimized. The nursery plot should be treated with bio-product containing *Trichoderma* or *Pseudomonas*. Red rot disease was noticed to biologically managed through *Trichoderma harzianum* and *Trichoderma viride* (20 kg/ha) application applied with sanitized decompose press mud/FYM (mixed with 200 kg FYM/press mud) at the time of planting to boost internal disease resistance, plant vigor, and higher yield (Solanki et al., 2023; Malviya et al., 2021, 2022). Because *Pseudomonas* spp. and *T. harzianum* produce the chitinase enzyme, they are effective at protecting crops from soil-borne red rot inoculum.

8.8.2 RESISTANT VARIETIES

Employing resistant cultivars, such as CoS 08272, Co 0118, Co 98014, UP 05125, CoS 13231, CoS 13235, CoS 18231, CoLk 14201, C0Lk 16202, Co 15023, etc., has proven to be the main approach of effectively controlling red rot. The breeding work in India primarily focuses on developing red rot-resistant cultivars. As the pathogen is more variable, therefore, even if a disease-resistant cultivars developed for the cultivation, it gets susceptible to red rot infection within nearly 10 years because of its recent development, most virulent races of the pathogen due to highly variable.

8.8.3 CHEMICAL TREATMENT

Regular watch of the healthy sugarcane crop and adjoining areas is necessary to save the crops from any disease (Wu et al., 2022; Xu et al., 2022; Huang et al., 2024). Generally, red rot infection appears in the month up to July. The initial symptom of red rot appears on the leaf mid-rib (Spindle infection). Later on, destruction and uprooting of the affected clumps without fail and broadcast of 10–20 g bleaching powder at the place of infected soil. Soil inoculum may also be destroyed by drenching 0.2% suspension of Thiophanate Methyl. Once red rot is detected in the earlier month, the rest of the crop may be sprayed with a systemic fungicide (0.2% Thiophenate Methyl or Carbendazime) at a monthly interval up to the monsoon. Suppose the infection of disease is expected to persist. In that case, Thiophanate Methyl (0.1%) can be applied as a spray of foliar at 3–4 months stage or can be delivered via micro irrigation if the system is developed in the farmers field (Viswanathan et al., 2024).

8.8.4 DISEASE MANAGEMENT IN RATOON CROP

To minimize disease development in the field in endemic areas, two sprays of Thiophanate Methyl (0.1%) may be applied to ratoon crops: one immediately following ratoon sprouting (30 days), and the second between 60 and 90 days later (Bhatt et al., 2023a, 2023b). Even if disease incidence is not observed in areas where the disease is prevalent, this suggestion may still be followed. It has been found that foliar infections through midrib lesions have been a major factor in the spread of the red rot epidemics in the cane variety Co 0238 (Viswanathan, 2023). Notwithstanding the crop's protection during the early crop stage, the midrib lesions inoculum caused the disease to spread throughout the command area, causing potentially high levels of crop loss. Therefore, it is preferable to apply fungicides topically to prevent fungal infections that occur through the foliage (Viswanathan et al., 2024).

8.8.5 QUARANTINE

Plant material from high-risk areas cannot be introduced without permission under quarantine restrictions. In our country, many sugarcane diseases have spread from one place to another with the active help of human agency. The unrestricted movements of seed cane from one district or state to another have been more responsible for the spreading of red rot in various areas/regions. Due to the introduction of highly sensitive species' seed material from the red rot endemic area, the disease has recently migrated to Karnataka and Maharashtra. Hence, there is a need to restrict cane transport from infected to disease-free zones. Seed material from research stations with valid phytosanitary certificates is the only source available for purchase outside of the state.

8.9 CONCLUSIONS AND FUTURE DIRECTIONS

To mitigate the severity of the disease, it is recommended to use certified seeds, healthy planting materials, crop rotation, field maintenance, proper drainage, and bundling of contaminated fields. The aforesaid agronomical cultural practices are not only to decrease the inoculum in the field, but also to reduce crop losses (Viswanathan, 2021b). Strict enforcement of a nursery program should be essential among sugar command area for preventing spread of the disease via cane seeds. However, endemic areas of any sugar mill command regions, availability of cane seeds free from infections and mixtures of sugarcane varieties must be ensured. Soaking of sugarcane setts with systemic fungicide and insecticide or micronutrients or microbial combination should be essential to producing disease-free, healthy sugarcane through the STD. The STD could produce disease-free, healthy seed nurseries with pesticides and micronutrients. The sugarcane seed should be treated with bio-products containing *Trichoderma* or *Pseudomonas* in endemic region of diseases. UP Council of Sugarcane Research, Shahjahanpur, India produces highest sugarcane seed materials in India, and this organization is being provided healthy and disease-free true to the types seed materials for cane growers. Integrated biotic management should be implemented as the availability of resistant varieties, precise and timely detection of the pathogens, the severity of their infection, and cost-effective control measures.

KEYWORDS

- **red rot disease**
- **smut and wilt disease**
- **pokkah boeng disease symptoms**
- **disease tolerance index**
- **disease severity**
- **sett treatment device**
- **ratoon crop**
- **management strategies**

REFERENCES

Agnihotri VP, Budhraja TR, Singh K. (1979) Role of diseased sett and soil in the annual recurrence of red rot in sugarcane. *Int. Sugar J.* 82, 263–265.

Agnihotri VP, Rao GP. (2002) A century status of sugarcane wilt in India. In: Singh SB, Rao GP Eswaramoorthy S. (Eds.), *Sugarcane Crop Management*, SciTech Publishing LLC: Houston, USA, pp. 145–160.

Agnihotri VP, Singh K. (1977) Seed-piece transmissible diseases of sugarcane and their chemical control. *Sugar News* 2, 90–95.

Agnihotri VP, Singh RP. (1989) Sugarcane wilt current status. In: Agnihotri VP, Singh N, Chaube HS, Singh US, Dwivedi TS. (Eds.), *Prospective in Plant Pathology*.

Agnihotri VP. (1990) *Diseases of Sugarcane and Sugarbeet*. Oxford and Ibh Publishing Company Private Limited: New Delhi, p. 483.

Agnihotri VP. (1996) Current sugarcane disease scenario and management strategies. *Ind. Phytopathol.* 49, 109–126.

Akalach M. (1994) First report of sugarcane smut in Morocco. *Plant Dis.* 78(5), 529.

Alexander KC, Viswanathan R. (2002) Diseases of sugarcane in India and its rapid diagnosis. In: Singh SB, Rao GP, Eswaramoorthy S. (Eds.), *Sugarcane Crop Management*, SCI TECH Publishing LLC: Houston, Texas, USA, pp. 10–51.

Banihashemi Z. (1995) The occurrence of sugarcane smut in Mazandarn Province. *Iran. J. Plant Pathol.* 31(1/4), 40–41.

Barber CA. (1901) Sugarcane diseases in Godawari and Ganjam District. *Madras Dept Land Records Agric. Bull.* 512, 181–194.

Bhatt R, Verma KK, Jamwal NS. (2023a) Can paddy-compost eliminate N fertilization for climate-smart sugarcane? *Sugar Tech.* 25, 1036–1046. https://doi.org/10.1007/s12355-023-01270-3

Bhatt R, Verma KK, Kumar R, Sanghera GS. (2023b) Foliar application of potassium salt of active phosphorus (PSAP) mitigates insect pests and improves yield along with sugarcane quality in response to agroclimatic conditions of Punjab. *Sugar Tech.* 25(3), 660–669.

Bolle PC. (1937) Over vatbaarheid en gevoeligheid voor Pokkah boeng en indirecte bestrijding. *Verh.* 25, 657–761.

Bourne BA. (1922) *Researches on the Root Diseases of Sugarcane*. Department of Agriculture: Barbados, pp. 1–17.

Butler EJ, Khan AH. (1913) Red rot of sugarcane. *Memoirs of Department of Agriculture, India, Botanical Series* 6, 151–178.

Butler EJ. (1906) Fungus disease of sugarcane in Bengal. *Memoirs of the Department of Agriculture, Indian Botanical Survey* 1, 2–24.

Chen G-L, Zheng FJ, Lin B, Lao SB, He J, Huang Z, Zeng Y, Sun J, Verma KK. (2020) Phenolic and volatile compounds in the production of sugarcane vinegar. *ACS Omega* 5, 30587–30595.

Chona BL, Bajaj BS. (1953) Occurrence in nature of *Physalospora tucumanensis Speg.*, the perfect stage of sugarcane red rot organism in India. *Ind. Phytopathol.* 6, 63–65.

Chona BL, Nariani TK. (1954) Investigations on the survival of the sugarcane red rot fungus in the compost. *Ind. Phytopathol.* 5, 151–157.

Chona BL, Srivastava DN. (1952) The perithecial stage of *Colletotrichum falcatum* Went in India. *Ind. Phytopathol.* 5, 158–160.

Chona BL. (1943) Sugarcane smut and its control. *Ind. Farm.* **4,** 401–404.

Davis MJ, Rott P, Warmuth CJ, Chatenet M, Baudin P. (1997) Intraspecific genomic variation within Xan thomonas albilineans, the sugarcane leaf scald pathogen. *Phytopathol.* 87(3), 316–324. https://doi.org/10.1094/ PHYTO.1997.87.3.316

Duttamajumder SK. (2008) *Red Rot of Sugarcane*. Indian Institute of Sugarcane Research: Lucknow, India.

Edgerton CW. (1955) *Sugarcane and its Diseases*. Louisiana Univ. Press: Baton Rouge, LA, p. 301.

Edgerton CW. (1959) *Sugarcane and Its Diseases*. Louisiana State University Press: Louisiana, USA.

Govender P, Mcfarlane SA, Rutherford RS. (2010) *Fusarium* species causing pokkah boeng and their effect on *eldana saccharina* walker (Lepidoptera: Pyralidae). *Proc. S. Afr. Sug. Technol. Ass.* 83, 267–270.

Gupta MR, Gupta SC. (1976) Assessment of losses caused by wilt disease of sugarcane in western U.P. *Ind. Sugar* 26, 151–152.

Hoy JW, Hollier CA, Fontenot DB, Grelen LB. (1986) Incidence of sugarcane smut in Louisiana and its effects on yield. *Plant Dis.* 70, 59–60.

Huang CM, Wu KC, Verma KK, Luo HB, Deng ZN, Shi GY, Cao HQ, Wu XJ, Ye LP, Wei YW, Xu L. (2024) Isolation and expression analysis of cellulose synthase 3 (*Ces3*) genes from sugarcane (*Saccharum officinarum* L.). *Chem. Biol. Technol. Agric.* 11, 71. https://doi.org/10.1186/s40538-024-00597-7

Khanna KL, Chacravarti AS. (1949) Studies on storage of gur. *Curr. Sci.* 18, 127–128.

Kirtikar, Singh GP, Shukla R. (1972) Role of seed material in carryover of wilt disease of sugarcane. *Ind. Sugar* 22, 89–90.

Lovick G. (1978) Smut of sugarcane—*Ustilago scitaminea. Rev. Plant Pathol.* 57(5), 181–188.

Luthra JC, Suttar A, Sandhu SS. (1940) Experiments on the control of smut of sugarcane. *Proc. Ind. Acad. Sci. Sec. B*. 12, 118–128.

Malathi P, Viswanathan R, Ramesh Sundar A, Elamathi E, Kaverinathan K, Manivannan K, Nithiyanantham R. (2022) Approaches endeavoured to manage red rot under field conditions. *J. Sugarcane Res*. 12, 128–139.

Malathi P, Viswanathan R, Ramesh Sundar A. (2017) Mechanized means of sett treatment: An effective way of delivering fungicides for the management of red rot in sugarcane. *Sugar Tech*. 19(2), 176–182.

Malviya MK, Li C-N, Lakshmanan P, Solanki MK, Wang Z, Solanki AC, Nong Q, Verma KK, Singh RK, Singh P, Sharma A, Guo D-J, Dessoky ES, Song X-P, Li Y-R. (2022) High-throughput sequencing-based analysis of rhizosphere and diazotrophic bacterial diversity among wild progenitor and closely related species of sugarcane (*Saccharum* spp. inter-specific hybrids). *Front. Plant Sci*. 13, 829337. https://doi.org/10.3389/fpls.2022.829337.

Malviya MK, Solanki MK, Li CN, Wang Z, Zeng Y, Verma KK, Singh RK, Singh P, Huang HR, Yang LT, Song XP, Li YR. (2021) Sugarcane-legume intercropping can enrich the soil microbiome and plant growth. *Front. Sustain. Food Syst*. 5, 606595. https://doi.org/10.3389/fsufs.2021.606595.

Martin JP, Han Loie Hong, Wismer CA. (1961) Pokkah boeng. In: Martin JP, Abbott EV, Hughes CG. (Eds.), *Sugarcane Diseases of the World*. Elsevier, Vol. 1, pp. 247–257.

Mohammadi A, Nejad RF, Mofrad NN. (2012) *Fusarium verticillioides* from sugarcane, vegetative compatibility groups and pathogenicity. *Plant Prot. Sci*. 48(2), 80–84.

Padmanaban P, Alexander KC, Shanmugan N. (1998) Effect of smut on growth and yield parameters of sugarcane. *Ind. Phytopath*. 41, 367–369.

Parthasarathy SV. (1972) *Sugarcane in India*. The KCP Ltd.: Madras, p. 6, 685.

Patil AS, Hapase DG. (1987) Studies on pokkah boeng disease in Maharashtra. *Ind. Phytopathol*. 40, 290.

Patil AS. (1995) Studies on pokkah boeng and pine apple disease of sugarcane in Maharashtra with their economic losses in yield and quality of sugarcane. *Final Project Report*, ICAR, VSI, Pune.

Que YX, Xu LP, Lin JW, Chen RK, Grisham MP. (2012) Molecular variation of *Sporisorium scitamineum* in Mainland China revealed by RAPD and SRAP markers. *Plant Dis*. 96, 1519–1525.

Ram B, Hemaprabha G. (2020) The sugarcane variety Co 0238—A reward to farmers and elixir to India's sugar sector. *Curr. Sci*. 118, 1643–1646.

Ram B. (2020) Co 0238—Its impact, management and what's next? *Ann. Convention Sugar Technol. Assoc. Ind*. 78, 167–175.

Rao GP, Viswanathan R, Singh SB. (2002) Current situation of sugarcane diseases in India. In: Singh SB, Rao GP, Easwaramoorthy S. (Eds.), *Sugarcane Crop Management*, SCI Tech Publishing LLC: Houstan, p. 734.

Rao GP. (2004) *Sugarcane Pathology*. Oxford and Ibh Publishing Company: UK, p. 2.

Rott P, Bailey RA, Comstock JC, Croft BJ, Saumtally AS. (2000) *A Guide to Sugarcane Diseases*. CIRAD/ISSCT: Montpellier, p. 340.

Rott P, Davis M. (2000) Leaf scald. In: Rott P, Bailey RA, Comstock JC, Croft BJ, Salem Saumtally A. (Eds.), *A Guide to Sugarcane Diseases*. CIRAD/ISSCT, La Librairie du Cirad: Montpellier, France, pp. 38–44.

Sarma MN. (1976) Wilt disease of sugarcane. *Sugarcane Pathol. Newslett.* 15/16, 30–33.
Satyavir. (2003) Red rot of sugarcane-current scenario. *Ind. Phytopathol.* 56, 245–254.
Sharma R, Tamta S. (2015) A review on red rot: The "Cancer" of sugarcane. J. Plant Pathol. Microbiol. S1, 003. https://doi.org/10.4172/2157-7471.S1-003
Sheldon JL. (1904) A corn mold (*Fusarium moniliforme* sp). *Ann. Rep. Nebraska Agric. Exp. Stat., 17, 23–32.*
Singh A, Isaac-Kirubakaran S, Sakthivel N. (2007) Heterologous expression of new antifungal chitinase from wheat. *Protein Expres. Purif.* 56, 100–109.
Singh K, Budhraja TR, Agnihotri VP. (1977) Survival of *Collatotrichum falcatum* in soil, its portals of entry and role of inoculum density in causing infection. *Int. Sugar J.* 79, 43–44.
Singh K, Singh RP. (1974) Involvement and pathogenicity of *Acremonium* in wilt syndrome of sugarcane. *Sugarcane Pathol. Newslett.* 11/12, 24–25.
Singh K, Singh RP. (1989) Red rot. In: Ricaud C, Egan BT, Gillaspie AG, Hughes CG. (Eds.), *Sugarcane Diseases—Major Diseases*. Elsevier: Amsterdam, pp. 169–188.
Singh O, Waraitch KS. (1981) Effect of wilt and red rot induced stress on quality deterioration of sugarcane. *Sugarcane Pathol. Newslett.* 27, 25–30.
Singh RK, Banerjee N, Khan MS, Yadav S, Kumar S, Duttamajumder SK, Lal Ram Ji, Patel JD, Guo H, Zhang D, Paterson AH. (2016) Identification of putative candidate genes for red rot resistance in sugarcane (*Saccharum* species hybrid) using LD-based association mapping. *Mol. Genet. Genomics.* https://doi.org/10.1007/s00438-016-1190-3
Singh RP, Lal S, Singh K. (1983) Development of dormant nodal infection of *Colletotrichum falcatum* Went In Sugarcane. *Ann. Convention Sugar Technol. Assoc. India* 47, 79–90.
Singh RP, Lal S. (1996) Air-Borne Propagules of *Colletotrichum falcatum* Went and their role in the epidemiology of sugarcane red rot. *Ind. Phytopathol.* 49(1), 89–91.
Singh RP, Lal S. (2000) Red rot. In: Rott P, Bailey RA, Comstock JC, Croft BJ, Salem Saumtally A. (Eds.), *A Guide to Sugarcane Diseases*. CIRAD/ISSCT, La Librairie du Cirad: Montpellier, France, pp. 153–158.
Solanki MK, Verma KK, Dastogeer KMG, Mora-Poblete F, Mundra S. (2023) Microbial resilience in plant nutrient management towards sustainable farming. *Front. Microbiol.* 14, 1280811. https://doi.org/10.3389/fmicb.2023.1280811
Song X-P, Mo F, Verma KK, Wei J, Zhang X, Yang L, Li YR. (2019) Effect of sugarcane smut (*Ustilago scitaminea* Syd.) on ultrastructure and biochemical indices of sugarcane. *Biomed J. Sci. Tech. Res. 17*(1), 002950. https://doi.org/10.26717/BJSTR.2019.17.002950.
Song XP, Mo F, Yan M, Zhang X, Zhang B, Huang X, Huang D, Pan Y, Verma KK, Li YR. (2022) Effect of smut infection on the photosynthetic physiological characteristics and related defense enzymes of sugarcane. *Life* 12, 1201. https://doi.org/10.3390/life12081201
Song XP, Verma KK, Tian DD, Zhang XQ, Liang YJ, Huang X, Li CN, Li YR. (2021) Exploration of silicon functions to integrate with biotic stress tolerance and crop improvement. *Biol. Res. 54*, 19. https://doi.org/10.1186/s40659-021-00344-4
Srinivasan KV, Alexander KC. (1966) A study of mode of nodal infection and occurrence of dormant infections of Red Rot. *Proc. Ind. Conf. Sugarcane Res. Develop. Workers* 5, 676–684.
Steib RJ, Chilton SJP. (1951) Infection of sugarcane stalks by the red rot fungus *Physalospora tucumanensis* Speg. *Phytopathol.* 41, 522–528.

Subba Raja KT, Natarajan S. (1972) Sugarcane wilt caused by *Cephalosporium sacchari* and Fusarium moniliforme in India. *Sugarcane Pathol. Newslett.* 8, 21–23.

Tian D, Qin L, Verma KK, Wei L, Li J, Li B, Zhou W, He Z, Wei D, Huang S, Long S, Huang Q, Li C, Wei S. (2023) Transcriptomic and metabolomic differences between banana varieties which are resistant or susceptible to *Fusarium wilt. PeerJ* 11, e16549. https://doi.org/10.7717/peerj.16549

Verma KP, Singh MP, Upadhyay UC, Singh HN. (1984) Knife cut, the first report of the incidence of this disease in India. *Sugarcane* 40, 817–820.

Vishwakarma SK, Kumar P, Nigam A, Singh A, Kumar A. (2013) Pokkah boeng: An emerging disease of sugarcane. *J. Plant Pathol. Microbiol.* 4, 170.

Viswanathan R, Goswami SK, Chandramani Raj, Singh S. (2023) Delivery of micropogation to improve settling vigour in nurseries. In: *Technical Report for 2022–23, Plant Pathology, All India Coordinated Research Project on Sugarcane (Indian Council of Agricultural Research)*, ICAR—Indian Institute of Sugarcane Research: Lucknow, India, p. 89.

Viswanathan R, Malathi P, Naik R. (2016) A sett treatment device for fungal disease management and healthy nursery programme in sugarcane. *Ann. Convention Sugar Technol. India* 77, 236–253.

Viswanathan R, Malathi P, Ramesh Sundar A, Poongothai M, Singh N. (2006) Current status of sugarcane wilt in India. *Sugar Cane Int.* 24(4), 1–7.

Viswanathan R, Padmanaban P, Mohanraj D, Nallathambi P. (1997) Occurrence of leaf scald disease in Tamil Nadu. *Ind. Phytopathol.* 50, 149.

Viswanathan R, Poongothai M, Malathi P, Sundar AR. (2012) Sugarcane wilt: New insights into pathogen identity, variability and pathogenicity. *Funct. Plant Sci. Biotechnol. Global Sci. Books* 6, 30–39.

Viswanathan R, Rao GP, Singh D, Solomon S. (2024) Strategies to contain red rot of sugarcane in subtropical India. *Indian Sugar, LXXIV* 11, 21–27.

Viswanathan R, Rao GP. (2011) Disease scenario and management of major sugarcane diseases in India. *Sugar Tech.* 13(4), 336–353.

Viswanathan R, Samiyappan R. (1999) Red rot disease of sugarcane: Major constraint for Indian sugar industry. *Sugar Cane* 5, 9–15.

Viswanathan R, Samiyappan R. (2000) Efficacy of *Pseudomonas* spp. strains against soil borne and sett borne inoculum of *Colletotrichum falcatum* causing red rot disease of sugarcane. *Sugar Tech* 2(3), 26–29.

Viswanathan R, Singh SP, Selvakumar R, Singh D, Bharti YP, Chhabra ML, Parameswari B, Sharma A, Minnatullah M. (2022) Varietal break down to red rot in the sugarcane variety Co 0238 mimics Vertifolia effect: Characterizing new *Colletotrichum falcatum* pathotype CF13. *Sugar Tech.* 24, 1479–1496.

Viswanathan R, Sundar AR, Malathi P, Padmanaban P. (2011) *Red Rot of Sugarcane*. In: Shanthy TR. (Ed.), Sugarcane Breeding Institute: Coimbatore.

Viswanathan R. (2010) *Plant Disease: Red Rot of Sugarcane*. Anmol Publishers: New Delhi, p. 301.

Viswanathan R. (2012) *Sugarcane Diseases and Their Management*. Sugarcane Breeding Institute: Coimbatore, p. 140.

Viswanathan R. (2013) Status of sugarcane wilt: One hundred years after its occurrence in India. *J. Sugarcane Res.* 3(2), 86–106.

Viswanathan R. (2021b) Red rot of sugarcane (*Colletotrichum falcatum* Went). *CAB Rev.* 16(023). https://doi.org/10.1079/PAVSNNR202116023

Viswanathan R. (2023) Severe red rot epidemics in sugarcane in Subtropical India: Role of aerial spread of the pathogen. *Sugar Tech.* 25, 1275–1277.

Went FAFC. (1893) Het root snot. *Arch. Jora-Suikerind* 1, 265–282.

Wu K-C, Huang C-M, Verma KK, Deng Z-N, Huang H-R, Pang T, Cao H-Q, Luo H-B, Jiang S-L, Xu L. (2022) Transcriptomic responses of *Saccharum spontaneum* roots in response to polyethylene glycol–6000 stimulated drought stress. *Front. Plant Sci.* 13, 992755. https://doi.org/10.3389/fpls.2022.992755

Xu L, Deng Z-N, Wu K-C, Malviya MK, Solanki MK, Verma KK, Pang T, Li Y-J, Liu X-Y, Kashyap BK, Dessoky ES, Wang W-Z and Huang H-R. (2022) Transcriptome analysis reveals a gene expression pattern that contributes to sugarcane bud propagation induced by indole-3-butyric *Acid. Front. Plant Sci.* 13, 852886. https://doi.org/10.3389/fpls.2022.852886

Xu LP, Chen RK, Chen PH. (2004) Analysis on infection index of smut caused by *Ustilago scitaminea* in sugarcane segregated population. *Chin. J. Trop. Crop.* 25, 33–36.

Zheng F-J, Lin B, Yang Y-X, Fang X-C, Verma KK, Chen G-L. (2023) Efficacy and functionality of sugarcane original vinegar on mice. *Front. Microbiol.* 14, 1224666. https://doi.org/10.3389/fmicb.2023.1224666

CHAPTER 9

Plant Biomass to Bioenergy: Use and Impact on Environment and Development

AKHTAR HUSSAIN[1], IRUM[2], GYANENDRA TRIPATHI[2], ALVINA FAROOQUI[2], and MOHAMMAD ASHFAQUE[1*]

[1]*Department of Biosciences, Integral University, Lucknow, Uttar Pradesh, India*

[2]*Department of Bioengineering, Integral University, Lucknow, Uttar Pradesh, India*

*Corresponding author

ABSTRACT

The exhaustion of fossil fuel reserves and increasing atmospheric environmental pollution and climate change have heightened the quest for sustainable alternatives. Within these options, biofuels from plant biomass have emerged as a better alternative. This chapter provides the accessibility of plant biomass to substitute fossil fuels with biofuels. The initial part delineates the environmental advantages of biofuels, including their ability to lessen greenhouse gas emissions and reduce reliance on limited fossil fuel reserves. Subsequently, the conversation delves into the various origins of plant biomass, spanning from specialized energy crops. The quantity and dispersion of raw materials are assessed, considering geographical differences, land accessibility, and farming efficiency. The technological progressions in biomass transformation, encompassing biochemical and thermochemical responses, are discussed. These processes include fermentation, pyrolysis, and gasification, which convert lignocellulosic

Sugarcane Production: Environmental Impacts and Socioeconomic Issues.
Krishan K. Verma, Xiu-Peng Song, Lin Xu, Rajan Bhatt, and Yang-Rui Li (Eds.)

DOI: 10.1201/9781779641137-9

biomass into biofuels such as ethanol, biodiesel, and renewable biodiesel. The chapter also tackles the obstacles associated with expanding biofuel production from plant biomass, including competition for land use with food crops, possible effects on biodiversity, and concerns regarding sustainability and land use practices. However, economic factors, policy structures, and the influence of government incentives in encouraging the shift to biofuels are discussed.

9.1 INTRODUCTION TO BIOFUELS

The worldwide energy environment is experiencing a significant shift due to the pressing necessity to address and reduce the impact of climate change and reliance on finite fossil fuel resources (Zohuri 2023; Huang et al., 2024). Central to this shift is exploring renewable energy sources, with biofuels emerging as a promising alternative to traditional fossil fuels (Rial, 2024). At the forefront of biofuel production is using plant biomass, offering a renewable and potentially sustainable solution to meet the world's energy demands (Verma et al., 2019a, 2019b; 2023a). Plant biomass, derived from various sources, including dedicated energy crops, forestry residues, and agricultural residues, holds significant promise as a feedstock for biofuel production (Mohammed et al., 2018; Chen et al., 2020, 2023). Unlike fossil fuels, which release carbon dioxide accumulated over millions of years into the atmosphere upon combustion, biofuels generated from plant biomass offer the potential to mitigate greenhouse gas (GHG) emissions by recycling atmospheric carbon through photosynthesis. This carbon-neutral cycle presents an opportunity to address energy security and environmental sustainability challenges (Verma et al., 2021a, 2021b; Bhatt et al., 2023a, 2023b). Biomass conversion technology advancements have recently enhanced biofuel production processes' efficiency and viability (Khan et al., 2022). From biochemical processes such as fermentation and anaerobic digestion to thermochemical techniques like pyrolysis and gasification, various methods exist to convert plant biomass into liquid fuels such as biodiesel, bio-oil, and ethanol (Verma et al., 2023b, 2023c; Zheng et al., 2023).

Moreover, ongoing research and development efforts continue to optimize these processes, driving down production costs and improving overall energy yields (Nanda et al., 2014). However, the widespread adoption of plant biomass for biofuel production faces several challenges and considerations. Competition for land between food and energy crops raises concerns about food security and land use change, necessitating sustainable practices and land management strategies (Popp et al., 2014; Joshi et al., 2022; Guo et al., 2022). Additionally, the environmental impact of intensive agricultural practices and the potential for biodiversity loss requires careful assessment and mitigation measures to ensure biofuel production systems' long-term viability and ecological sustainability (Correa et al., 2019). The availability of plant biomass as a renewable resource offers a compelling pathway toward reducing and enhancing energy security and promoting rural development (Sikkema et al., 2021). By leveraging the inherent energy potential of biomass, coupled with ongoing technological innovation and sustainable land management practices, the transition from fossil fuels to biofuels holds the promise of a more resilient and sustainable energy future (Strezov et al., 2019; Verma et al., 2020a, 2020b; Huang et al., 2022).

Biofuels are pivotal in renewable energy, offering a sustainable substitute to conventional fossil fuels (Rial, 2024). They derive from organic sources such as plant biomass, animal fats, and algae, harnessing the stored energy through various conversion processes. The primary reason for their significance lies in their ability to reduce GHG emissions. Unlike fossil fuels, which release CO_2 from deep within the Earth's crust, biofuels recycle atmospheric CO_2 through the growth of plants used in their production (Schobert, 2013; Verma et al., 2021c, 2022a, 2022b; Joshi et al., 2023). This characteristic contributes to mitigating climate change, aligning with efforts to transition toward low-carbon energy systems (Van Sluisveld et al., 2020; Deng et al., 2023).

Moreover, biofuels bolster energy security by providing a domestically produced alternative to imported fossil fuels, diminishing reliance on geopolitically unstable regions. As renewable resources, plant biomass used for biofuels can be sustainably harvested and replenished, ensuring a reliable energy source in the long term (López-Bellido et al., 2014; Guo et al., 2021, 2023). Additionally, biofuels integrate into existing infrastructure, requiring minimal modifications in transportation, heating, and electricity generation sectors. This compatibility expedites adopting of renewable energy solutions (Jayachandran et al., 2022). Economically, biofuels drive growth and employment creation across various sectors, from agriculture to fuel processing, contributing to the development of a robust green economy. Their versatility encompasses a range of products such as ethanol, biodiesel, biogas, and bio-oil, catering to diverse energy needs across different sectors (Gada et al., 2022). In essence, biofuels are critical in advancing toward a cleaner, more sustainable energy future by offering a renewable, low-carbon alternative to fossil fuels, promoting environmental protection, energy security, economic development, and technological innovation (Cantarero, 2020).

Biofuels encompass a variety of types, each originating from different organic sources and processed through distinct methods. Ethanol, derived from fermenting sugars in crops such as corn, sugarcane, and switch grass, is widely recognized (Anastassiadis, 2016). It can be blended with gasoline to reduce emissions in conventional vehicles or utilized in flex-fuel vehicles. Biodiesel, produced from vegetable oils, animal fats, or recycled cooking grease via transesterification, reduces emissions of particulate matter and hydrocarbons in diesel machines. Biogas, generated through anaerobic digestion of biological matter such as agricultural residues or food waste, predominantly consists of methane and finds application in electricity generation, heating, or vehicle fuel postpurification (Lemmer et al., 2017). Bio-oil, obtained from fast pyrolysis of biomass like wood chips or algae, can be refined into transportation fuels or serve as a chemical feedstock. Additionally, second-generation bioethanol, derived from lignocellulosic biomass (LCB) such as agricultural or forestry residues, presents a nonfood alternative with potentially higher energy yields (Guo et al., 2015; Verma et al., 2022c, 2022d; Wu et al., 2022). Each biofuel type contributes uniquely to renewable energy diversification, playing a role in reducing GHG emissions across various sectors (Algarni et al., 2023).

One significant lacuna in the current discourse surrounding biofuels is the need for comprehensive sustainability assessments across the entire lifecycle of biofuel production and utilization (Mutopo et al., 2011). While biofuels propose the potential to mitigate GHG emissions and reduce reliance on fossil fuels, their environmental and social impacts can vary widely depending on feedstock selection, land use practices, and production techniques. Further research is required to evaluate the environmental footprint of different biofuel pathways, including their impacts on water consumption, air quality, soil health, land use change, and biodiversity (Qin et al., 2018). Social considerations such as impacts on food security, land tenure rights, rural livelihoods, and human health must be integrated into sustainability assessments to ensure that biofuel development benefits (Xu et al., 2022; Das et al., 2024).

Addressing this lacuna will require interdisciplinary collaboration among scientists, policymakers, industry stakeholders, and local communities to develop robust sustainability criteria and standards for biofuel production and certification (Afionis and Stringer, 2020). By incorporating holistic sustainability assessments into biofuel policies and decision-making processes, we can ensure that biofuels contribute to a sustainable energy future while avoiding unintended negative consequences on the environment and society (Jeswani et al., 2020). Hence, this chapter discussed the availability of plant biomass for replacing fossil fuels with biofuels for a green environment.

9.2 PLANT BIOMASS AS SECOND-GENERATION BIOMASS FOR BIOFUEL PRODUCTION

Plant biomass, comprising organic matter derived from plants, represents a critical renewable resource for biofuel production. This biomass encompasses a wide array of plant materials, including agricultural residues, energy crops, and forestry residues. The significance of plant biomass lies in its abundance, renewability, and possibility to mitigate by substituting for fossil fuels in the transportation and energy sectors (Prasad and Ingle, 2019). Second-generation biomass, such as plant biomass, is distinguished from first-generation feedstocks such as sugarcane by its noncompetition with food crops and utilization of nonedible parts of plants (Oladunni and Ishola, 2023).

This characteristic addresses concerns related to food security and making a second-generation biomass substitute for biofuel production (Havlík et al., 2011). Using agricultural residues and energy crops, second-generation biomass maximizes resource efficiency and minimizes environmental impacts associated with land-use change and food production. The utilization of plant biomass for the production of biofuel offers numerous environmental benefits. During growth, plants sequester CO_2 through photosynthesis, effectively reducing GHGs from the atmosphere (Shen, 2014; Verma et al., 2022e, 2022f; Li et al., 2023). When converted into biofuels, the carbon stored in plant biomass is temporarily removed from the carbon cycle, reducing net carbon emissions related to fossil fuels (Searchinger, 2010). Furthermore, farming energy crops and using agricultural residues can improve soil health, enhance biodiversity, and reduce erosion, contributing to sustainable land management practices (Zuazo et al., 2011; Malviya et al., 2021, 2022; Solanki et al., 2023).

Advancements in biorefinery technologies have played a crucial role in increasing the productivity and viability of converting plant biomass into biofuels (Yamakawa et al., 2018). Integrated biorefineries employ various conversion pathways, including biochemical, such as fermentation and enzymatic hydrolysis, and thermochemical processes, such as pyrolysis and gasification, to extract sugars, lignin, and other valuable components from biomass feedstocks (Calvo-Flores and Martin-Martinez, 2022). These processes permit the production of a diverse variety of biofuels, like bioethanol, biodiesel, and renewable diesel, thereby diversifying the renewable energy portfolio and reducing dependence on fossil fuels. Furthermore, using plant biomass for biofuel production can stimulate rural economic development (Verma et al., 2021d, 2021e; Song et al., 2019, 2021). Fuel production can revitalize agricultural communities and enhance rural livelihoods by fostering the growth of energy crop industries and supporting biomass supply chains (Peskett et al., 2007; Verma et al., 2022g, 2022h).

Moreover, establishing biorefineries and associated infrastructure can attract investment and stimulate innovation in renewable energy technologies, further driving economic growth and sustainability. Plant biomass, as a second-generation feedstock for biofuel production, holds significant promise in addressing the environmental, economic, and social challenges associated with fossil fuel dependence (Anas et al., 2021a, 2021b). Through sustainable cultivation practices, technological advancements, and supportive policies, plant biomass can play a pivotal role in advancing the transition toward a more sustainable and renewable energy future (Antar et al., 2021). Plant biomass is considered as a renewable resource primarily due to its regenerative nature and the sustainable practices in its cultivation and utilization. Plants can grow and reproduce, making them renewable over time (Kalak, 2023). Through photosynthesis, plants utilize solar energy and convert CO_2 and H_2O into organic compounds, resulting in biomass accumulation (Shen, 2014; Sharma et al., 2021a, 2022). This ongoing process can be sustained through proper management practices, ensuring a continuous supply of plant biomass for biofuel production (Sims et al., 2010).

Moreover, plant biomass is renewable because it can be grown and harvested on a cyclical basis without depleting finite resources. Unlike fossil fuels, which are nonrenewable and formed over millions of years through geological processes, plant biomass can be replenished within a short timeframe through agricultural activities (Singh et al., 2021; Adeoye et al., 2023). By implementing sustainable land management practices and crop rotation techniques, farmers can maintain the productivity of biomass feedstocks while preserving soil health and biodiversity. Furthermore, the carbon cycle plays a vital role in the renewability of plant biomass. During growth, plants absorb CO_2 from the atmosphere and incorporate it into their tissues (Gayathri et al., 2021). When biomass is converted into biofuels and used for energy, the carbon stored in plants is released back into the atmosphere as CO_2. However, since this carbon was recently captured from the atmosphere, the net carbon emissions from biofuel combustion are considered neutral over the long term if the rate of biomass regrowth matches or exceeds the biomass consumption rate (Ahamer, 2022).

Additionally, plant biomass can be considered renewable when managed sustainably, balancing resource utilization with ecosystem conservation (Omer, 2008). Sustainable practices include minimizing inputs such as water, fertilizers, and pesticides, optimizing land use to maximize biomass productivity, and protecting natural habitats and biodiversity (Shah and Wu, 2019; Krishna et al., 2022; Verma et al., 2022i). By promoting sustainable biomass production, stakeholders can ensure the long-term viability of biofuel feedstocks while minimizing negative environmental impacts (Baudry et al., 2017). Moreover, advancements in agricultural technologies and biorefinery processes have further enhanced plant biomass renewability for biofuel production (Chen and Zhang, 2015). Improved crop genetics, precision farming techniques, and irrigation technologies have increased biomass yields and resource use efficiency, allowing for more sustainable biomass production. Additionally, innovations in biorefinery processes, such as enzyme optimization and integrated biorefinery systems, have improved biomass conversion efficiency into biofuels, reducing waste and enhancing resource utilization (De Bhowmick et al., 2018; Verma et al., 2022j, 2022k). Moreover, policies and regulations promoting renewable energy and sustainable agriculture play a critical role in ensuring the renewability of plant biomass (Saleem, 2022). Incentives such as renewable energy mandates, tax credits, and subsidies inspire investment in biofuel production and support the development of sustainable biomass supply chains. Additionally, environmental regulations and certification schemes promote responsible land management practices and biodiversity conservation, further enhancing the renewability of plant biomass resources (Song et al., 2022; Oliveira et al., 2022).

9.3 TYPES OF PLANT BIOMASS

The fundamental necessities for human beings encompass food, shelter, clothing, healthcare, environment, and energy, which are considered as secondary requisites. Fossil fuels are a primary energy source, yet their reserves are progressively diminishing, prompting researchers to explore alternative renewable energy sources (Banerjee et al., 2010). Indeed, numerous investigations highlight biomass as a significant reservoir from which renewable energy could be harnessed (Lund, 2007). Biomass properties encompass agro-industrial residues and agricultural, animal byproducts, woody/hard biomass, aquatic biomass, industrial waste, and domestic, nonedible plant parts, as well as residual commercial materials. This section discusses various reservoirs for biofuel production. Figure 9.1 shows the variety of plant parts used for the production of biofuel.

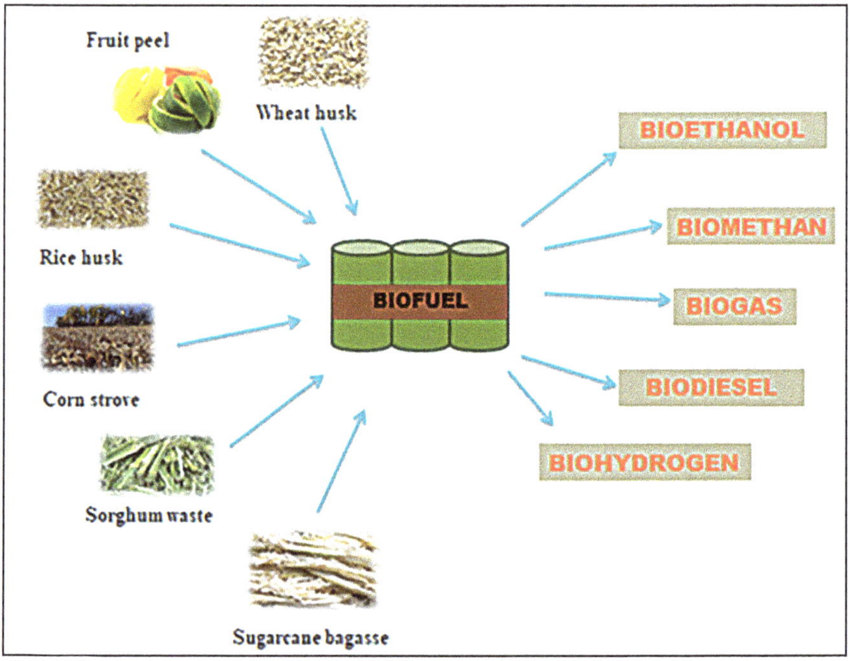

FIGURE 9.1 Plant biomass producing different types of biofuels.

9.3.1 BIOFUELS CROPS

Energy crops provide a substantial opportunity to meet the future energy demands of a consistently growing population (Verma et al., 2020c, 2020d; Singh et al., 2020). The selection of the appropriate biofuel crop plays a vital role in the ensuing biofuel transformation and energy yield (Jørgensen, 2011). To achieve the national biofuels target command by 2022, biofuel crops like *Miscanthus* and Switchgrass must be grown (Zhuang et al., 2013). Sweet sorghum, *Miscanthus*, and Switchgrass, known as C_4 crops, are beneficial to cultivate due to their ability to thrive on poor soil and produce larger amounts of biomass. Additionally, they are resilient to dry conditions, exhibit other favorable traits such as high photosynthetic efficiency, and capture CO_2 faster than C_3 crops (Koçar and Civaş, 2013).

Miscanthus, originating from Eastern Asia, is a highly productive biomass crop with low nutrient requirements. It is cold-tolerant and can be utilized for heat, electricity, and biofuel production (Stewart et al., 2009). With its efficient C_4 photosynthetic system, sweet sorghum is rich in sugars and ideal for bioethanol production (De Vrije et al., 2009; Verma et al., 2023d). It is adaptable to various conditions and can significantly contribute to reducing GHG emissions (Kumari et al., 2022, 2023a, 2023b). Switchgrass, native to North America, is versatile and productive, suitable for direct burning, conversion into energy-rich forms, or cofiring with coal for electricity generation (Wang et al., 2008). Other crops like barley, corn, oats, rye, potatoes, sugar beets, sugarcane, and sweet potatoes also hold potential for ethanol production or solid fuel generation, contributing to renewable energy sources (Sims et al., 2006).

Sugarcane stands out as the primary sugar crop for fuel ethanol worldwide (Verma et al., 2020e: Singh et al., 2023), favored for its high yields and sucrose-rich composition, particularly in Brazil, which leads global production. Brazil's long-standing policies promoting ethanol blending have entrenched sugarcane ethanol as a conventional industry despite market variations and fossil-fuel competition (Nogueira and Capaz, 2013; Hussain et al., 2024). Sugarcane ethanol ranks second globally after US corn ethanol, offering GHGs reduction, ethanol yield, and production cost advantages. Argentina also contributes to bioethanol production, with a significant portion derived from sugarcane.

Sugar beet crop, prevalent in temperate regions, notably Europe, represents another source of bioethanol production. While less energy-efficient than sugarcane, sugar beet integration into biorefinery processes, including crystallized sugar production for food, adds value. About 45% of European ethanol originates from sugar beet processing, primarily from sugar beet agro-industries' molasses rather than direct sugar beet utilization (Flach et al., 2019). Similar to sugarcane in botanical characteristics and sucrose accumulation, sweet sorghum has gained more attention for bioethanol production in temperate climates. While global statistics on sweet sorghum cultivation and bioethanol production are limited, commercialization efforts have been noted in Brazil, China, and India, with China specifically promoting ethanol production from cassava, sweet sorghum, and other nonfood grain feedstocks (Macke, 2017).

9.3.2 AGRICULTURAL WASTE

The global population of humans is on the rise every day, generating larger amounts of waste. As waste generation rises, the issue of disposal becomes more prominent (Tian et al., 2023). Instead of investing energy in waste disposal, using waste to produce energy would be more economically viable, as suggested by Kumari and Singh (2018). Research has shown that about 5–7% of biomass energy comes from agricultural waste, including beets, corn, fruits, and sugarcane (Deshmukh et al., 2008). This biomass comprises food-like portions of crops like orchard trimmings, stalks, rice straw, leaves, rice husk, and cobs of corn stover, along with food-based portions (Sims, 2004). The main components of agricultural waste are ash, cellulose (C6), hemicelluloses (C5), lignin, and protein. LCB undergoes several pretreatment methods, such as biological or combined pretreatment, as well as physicochemical. Physical methods like grinding, freezing, chipping, milling, and radiation are used to reduce the particle size and increase surface area (Kumari and Singh, 2018).

Chemical pretreatment involves treatment with acids like H_2SO_4 and HCl to enhance enzymatic hydrolysis and oxidizing agents like H_2O_2. Alkalis such as $Ca(OH)_2$, KOH, NH_3, and NaOH are also used for pretreatment (Sheikh et al., 2015). Other methods include ozone treatment, ionic liquid pretreatment, and treatment with organic solvents. Physicochemical methods like liquid hot water, ultrasonication, ammonia fiber explosion, CO_2 explosion, steam explosion, and wet oxidation are employed. Biological pretreatment involve microbes and enzymatic treatment, while combined methods include electron beam irradiation and alkali, alkali and photocatalysis, ionic liquid and alkali, dilute acid and biological, steam explosion and biological, dilute acid and steam explosion, dilute acid and microwave, organ solvent and ionic liquid and ultrasonic, enzyme hydrolysis, superfine grinding with steam, and biological, SO_2 and steam explosion, supercritical CO_2 and microwave-assisted acid and, alkali pretreatment, steam explosion. Figure 9.2 shows that these procedures aim to produce various biofuels using agricultural wastes (Kumari and Singh, 2018).

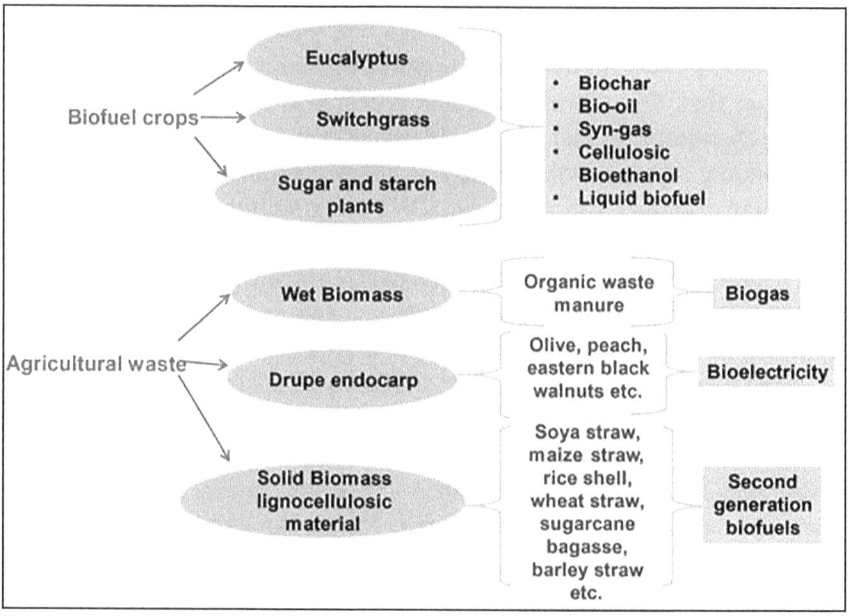

FIGURE 9.2 Resources of biomass and use in biofuel production.

Biohydrogen is a safe, clean, and nontoxic biofuel with immense potential for future energy needs (Show et al., 2011). Unlike traditional fuels, its combustion results in water without carbon emissions or pollutants. However, the predominant method of H_2 production involves physicochemical processes reliant on fossil fuels, contributing to pollution and GHGs (Yun et al., 2018). This underscores the urgency of exploring alternative, sustainable sources for production. Biological processes offer a promising source, being more biologically friendly and energy-efficient compared to physicochemical methods. These processes encompass dark fermentation, indirect and direct biophotolysis, and photo-fermentation (Yun et al., 2018). Among these, dark fermentation emerges as the most practical, given its ability to degrade organic waste effectively and achieve optimum yield. Dark fermentation relies on mixed bacterial cultures, including *Selenomonas* sp., *Megasphaera* sp., *Lactobacillus* sp., *Prevotella* sp., and Clostridium sp., *Enterobacter* sp. (Lopez-Hidalgo et al., 2018). Additionally, LCB wastes like rice straw, wheat straw hydrolysate, corn stover, bean husk, corn stalk, and vegetable waste are viable substrates for biohydrogen production (Lopez-Hidalgo et al., 2018). Pretreatment of LCB with NaOH or H_2SO_4 before heating has been identified as advantageous for enhancing biohydrogen production yields.

Ethanol, known for its diverse applications, including as an antifreeze, beverage, fuel, and solvent, is gaining prominence due to worries over climate change and global warming derived from renewable sources and is experiencing rapid growth globally as countries transition toward renewable energy sources (Sarkar et al., 2012). Agricultural waste presents a viable and environmentally friendly feedstock for bioethanol production, with residues such as corn stalk, citrus peel, corn stover, bagasse, rapeseed waste, corncob, cornhusk, straws, leaves, and sugarcane bark offering significant potential (Kumari and Singh, 2018).

Biomethanol emerges as a potent biofuel with enhancement in fuel cell-powered vehicles and as a hydrogen storage compound. Utilizing lignocellulosic wastes for methanol production offers economic and environmental advantages with agricultural residues like straw, banana peel, husk, rice bran, and herbal biomass being viable feedstocks (Arteaga-Perez et al., 2016). Although less explored, further research is warranted to fully utilize agricultural wastes for biomethanol production. Butanol, a superior biofuel with favorable fuel characteristics compared to ethanol, holds promise as a renewable transportation fuel (Green, 2011). Agricultural wastes such as corn stover, rice straw, corn cobs, wheat straw, sugarcane bagasse, cassava waste, barley straw, wheat bran, vegetable waste, and fruit residues serve as potential feedstocks for biobutanol production (Huzir et al., 2018).

9.3.3 LIGNOCELLULOSIC CROPS

Over the past few decades, significant research endeavors have been dedicated to advancing a novel approach to agricultural biomass production centered on lab crops. These crops are specifically cultivated as dedicated energy sources for generating solid biomass or advanced biofuels, focusing on solid biomass, liquid biofuels for power/energy, and heat and biorefineries. The rise of lignocellulosic species stems from several factors (Table 9.1). Unlike sugar or oil plant species, the diversity of lignocellulosic plants in nature is notably broader, offering the potential for their cultivation across diverse geographical locations. Moreover, lignocellulosic species exhibit greater resilience than conservative energy crops. Additionally, the energy expenditure involved in producing C6 or C5 in plants is roughly half that required for lipid production (Bentsen and Moller, 2017). The abundance of cellulose as the most dominant biopolymer on Earth has spurred the exploration of advanced methodologies for the production of biofuels (Carere et al., 2008).

TABLE 9.1 Summary of the Contents (wt%) of C6, C5, and Lignin in Common Plant Biomass Feedstocks

Plant Biomass	Type of Biomass	C6	C5	Lignin	References
Crop residue	Corn stover	31–46	16–36	8–23	Wang et al. (2020)
	Sugarcane bagasse	39–45.5	18–34	5–34	Hussain et al. (2024)
	Sorghum bagasse	28–45	14–36	15–26	Usmani et al. (2020)
	Rice husk	33–41.5	14–28.8	34	Irum et al. (2024)
	Wheat straw	44	20–45	19-34	Chandra et al. (2012)
	Tobacco residue	43–45	29–42	15–27	González-García (2018)
	Corn cob	40–53	24–36	13–20	Sharma et al. (2020)
	Corn straw	52	31	18	Demirbas (2001)
	Corn stalk	37	30	14	Hassan et al. (2016)
Hardwood	Poplar	48	15–22	22–30	Sharma et al. (2020)
	Willow	35–38	23	21	Sharma et al. (2020)
	Eucalyptus	35-52	9-31	22-40	Penín et al. (2020)
Grasses	Bamboo	38–46	17–19	26-40	Yang et al. (2020)
	Miscanthus	41–54	27	27	Kang et al. (2013)
	Switch-grass	40–45	21–31.6	13–21	Ussiri et al. (2015)
	Agave bagasse	39–48	13–24	11–15	Alemán-Nava et al. (2018)
	Agave leaf	47–80	16–31	5–12	Li et al. (2012)
Fruit waste	Walnut shell	41	21	19	González-García (2018)
	Cocoa shell	12.8	11	14	González-García (2018)
	Peanut shell	25	40	34	González-García (2018)
	Banana peel	53	10	12	González-García (2018)
	Olive leaves	16	10	18	Lama-Muñoz et al. (2020)
	Olive stone	31	18	33	González-García (2018)

TABLE 9.1 *(Continued)*

Plant Biomass	Type of Biomass	C6	C5	Lignin	References
Softwood	Softwood stems	45–51	24-36	36	Ussiri and Lal (2014)
	Spruce	25–48	11–22	32–36	Usmani et al. (2020)
	Pine	35–46	21–35	25–33	Usmani et al. (2020)
	Japanese larch	59	12.9	21	Park et al. (2008)
	Beech	52	32	20–23	Hassan et al. (2016)

In the late 1970s, the United States initiated a dedicated research and development program to identify energy crops suitable for solid biomass production. This effort led to the recognition of switchgrass as a prime candidate among poplar and perennial herbs and willow as exemplary crops for Short Rotation Coppice, sparking further exploration in research avenues (Kszos et al., 2000). Remarkable progress has been achieved in plant production, particularly in enhancing the agronomy of perennial grasses and short-rotation woody species, alongside advancements in lignocellulose conversion technologies (Barth et al., 2016; Aditiya et al., 2016). Present trends in lignocellulosic energy crops are geared toward assessing their suitability for cultivation in marginal lands (Mehmood et al., 2017).

Plant biomass is the total amount of living or recently dead organic material derived from plants, including leaves, stems, roots, and other plant parts. It is a significant renewable resource with diverse characteristics and availability depending on the type of plant and its geographical location (Chaudhary et al., 2023; Jagadesh et al., 2024). Understanding the different types of plant biomass is crucial for harnessing their potential sustainably (Norfarhana et al., 2023). First, agricultural biomass comprises crop residues, such as sugarcane bagasse, wheat straw, and corn stover, readily available from farming activities. These residues are abundant in agricultural regions, providing a consistent and reliable source of biomass feedstock. Energy crops like switchgrass and miscanthus are specifically grown for biomass production, offering high yields and energy content suitable for bioenergy applications (Smith et al., 2016).

Second, forest biomass encompasses woody materials derived from trees and shrubs, including logging residues, sawdust, and forest thinning. Forest biomass is characterized by its high energy density and versatile applications, ranging from heat and power generation to bio-based products like wood pellets and biochar. However, sustainable management practices are essential to maintain forest ecosystems and prevent the depletion of valuable habitats (Cherubini and Strømman, 2011). Third, aquatic biomass refers to plant matter derived from aquatic environments, such as seaweeds and freshwater macrophytes like water hyacinth and duckweed. These aquatic plants exhibit rapid growth rates and high biomass productivity, making them promising sources for biofuels, bioplastics, and wastewater treatment. However, harvesting, processing, and environmental challenges need to be addressed for sustainable utilization (Mata et al., 2010).

Moreover, urban biomass comprises green waste, landscaping trimmings, and municipal solid waste from urban areas. Recycling organic waste streams into bioenergy and compost can mitigate landfill emissions and contribute to circular economy initiatives. Innovative technologies like anaerobic digestion and thermal conversion can convert urban biomass into valuable products while reducing GHG emissions and landfill pressures. Also, algae biomass represents diverse photosynthetic microorganisms found in various aquatic environments. Microalgae, such as *Chlorella and Spirulina*, are rich in proteins, lipids, and carbohydrates, making them suitable for food, feed, and biofuel production. Macroalgae offer bioactive compounds, biopolymers, and biofuels with minimal land and freshwater requirements (Mata et al., 2010). Furthermore, waste biomass encompasses organic residues from industrial processes, such as agricultural residues, food processing byproducts, and lignocellulosic waste streams. These waste materials can be valorized through biorefinery approaches, converting them into biofuels, biochemicals, and bioproducts. Integrated biorefinery concepts maximize resource efficiency and economic viability while reducing environmental impacts associated with waste disposal (Ragauskas et al., 2014).

9.4 GLOBAL AVAILABILITY OF PLANT BIOMASS

The potential for production and its capacity to generate renewable and sustainable bioenergy differ significantly among nations due to resource availability, technology, economy, geography, and biodiversity. Plans suggest that by 2050, biomass could contribute 3000 terawatt-hours (TWh) of power and mitigate 1.4 billion tons of CO_2 comparable to emissions annually.

However, it is noteworthy that for each terawatt-hour of energy produced, approximately 472.89 kilotons of CO_2 are generated (Gielen et al., 2019). Worldwide, production of renewable energy reached 19% by 2014, with a 14.5% increase in 2018, slightly below the average growth rate of the past decade, with the Asia-Pacific region leading with 39–41% of worldwide renewable energy generation, followed by North America and Europe. Notably, China is a significant contributor, producing 634.2 TWh of renewable energy, followed by the US and Germany. Over the past decade, there has been a remarkable shift in renewable energy consumption patterns worldwide, particularly in China, where consumption has increased 20-fold since 2008. However, significant albeit lower increases have also been observed in the US, Germany, Canada, and India. The growing demand for biofuels as an alternative to conventional fuels has led to a global increase in biofuel production.

The US, Brazil, and Germany are among the top contributors to biofuel production. However, the expansion of biofuel production may affect agricultural commodity prices, as biofuel production relies on feedstocks such as soybean, rapeseed, wheat, sugarcane, corn, and sunflower, which are also used in the nutrition industry. Some concerns are that increased utilization of these feedstocks for biofuel production could lead to higher feedstock and nutrition prices due to improved demand and condensed supply in the global market (Zhang et al., 2010). The rise in food and fuel commodity prices in the early 2000s, peaking in 2008, was partly attributed to the biofuel industry's dramatic expansion during that period. Ethanol and biodiesel production surged, leading to increased prices for major feedstocks such as corn and soybeans (Mueller et al., 2011). However, other factors such as global economic growth, population growth, energy price inflation, exchange rate fluctuations, adverse weather, trade policies, government incentives, and feedstock production are all factors that impact various aspects of the economy and also contribute to the rise in feedstock prices. The food price index before and after the biofuel boom of the 2000s, using empirical information, suggests that while biofuel demand and production played a significant role, several other factors were also important (Koh and Ghazoul, 2008).

Biofuels have been heralded as a solution to the energy calamity and a step toward sustainability. Still, they have faced criticism, particularly regarding their impact on global nutrition prices, the change of nonagricultural lands into agricultural ones, and altering nutrition crops for biofuel production (Fischer et al., 2009). Critics argue that biofuel production and crop diversion exacerbate hunger, disproportionately affecting the 2 billion people worldwide who spend a significant portion of their income on food (Von Braun, 2008). While biofuel production is vital for addressing energy challenges, the potential burden on the food supply must be considered. Strategies must be implemented to strike a balance between bioenergy and food availability. One approach involves reducing the use of food crops for biofuel and exploring techniques to extract translatable sugar from nongrain parts of crops, such as corn stalks and husks, for ethanol production.

Additionally, utilizing nonfood crops like switchgrass and miscanthus for biofuel can alleviate pressure on food crops. However, intensive monocropping and agricultural practices may impact biodiversity and environmental sustainability. Moreover, modern techniques like genetics and microbiota can enhance biomass production per capita, but further research is needed to implement these tools effectively. The European Parliament has taken steps to restrict the use of first-generation biomass sources, favoring agricultural residues and promoting advanced biofuels like straw. These legislative efforts encourage sustainable biofuel production while mitigating adverse effects on the food supply and the environment (Dudley, 2018).

9.5 CURRENT UTILIZATION OF PLANT BIOMASS FOR THE PRODUCTION OF BIOFUEL

The energy markets of the world have long depended heavily on oil, natural gas, and coal, collectively referred to as fossil fuels. Emissions from the combustion of fossil fuels consist of various harmful air impurities and CO_2, thereby posing a significant risk to human health and well-being while also playing a substantial role in global warming and environmental decline. The onset of coal mining in the 18th century marked coal as the primary fuel of the Industrial Revolution; before this, energy needs were entirely fulfilled by charcoal and wood, both of which fall under biomass (Abbasi and Abbasi, 2010). Biomass stands out as the sole recognized organic, renewable carbon source of a sizeable magnitude that can potentially substitute fossil fuels. Biomass energy, categorized as a form of renewable energy, can be harnessed directly or, following change into other systems, consequently releasing the CO_2 recently absorbed from the atmosphere during its growth. This characteristic renders biomass "carbon neutral," sparking global interest in utilizing biomass as an alternative to energy sources derived from fossil fuels, particularly in producing liquid fuels and chemicals like biodiesel, methanol, and ethanol (Abbasi and Abbasi, 2010). Presently, a diverse array of sectors derives benefits from biomass, whether as food/feed for generating heat, electricity, and fuels or as materials and properties in industrial sectors, primarily within chemical industries and the wood and paper.

9.5.1 BIOENERGY

The current state of affairs within the energy area is seen as economically, ecologically, and communally unmanageable. Over the past decades, global GHG emissions have doubled since 1970, and the escalating demand for oil has raised worries regarding the stability of supplies, indicative of a pressing need for action. One likely solution to address these challenges is bioenergy. It stands out as the oldest and largest renewable energy source, primarily resulting from waste materials such as lab residues from forestry, municipal solid wastes, and agriculture. Additionally, bioenergy encompasses cultivated crops, virgin lab, and algae as raw materials for the production of energy (Hanssen et al., 2020). In contemporary times, bioenergy has become a versatile source capable of providing heat, power, and transportation fuels. It currently contributes around 10% to the global primary energy supply, with this proportion steadily growing by approximately 2.6% annually worldwide (Edrisi and Abhilash, 2016). The efficacy of bioenergy production largely depends on the kind of feedstock utilized. For instance, /hard/ woody biomass, characterized by higher lignin content (24–31.0%) and greater density (360–480 kg/m^3) compared to agricultural biomass (107–129 kg/m^3), exhibits enhanced resistance to microbial and enzymatic degradation, thereby influencing the amount of bioenergy released from bioresources (Zou et al., 2018).

In bioenergy production, particularly biofuel, a higher concentration of cellulose within biomass is essential for generating increased volatiles (Cao et al., 2019), which are crucial for facilitating ignition and oxidation during the combustion process (Wijayanta et al., 2015). Conversely, a lower content of C5 and lignin is advantageous in feedstock selection. This is because lignin, functioning as a binding agent, hinders the separation of C6 from C5 and prevents the formation of unstructured C6 from crystalline C6. Additionally, lignin obstructs the access of cellulase enzymes to C6, thereby impeding the biological conversion of biomass (Yoo et al., 2020). In less economically well-developed countries, bioenergy is primarily used for heating and cooking (Lynd et al., 2015). However, a major issue with this type of bioenergy is that the biomass used is often collected weakly, leading to deforestation. Additionally, burning biomass for heat and cooking is not very efficient, typically between 10% and 20%, and it also leads to significant indoor air pollution (Fullerton et al., 2008). In contrast, large-scale combustion plants are highly efficient in producing heat and can compete with fossil fuels. These plants use various biomass sources, including timber, wood chips, or wood pellets, to generate heat, with some systems reaching capacities of up to 10 MW, mainly for industrial use.

Biomass can also be converted into electricity through different procedures, but most bioenergy plants employ fiery biomass through combustion techniques to produce higher-pressure steam that powers turbine generators. The amount of electricity a steam turbine generates depends on its capacity, with typical electric revenue rates falling between 14% and 36%. The quantity of carbon dioxide emitted during bioenergy/electricity generation depends on the efficiency of the changing technology and the amount of fossil fuel utilized in biomass production. Effective agricultural and forestry management practices play a crucial role in this regard. A study conducted in the US assessed the production costs and carbon reduction per unit of electricity generated for various feedstocks, revealing that bioelectricity production has significantly lower GHG emissions than fossil fuels. Cotton stalks were the most GHG gas-intensive, emitting 325 kg of CO_2 equivalent per megawatt-hour (MWh), followed by corn stoves at 235 kgs of CO_2 equivalent per MWh. In contrast, pine chips emitted the least amount of GHGs, at 134 kg of CO_2 equivalent per MWh, for a similar electricity output (Masum et al., 2020). The global electricity demand is projected to increase from approximately 20,000 TWh in 2009 to 41,000 TWh by 2050, driven by the expansion of sustainable energy economies. Renewable power is projected to rise from 20% in 2009 to nearly 55–62% in 2050 (Eisentraut and Brown, 2012).

9.5.2 BIOFUEL

There is a growing energy demand and associated setups to support social, societal, and economic progress, as well as health and welfare. Embracing renewable sources to address climate change challenges presents a viable approach to meeting the energy needs of future generations (Owusu and Asumadu-Sarkodie, 2016). Biofuels, derivatives from biotic sources, including oil crops, starch, grains, sugar crops, cellulosic materials such as grasses and trees, as well as biological waste, are recognized as a category of fuels (De Fraiture et al., 2008). Biofuels are classified into three primary categories: (solid, liquid, and gaseous) phases. Solid biofuels, including (1) charcoal, wood chips, firewood, and wood pellets, have been utilized used since early human civilization for heating and cooking. While historically prevalent, these materials are now predominantly consumed in developing nations. (2) Biogas, conversely, is a type of biofuel generated through biomass for anaerobic digestion. It represents a renewable gaseous (GHGs) fuel with the possibility to serve as a substitute for natural gas (Nayono, 2010).

Given the availability of domestic wastes and agro-waste cultural feedstock, it is estimated that over 999–1000 billion normal cubic meters (B Nm^3) of biomethane can be generated annually for heat and electricity production (World Bioenergy Association, 2013). Switchgrass, a tall warm-season constant grass, shows the potential to yield more biomethane compared to other crops through anaerobic digestion (Zhang et al., 2015), an environmentally friendly and cost-effective process wherein microbes change raw biomass into CO_2 and CH_4 (Zou et al., 2018). However, the complete breakdown of lignin solely by microbes does not result in optimal biogas production, microbial, mechanical, or thermal pretreatments, and necessitating chemicals (Brodowska et al., 2018). The methane production from most grasses is increased when the harvest is delayed, although the impact of collecting time in moist and cool climates, such as Eastern Canada, varies. Switchgrass harvested in mid-summer or early fall yields the highest CH_4 yield, with approximately 20–26% more methane found using a two-cut policy than a single late-seasonal harvest (Masse et al., 2010).

Syn-gas, another form of gaseous biofuel, is produced through pyrolysis or gasification of feedstock. Following purification, syn-gas serves as a raw material for producing transportation fuel, methanol, ethanol, CH_4, C_2H_6O, and other products (McDowall et al., 2012). (3) Biodiesel, used as a diesel substitute, and bioethanol, serving as a replacement for petrol/Diesel, represent the two most commonly employed liquid biofuels (Hassan and Kalam, 2013). Projections from the latest US Energy Information Administration indicate a 38% increase in the global consumption of petroleum and other liquid fuels from 2014 to 2040, with the transportation and industrial sectors projected to account for 90–92% of worldwide liquid gasoline consumption by 2038–2040 (Sieminski, 2014). This escalating demand underscores the need to explore various biomass feedstocks for fuel production to replace fossil fuels. While the combustion of biofuels in machines results in CO_2 emissions, the emissions produced are balanced out through the photosynthesis carried out by the plants that serve as the source for the biofuels.

However, GHG emissions can remain substantial due to the use of fossil fuels for agricultural machinery and transportation. Moreover, biomass production must also apply pesticides, fertilizers, and herbicides. Hence, conducting a life cycle analysis (LCA) that evaluates land use changes, raw materials, biofuel production, and final utilization for each biofuel category is imperative to assess their efficiency compared to fossil fuels (Niculescu et al., 2019). For instance, an LCA of soy biodiesel in the US revealed an 80% reduction in fossil energy consumption and a 60–75% decline in total GHG emissions compared to fossil diesel. The agricultural process significantly influences GHG emissions of biofuels; for example, canola biodiesel emits considerably more GHG during biomass production compared to soy biodiesel, with emissions of 18.5 g CO_2 e/MJ for canola versus 9.2 g CO_2 e/MJ for soybeans. This discrepancy is primarily attributed to nitrogen fertilizer application, as canola lacks the ability to fix atmospheric nitrogen, unlike soybeans (Chen et al., 2018). While first- and second-generation biofuels are widely adopted by industry, third-generation biofuels are unmovingly emerging but hold promising prospects for the future (Sharma et al., 2019).

Biofuels that extend beyond the first generation are commonly grouped under the term "advanced biofuels." The initial biofuel generation, like ethanol, is predominant and is derived from food crops such as soybean oil, corn, beet, sugarcane, and wheat in Europe, Brazil, and the US, and oilseed rape in Germany and France, as well as palm oil in Asia and Africa. Most commercially produced biofuels originate from these sources, while converting lignocellulosic materials, such as C6, C5, and lignin, yields second generation or advanced biofuels (Hayes, 2009). Third-generation biofuels rely on marine plants and algae (seaweeds). Algae-produced biofuels show significant potential as a substitute for fossil fuels since they can thrive in wastewater, saltwater, and sewage growing in regions unsuitable for crops used in second-generation biofuels (Demirbas and Demirbas, 2010). However, the working processes are expensive and require further study and development to ensure the efficiency of the third generation of clean energy. These feedstocks can grow three or four times earlier than earthly plants and possess comparable C6 content to agricultural or forest remains, with lower lignin and C5 contents (Panahi et al., 2019).

Untreated macroalgae demonstrate a C6 conversion rate of approximately 76%, whereas treating the feedstock with CH_3CO_3H followed by ionic liquid results in near-complete cellulose conversion. If biomass waste-to-biofuel conversion methods can operate without excessive energy requirements, biofuel production may offer significant advantages over crude oil and coal. Numerous technologies and procedures are available for biomass energy conversion, with thermochemical and biochemical transformations being two prevalent approaches. Additionally, transesterification is utilized to produce 1st generation biodiesel (Lee et al., 2019). The first technique utilizes heat to decompose biomass (Won and Maravelias, 2017), whereas the latter involves using microorganisms or enzymes to convert biomass into biofuels (Hossain et al., 2017). In summary, the typical biofuel production process consists of two main steps.

1. Reducing the O_2 content of the feedstocks to improve the density of energy.
2. Formation of C–C bonds between intermediates sourced from biomass to elevate the molecular weight of the end product (Alonso et al., 2010).

The global production of biofuels has experienced a significant surge, rising from 9.2 million metric tons (Mt) of oil equivalent in 2000 to 96 Mt of oil equivalent in 2018 (Wang et al., 2019). This substantial expansion has been propelled by policies promoting the adoption and manufacture of biofuels, driven by the belief that they could enhance energy safety and mitigate GHG emissions within pertinent areas.

9.5.3 BIOFUEL FROM CANADIAN PERSPECTIVE

Agricultural land, Canadian forests, and municipal wastes collectively provide 60%, 40%, and 4%, respectively, of the projected net energy content (6 EJ yr^{-1})) from biomass (Littlejohns et al., 2018). It is important to highlight that the residues from harvesting municipal waste, agricultural fields, and forests remain mainly untapped. Consequently, Canada's biofuel sector faces no shortage of feedstock. It is estimated that nearly 176,000 million liters (ML) of biofuels can be sustainably produced in Canada annually from all biomass sources. Even if only 8–10% of the net biomass from forest harvest residue, agriculture, and municipal waste is utilized for biofuel production, approximately 18 ML of biofuels could be generated, whereas the present production stands at around 2.2 ML (Littlejohns et al., 2018). Leveraging biomass resources could significantly reduce GHG emissions. Government data reveal a notable increase in ethanol utilization from approximately 1.7 billion liters in 2010 to 3100 ML in 2017, and biodiesel consumption rose from around 125 ML in 2010 to 380 ML in 2017, resulting in a reduction of 34.3 million tons of CO_2 emissions between 2010 and 2017.

9.5.4 BIOPRODUCTS

In addition to biofuels and bioenergy, biomass is utilized to produce specific chemicals derived from biomass, known as bioproducts. Approximately, 50 million tons of polymer materials and bio-based materials are manufactured globally, signaling the potential for economic sustainability (Davis et al., 2013) and compact reliance on fossil fuels, leading to significant reductions in GHGs emissions. Despite this, gas and oil continue to be extensively utilized for polymer and organic chemical production, which contradicts the growing movement toward minimizing environmental degradation, pollution, and resource depletion. Various biochemicals, from long-established products like citric acid to newly introduced materials such as $C_3H_8O_2$, and those showing important growth potential, are derived from biomass (de Jong et al., 2012).

Most of these chemicals are derived from specific structure blocks, all of which could potentially be substituted with their counterparts from biomass, known as bio-derived building blocks, including $C_5H_{12}O_5$, $C_6H_{14}O_6$, $C_6H_{14}N_2O_2$, HCOOH, $C_3H_6O_3$, glycol, C_3H_6, $C_4H_6O_4$, $C_5H_4O_2$, and $C_2H_6O_2$. Following the principle of cascade utilization in the chemical industry, a single chemical precursor can be transformed into multiple derivatives, thereby connecting chemicals along the value chain (Kannegiesser, 2008).

Consequently, a diverse range of value-added materials or chemicals can be produced from biomass, either as final products such as pest control agents, paints and coatings, adhesives, lubricants, pigments and inks, or as intermediates for further processing, such as the manufacturing of bioplastics (Karan et al., 2019), biocomposites (Haque et al., 2017), and ingredients for cosmetics, pharmaceuticals, medical devices, fragrances, and foods.

9.5.5 BIOCHAR

Biomass with a defined chemical structure can undergo various change procedures, like pyrolysis, to produce multifunctional products like biochar (Jahirul et al., 2012). The International Biochar Initiative defines biochar as a carbonaceous substance with a higher carbon content that is resistant to decay, derived from the thermochemical breakdown of bio-based feedstock through gasification/pyrolysis in the absence of added oxygen (Arbestain et al., 2015). Biochar finds wide-ranging applications in agriculture, industry, and the natural atmosphere. It can serve as a soil amendment, enhancing soil fertility and crop yields, as well as a water treatment agent (Malinska, 2015). Biochar has been recognized as a potentially effective method for reducing CO_2 concentration, as it sequesters soil carbon and mitigates the emissions of other GHGs like N_2O and CH_4 (Rondon et al., 2006). This is attributed to its ability to increase the adsorption and retention of ammonium ions in soils, thereby reducing the availability of N_2 for denitrification.

Furthermore, biochar can immobilize contaminants in soil and sewage by providing binding sites for impurities due to its high adsorption capacity and resistance to microbiological degradation (Mohan et al., 2014). Marketed to initiate a "doubly green revolution," biochar aims to address GHG emissions from soil organic matter while maintaining food security (Awasthi et al., 2017). Additionally, it positively impacts soil physical properties such as texture, structure, porosity, and density, as well as chemical characteristics including cation exchange capacity, pH, soil carbon content, nutrient cycling, and water retention (Głąb et al., 2016; Laghari et al., 2016). Moreover, the application of biochar elicits various effects on soil biota by serving as a carbon substrate (Smith et al., 2010), releasing or absorbing substances that either benefit (Kasozi et al., 2010) or impede microorganisms (Spokas et al., 2010), or expanding niche spaces for microorganisms by providing habitat. Consequently, biochar holds promise as a valuable tool in fostering sustainable agriculture, especially in tropical regions where soil organic matter is depleted due to rapid degradation caused by climatic conditions (Jeffery et al., 2017). For example, a meta-analysis of data from 15 to 16 fields and pots revealed yield rises of up to 25–30% when biochar amendments raised soil pH by 2.0 in acidic, medium-, and coarse-textured soils.

Similarly, biochar soil amendments enhanced corn yield by improving soil water retention, modifying root system development, and nutrient availability (Major et al., 2010). Furthermore, beyond tropical regions, there are indications that biochar could also offer benefits in temperate areas. Its use in soil enhances nitrogen (N) availability for plants by increasing N retention, thereby reducing the discharge of plant-available ammonium (NH_4^+) and nitrate (NO_3^-), which is particularly advantageous for non-N2-fixing crops. Additionally, biochar alters the nitrogen cycle through its impact on soil microbes. For example, in China, rice yields were observed to be 10–20% higher after biochar application during the first and second cropping cycles, attributed to increased soil, total nitrogen, pH, and soil organic carbon, along with enhancements in soil bulk density (Zimmerman et al., 2011).

Similarly, a study in southern Quebec evaluated the responses of switchgrass, soybean, and corn to pine wood/hard biochar applied at 5, 10, 15, and 20 Mg ha^{-1} in loamy sand and sandy clay loam soils. Applying 20 Mg biochar ha^{-1} on loamy sand resulted in 14.2% improvement in corn yields associated with the control, while no significant effect was observed on sandy clay loam soil. Biochar did not impact nutrient availability or switchgrass yield for soybean plots on either soil type. Consistent with these findings, the effects of using biochar depend on factors such as soil texture and the type of crop being cultivated (Backer et al., 2016). Beyond its agricultural advantages, biochar can also serve as an auxiliary material in NH_4 fermented and composting processes (Malińska et al., 2014). Furthermore, biochar has been explored for various applications, including its potential use as a substrate in hydrogen production, as a filtering agent in gasification/pyrolysis processes, and even as a fuel when pelletized.

9.6 BIOMASS CONVERSION TECHNOLOGIES

In recent years, numerous researchers have explored various methods for converting biomass into valuable forms of energy. These methods encompass thermochemical and biochemical processes, chosen based on property size, biomass type, energy requirements, commercial considerations, environmental concerns, and project-specific requirements (Tripathi et al., 2023). Biomass conversion typically yields three primary products: transportation fuels, chemical feedstock, and heat generation (Saxena et al., 2009). While thermochemical and biochemical processes have been extensively studied, there is limited research on physical processes and hybrid systems. Physical conversion methods, often applied in rural settings for domestic uses like heating and cooking, have been less documented, while hybrid systems remain in the research phase (Adams et al., 2018).

In this chapter, we offer an overview of various technologies used for biomass conversion applied for, encompassing thermochemical, bioenergy production, physical, biological, biochemical, and combined systems. The content includes the production of diverse bioenergies such as fuel, bio-oil, and biodiesel, along with insights into the operational processes. Additionally, we explore the potential environmental benefits, challenges, and impacts associated with the technologies mentioned above. Furthermore, we examine the economic viability and the prospects of biomass feedstock-driven bioenergy.

9.6.1 THERMOCHEMICAL CONVERSION

It involves converting biomass into efficient fuel or syngas rich in H_2, often utilized for chemical production via Fischer-Tropsch synthesis (Yohanes, 2021). This process encompasses several key phases, including pyrolysis, drying, heterogeneous char formation, and homogeneous reforming reactions. Commercially, thermochemical processes offer enhanced product selectivity, flexibility, quicker conversion rates, efficient processes, and diverse byproduct markets, rendering them more attractive (Muh et al., 2021). However, these technologies typically necessitate expensive resources in terms of process conditions and equipment, involving high temperatures and pressures. One of the primary challenges associated with thermochemical processes is the substantial initial investment required for infrastructure setup, which contrasts with the relatively low cost of biomass feedstock (Chan et al., 2019). Therefore, achieving cost-effectiveness necessitates further research and development efforts within the industry.

Current thermochemical biomass conversion methods predominantly include hydrothermal liquefaction (HTL), pyrolysis, gasification, torrefaction, incineration, and combustion.

9.6.1.1 HYDROTHERMAL LIQUEFACTION

HTL is a method that converts biomass into biochar (hydrothermal) within a closed system, using water as primary medium and specific pressures and temperatures (Zhuang et al., 2022). This process is effective in increasing the energy density of biomass. HTL involves complex reactions and changes in subcritical water. It starts with the hydrolysis of biopolymers into water-soluble peptides, followed by the breakdown of hydrogen linkages within and between molecules into simpler monomers such as glucose, as well as other compounds like furfural compounds, acetic acid, and acetaldehyde (Lachos-Perez et al., 2022). Compared to other thermochemical techniques, HTL offers several advantages. It eliminates the need for prior thermal drying of the biomass, making it more cost-efficient for wet biomass. It employs pressurized water as a reactant as well as reaction medium, making it environmentally friendly and versatile (Shahbeik et al., 2024). The bio-oil produced through HTL has higher yield and quality and lower oxygen and water content. However, the high-pressure requirements of this process may lead to expensive equipment for large-scale industrial operations (Sharma et al., 2021b).

9.6.1.2 PYROLYSIS

It is a thermal decomposition process of biomass at temperatures ranging from 500 to 1000 K in the absence of oxygen, producing liquid, gas, and solid residue (Dhyani and Bhaskar, 2019; Arbestain et al., 2015). The solid residue, commonly referred to as biochar, is a carbon-rich material having a porous structure. During pyrolysis, a diverse mixture of products is formed, including liquid bio-oil, solid char, and gases such as hydrogen (H_2), carbon monoxide (CO), carbon dioxide (CO_2), and methane (CH_4). Pyrolysis oils contain aromatic compounds, solid particles, and oxygenated aliphatics and typically have a high concentration of water between 15 and 30 wt%, along with various nitrogen and oxygenate compounds, necessitating further treatment before being used as a fuel for transportation (Abou Rjeily et al., 2021; Tripathi et al., 2024).

Several factors, including temperature, retention time, and catalysts, influence the pyrolysis process. It involves the thermal breakdown of lignocellulosic feedstock at approximately 773 K, often in a fluidized bed reactor with rapid cooling to produce bio-oil. Optimal production of bio-oil typically requires residence times of 0.5–3 s and thermal treatment temperatures between 400 and 600°C, which can be tailored to enhance the yield of hydrocarbon liquid fuels (Kumar et al., 2019). Among the different pyrolysis types, fast pyrolysis yields the highest bio-oil percentage, typically ranging from 50 to 75%, while approximately 10–20% is converted into char. Fast pyrolysis is characterized by the rapid cooling of pyrolysis vapors, high heating, high heat transfer rates, and controlled reaction temperatures (Chandran et al., 2020).

9.6.1.3 GASIFICATION

It represents an indirect technique of combustion applied to municipal plastic waste, which is converted into synthetic fuel or gases through partial oxidation in the presence of oxidants (Ciuta et al., 2017). This process involves transforming a mixture of solid or liquid organic materials into both solid and gaseous phases. The resulting gas phase possesses a high heating capacity and can be utilized for biofuel production or power generation (Solarte-Toro et al., 2018). The gas product mixture obtained from gasification typically consists of carbon dioxide (CO_2), nitrogen (N_2), methane (CH_4), hydrogen (H_2), and carbon monoxide (CO) (Sebothoma, 2021). Key factors influencing gasification include the presence of carbon dioxide, oxygen, air, and steam, with steam and air being commonly employed. However, each of these factors has its own set of advantages and limitations.

Biomass gasification can be categorized into low-temperature gasification (LTG) and high-temperature gasification based on operating temperature. LTG is particularly suited for obtaining gas and hydrogen from biomass with low calorific value. This method offers significant advantages such as simplicity, operational efficiency, and avoidance of ash-related issues (Mishra and Upadhyay, 2021). Various technologies are available for gasification, including fixed bed gasifiers, entrained flow gasifiers, and fluidized bed gasifiers. Recent advancements in technologies aiming at gasification aim to reduce tar content, increase the concentration of hydrogen yield in the initial producer gas, and enhance the energy efficiency of biomass gasification (Shamsi et al., 2022). Biomass gasification is a versatile technology that can be utilized for electricity generation or as a feedstock for chemical processes. However, it does have limitations that impact its flexibility and efficiency. Successful operation of a gasification installation requires an adequate and steady biomass supply, as well as careful consideration of the type of biomass used (González and Sandoval, 2020).

9.6.1.4 TORREFACTION

It is also known as roasting or high-temperature drying. It is a thermal method of treatment where biomass is heated within a temperature range of 200–300°C for a brief period under inert conditions (Chen et al., 2021). Like mild pyrolysis, this process involves several stages, including warming, pre and postdrying, transitional heating, roasting, and cooling of biomass. Compared to pyrolysis, torrefaction operates at lower temperatures and shorter durations, making it more suitable for enhancing the quality of solid char (Abdulyekeen et al., 2021).

During torrefaction, biomass properties are altered to yield a higher characteristic fuel suitable for applications in combustion and gasification. Laboratory-scale torrefaction is typically conducted using thermogravimetric analysis, small-scale tube reactors, or ovens. Torrefied biomass offers numerous advantages, including hydrophobic characteristics similar to coal, improved grindability, and ease of crushing. The process yields various products such as condensable gases like furfural, water vapor, phenol, methanol, lactic acid, acetic acid, oxygenates, and formic acid. Noncondensable gases such as carbon monoxide, carbon dioxide, and small traces of methane and hydrogen are also produced, with the combustion heat increasing with higher torrefaction temperatures. Despite the significant benefits of torrefaction technology, challenges remain in terms of economic and technical viability for full commercialization. Additionally, further scientific investigation is needed to better understand the impact of reaction conditions on the process and its products (Sarker et al., 2021).

9.6.1.5 COMBUSTION

Biomass combustion technology represents one of the oldest and most widely utilized thermochemical conversion processes. It is primarily employed to generate electric and heat power. The combustion process involves three main steps: drying, pyrolysis, and reduction, followed by the combustion of unstable gases and solid char (Shahbaz et al., 2021). Over 70% of the total heat generated comes from the combustion of unstable gases. Effective biomass combustion requires moisture levels in the biomass to be below 50%.

Biomass combustion is highly proficient in heat production, rendering it economically feasible. However, the process is complex, involving continuous solid–solid and solid–gas reactions. Successful combustion necessitates the presence of heat, fuel, and air. If any requirements are absent, combustion ceases. Conversely, when three elements are present in an appropriate amount, combustion becomes self-sustaining because of the additional heat coming out of fuel, which further fuels the burning process (Fiore et al., 2020).

9.6.1.6 INCINERATION

It involves aerobic oxidative combustion conducted at high temperatures, typically 900–1100°C. The primary objectives of incineration are energy recovery and volume reduction. The end products of the incineration process include solids, combustion gas, and ash, which can be utilized as heat sources. Key phases in this technique include evaporation and degassing, pyrolysis, and gasification (Zakaria et al., 2021). Incineration generates a plethora of harmful gases, GHGs, and chemicals, leading to water and air pollution. Therefore, the establishment of incineration facilities in any given area requires effective emission-cleaning technologies and approval from the local community (Karmakar et al., 2023; Irum et al., 2024).

9.6.2 BIOCHEMICAL

Biochemical biomass conversion technologies involve utilizing biological pretreatments to convert biomass into various products and intermediates by employing different microorganisms or enzymes (Verma et al., 2024a, 2024b). This process serves as a platform for obtaining fuels and chemicals, including biogas, hydrogen, ethanol, butanol, acetone, and diverse organic acids. These technologies primarily aim to produce products capable of replacing those derived from petroleum and grains. Compared to other conversion methods, biomass biochemical conversion technologies are characterized by their cleanliness, purity, and efficiency (Muh et al., 2021).

9.6.2.1 TRANSESTERIFICATION

It is a series of reversible reactions where lipids or triglycerides react with alcohol, typically methanol, to produce diglycerides, monoglycerides, and glycerol (Farooqui et al., 2021; Salvi, 2022). The efficiency of the transesterification process is influenced by various factors, including temperature, alcohol-to-oil ratio, moisture content, stirring speed, cell wall structure, reaction duration, and the type of catalyst employed. Different catalysts, including acidic or alkaline catalysts (heterogeneous and homogeneous), ionic liquids, carbon-based catalysts, and enzymes, can accelerate this process. Catalysis of homogeneous alkaline, in particular, is widely used for biodiesel production due to its ability to expedite the reaction at low temperatures and atmospheric pressure, resulting in rapid and significant conversion yields (Pan et al., 2022). Sodium hydroxide (NaOH) and potassium hydroxide (KOH) are commonly used alkaline catalysts in this method. However, the applicability of alkaline catalysts can lead to the formation of soap and free fatty acids in oils, which is problematic for microalgal biodiesel production due to the high concentration of free fatty acids in microalgal oils. To address this issue, acid catalysts are typically employed when the free fatty acid concentration exceeds 1%. Currently, commercial methods rely on triglyceride transesterification using homogeneous alkaline catalysts and methanol (Faruque et al., 2020).

9.6.3 PHYSICAL CONVERSION

9.6.3.1 BRIQUETTING

It is a process that transforms forestry and agricultural waste into briquettes of biomass or bio-coal, which can be used as fuel in traditional burners. This technology holds significant potential for environmental and socio-economic benefits, particularly in developing countries where traditional biomass is the primary energy source (Kuthe et al., 2021). Biomass briquettes not only contribute to the economic progress of developing countries but also aid in environmental preservation. Briquettes find various applications in both household and small industrial settings, including heating water, facilitating productive operations such as drying fruits, tobacco, tea, and heating clay and ceramics products like pottery, bricks, and cook stoves, serving as fuel for operating boilers for steam generation and electricity-generating gasifiers. One of the key advantages of biomass briquettes is their ability to be transported over long distances with reliable energy storage options, ensuring proper size and density, and being cost-effective (Obi et al., 2022).

In terms of production, briquettes can be made from various organic waste materials. However, implementing this technology can be costly due to the high pressure required by briquetting presses, excessive energy consumption, and the need for skilled personnel. The choice of briquette binder significantly impacts the briquetting process, with various binders such as sewage sludge, cow dung, microalgae, molasses, starch, waste-paper pulp, and okra stem gum having been tested. Briquettes' burning and quality efficiency are influenced by both the binders used and the type of agricultural biomass selected (Sanchez et al., 2022).

9.6.3.2 DISTILLATION

It begins with the oilseeds being crushed to extract oil, followed by steam distillation of the oil mixture and vaporization of essential oils (Kaur et al., 2022). The resulting vaporized oils are then collected, condensed, or cooled to obtain bio-oil. However, one limitation of the distillation method is its high energy requirement. This process consumes substantial energy to heat the liquid, convert it into vapor, and then condense it into liquid at the condenser (Dirker et al., 2019).

9.6.3.3 EXTRACTION

Extraction is a mechanical process to extract oil from various biomass crops such as cotton and groundnuts. This method yields oil and produces excess solid residue or "cake," which can be used as livestock feed (Sarkar et al., 2021). Effective oil extraction requires breaking down the cell walls to ensure complete extraction, which can be achieved using various mechanical and nonmechanical methods. The efficiency of the oil recovery process depends on optimizing extraction parameters, which are determined by factors such as the oil content in physical and chemical properties and vegetable fractions of the oil and its intended application (Rangaswamy et al., 2021).

9.7 FUTURE PROSPECTS

The prospects of utilizing plant biomass to substitute fossil fuels with biofuels are highly promising, supported by advancements and innovations across multiple domains. Particularly noteworthy is the continuous enhancement of biomass conversion technologies, which are anticipated to improve the efficiency and cost-effectiveness of biofuel production processes. Enzymatic hydrolysis, advanced pyrolysis, and synthetic biology are at the forefront of reshaping the conversion of plant biomass into valuable biofuels, presenting pathways that diminish environmental impact and boost scalability. Moreover, genetic engineering can revolutionize the development of energy crops by enhancing biomass yields, resilience to environmental stressors, and optimizing biochemical composition, potentially leading to the creation of energy crops that significantly increase biomass availability while reducing land and resource demands. This genetic manipulation may initiate a new era of sustainable biofuel production.

However, incorporating biorefinery concepts is poised to significantly contribute to maximizing the potential of plant biomass. These refineries will not solely focus on biofuel production but will also generate biochemicals, bioplastics, and other value-added products from biomass, diversifying income streams and improving economic sustainability. Sustainable land management practices, including agroforestry and conservation agriculture, will persist in being pivotal for guaranteeing the sustained availability of biomass without compromising food security or environmental preservation. Circular bioeconomy initiatives are also gaining momentum, highlighting the importance of resource efficiency and waste valorization in biofuel production. Through the valorization of agricultural residues, organic waste streams, and byproducts from biofuel production, circular bioeconomy strategies have the potential to mitigate waste generation and optimize resource utilization, thereby further enhancing the sustainability of biofuels. In conclusion, the future utilization of plant biomass for biofuels shows great promise, driven by technological advancements, sustainable methodologies, and an increasing acknowledgment of the necessity to transition toward renewable energy sources. Embracing these forthcoming elements and promoting interdisciplinary cooperation can unlock the complete potential of biofuels as a fundamental element of a sustainable energy future.

9.8 CONCLUSION

In conclusion, plant biomass offers a promising avenue for replacing fossil fuels with biofuels, providing a renewable and potentially sustainable solution to address the dual challenges of energy security and climate change. Plant biomass, derived from agricultural residues, dedicated energy crops, and forestry residues, demonstrates substantial promise as a raw material for biofuel production, capitalizing on the carbon-neutral process of photosynthesis to alleviate GHG emissions. Notwithstanding obstacles like land competition and environmental repercussions, continuous progress in biomass conversion technologies and sustainable land management strategies present prospects to amplify the effectiveness and feasibility of biofuel manufacturing procedures. By exploiting the intrinsic energy capacity of biomass and encouraging ongoing exploration and novelty, the shift from fossil fuels to biofuels holds the potential of a more robust and sustainable energy prospect, aiding global endeavors to combat climate change and advocate for environmental sustainability.

KEYWORDS

- **biofuel**
- **fossil fuel**
- **plant biomass**
- **agricultural waste**
- **lignocellulosic biomass**
- **productivity**

REFERENCES

Abbasi T, Abbasi SA. (2010) Biomass energy and the environmental impacts associated with its production and utilization. *Renew. Sust. Energy Rev.* 14(3), 919–937.

Abdulyekeen KA, Umar AA, Patah MFA, Daud WMAW. (2021) Torrefaction of biomass: Production of enhanced solid biofuel from municipal solid waste and other types of biomass. *Renew. Sust. Energy Rev.* 150, 111436.

Abou Rjeily M, Gennequin C, Pron H, Abi-Aad E, Randrianalisoa JH. (2021) Pyrolysis-catalytic upgrading of bio-oil and pyrolysis-catalytic steam reforming of biogas: A review. *Environ. Chem. Lett.* 19(4), 2825–2872.

Adams P, Bridgwater T, Lea-Langton A, Ross A, Watson I. (2018) Biomass conversion technologies. In: *Greenhouse Gas Balances of Bioenergy Systems.* Academic Press, pp. 107–139.

Adeoye AO, Lawal OS, Quadri RO, Malomo D, Aliyu MT, Dang GE, Emojevu EO, Maikato MJ, Yahaya MG, Omonije OO, Edidem VG. (2023) Sustainable energy via thermochemical and biochemical conversion of biomass wastes for biofuel production. In: *Transportation Energy and Dynamics.* Springer: Singapore, pp. 245–306.

Aditiya HB, Mahlia TMI, Chong WT, Nur H, Sebayang AH. (2016) Second generation bioethanol production: A critical review. *Renew. Sust. Energy Rev.* 66, 631–653.

Afionis S, Stringer LC. (2020) Fuelling friendships or driving divergence? Legitimacy, coherence, and negotiation in Brazilian perceptions of European and American biofuels governance. *Energy Res. Soc. Sci.* 67, 101487.

Ahamer G. (2022) Why biomass fuels are principally not carbon neutral. *Energies* 15(24), 9619.

Alemán-Nava GS, Gatti IA, Parra-Saldivar R, Dallemand JF, Rittmann BE, Iqbal HM. (2018) Biotechnological revalorization of Tequila waste and by-product streams for cleaner production—A review from bio-refinery perspective. *J. Cleaner Prod.* 172, 3713–3720.

Algarni S, Tirth V, Alqahtani T, Alshehery S, Kshirsagar P. (2023) Contribution of renewable energy sources to the environmental impacts and economic benefits for sustainable development. *Sust. Energy Technol. Assess.* 56, 103098.

Alonso DM, Bond JQ, Dumesic JA. (2010) Catalytic conversion of biomass to biofuels. *Green Chem.* 12(9), 1493–1513.

Anas M, Verma KK, Haq IU, Naeem M, Li Q, Liao F, Xiang YX, Li YR. (2021a) Characterization of exotic, native and wild-type genotypes of sugarcane (*Saccharum* spp. Hybrids) for internal nitrogen use efficiency under different nitrogen levels. *Sugar Tech.* 23, 1258–1267. https://doi.org/10.1007/s12355-021-01014-1

Anas M, Verma KK, Riaz M, Qiang L, Liao F, Liu Y, Li YR. (2021b) Physio-morphological and biochemical mechanism of nitrogen use efficiency in sugarcane (*Saccharum* spp.) genotypes under different growth stages and nitrogen levels. *J. Plant Interact.* 16, 332–343.

Anastassiadis SG. (2016) Carbon sources for biomass, food, fossils, biofuels and biotechnology-review article. *World J. Biol. Biotechnol.* 1(1), 1–32.

Antar M, Lyu D, Nazari M, Shah A, Zhou X, Smith DL. (2021) Biomass for a sustainable bioeconomy: An overview of world biomass production and utilization. *Renew. Sust. Energy Rev.* 139, 110691.

Arbestain MC, Amonette JE, Singh B, Wang T, Schmidt HP. (2015) A biochar classification system and associated test methods. In: Lehmann J, Joseph S., (Eds.) *Biochar for Environmental Management.* Routledge, pp. 209–248.

Arteaga-Perez LE, Gomez-Capiro O, Karelovic A, Jimenez R. (2016) A modelling approach to the techno-economics of biomass-to-SNG/methanol systems: Standalone vs integrated topologies. *Chem. Eng. J.* 286, 663–678.

Awasthi MK, Wang M, Chen H, Wang Q, Zhao J, Ren X, Li DS, Awasthi SK, Shen F, Li R, Zhang Z. (2017) Heterogeneity of biochar amendment to improve the carbon and nitrogen sequestration through reduce the greenhouse gases emissions during sewage sludge composting. *Bioresour. Technol.* 224, 428–438.

Backer RG, Schwinghamer TD, Whalen JK, Seguin P, Smith DL. (2016) Crop yield and SOC responses to biochar application were dependent on soil texture and crop type in southern Quebec, Canada. *J. Plant Nutr. Soil Sci.* 179(3), 399–408.

Banerjee S, Mudliar S, Sen R, Giri B, Satpute D, Chakrabarti T, Pandey RA. (2010) Commercializing lignocellulosic bioethanol: Technology bottlenecks and possible remedies. *Biofuels, Bioprod. Bioref.* 4(1), 77–93.

Barth S, Murphy-Bokern D, Kalinina O, Taylor G, Jones M. (Eds.), (2016) *Perennial Biomass Crops for a Resource-Constrained World.* Springer Nature.

Baudry G, Delrue F, Legrand J, Pruvost J, Vallée T. (2017) The challenge of measuring biofuel sustainability: A stakeholder-driven approach applied to the French case. *Renew. Sust. Energy Rev.* 69, 933–947.

Bentsen NS, Moller IM. (2017) Solar energy conserved in biomass: Sustainable bioenergy use and reduction of land use change. *Renew. Sust. Energy Rev.* 71, 954–958.

Bhatt R, Verma KK, Jamwal NS. (2023a) Can paddy-compost eliminate N fertilization for climate-smart sugarcane? *Sugar Tech.* 25, 1036–1046. https://doi.org/10.1007/s12355-023-01270-3

Bhatt R, Verma KK, Kumar R, Sanghera GS. (2023b) Foliar application of potassium salt of active phosphorus (PSAP) mitigates insect pests and improves yield along with sugarcane quality in response to agroclimatic conditions of Punjab. *Sugar Tech.* 25(3), 660–669.

Brodowska MS, Muszyński P, Haliniarz M, Brodowski R, Kowalczykjusko A, Sekutowski T, Kurzyna-Szklarek M. (2018) Agronomic aspects of switchgrass cultivation and use for energy purposes. *Appl. Ecol. Environ. Res.* 16(5).

Calvo-Flores FG, Martin-Martinez FJ. (2022) Biorefineries: Achievements and challenges for a bio-based economy. *Front. Chem.* 10, 973417.

Cantarero MMV. (2020) Of renewable energy, energy democracy, and sustainable development: A roadmap to accelerate the energy transition in developing countries. *Energy Res. Social Sci.* 70, 101716.

Cao W, Li J, Martí-Rosselló T, Zhang X. (2019) Experimental study on the ignition characteristics of cellulose, hemicellulose, lignin and their mixtures. *J. Energy Inst.* 92(5), 1303–1312.

Carere CR, Sparling R, Cicek N, Levin DB. (2008) Third generation biofuels via direct cellulose fermentation. *Int. J. Mol. Sci.* 9(7), 1342–1360.

Chan YH, Cheah KW, How BS, Loy ACM, Shahbaz M, Singh HKG, Shuhaili AFA, Yusup S, Ghani WAWAK, Rambli J, Kansha Y. (2019) An overview of biomass thermochemical conversion technologies in Malaysia. *Sci. Total Environ.* 680, 105–123.

Chandra R, Takeuchi H, Hasegawa T. (2012) Methane production from lignocellulosic agricultural crop wastes: A review in context to second generation of biofuel production. *Renew. Sust. Energy Rev.* 16(3), 1462–1476.

Chandran R, Kaliaperumal R, Balakrishnan S, Britten AJ, MacInnis J, Mkandawire M. (2020) Characteristics of bio-oil from continuous fast pyrolysis of Prosopis juliflora. *Energy* 190, 116387.

Chaudhary P, Xu M, Ahamad L, Chaudhary A, Kumar G, Adeleke BS, Verma KK, Hu D-M, Siric I, Kumar P, Popescu SM, Fayssal SA. (2023) Application of synthetic consortia for improvement of soil fertility, pollution remediation, and agricultural productivity: A review. *Agronomy* 13, 643. https://doi.org/10.3390/agronomy13030643.

Chen GL, Zheng FJ, Lin B, Lao SB, He J, Huang Z, Zeng Y, Sun J, Verma KK. (2020) Phenolic and volatile compounds in the production of sugarcane vinegar. *ACS Omega* 5, 30587–30595.

Chen G-L, Zheng F-J, Lin B, Yang Y-X, Fang X-C, Verma KK, Yang L-F. (2023) Vinegar: A potential source of healthy and functional food with special reference to sugarcane vinegar. *Front. Nutr.* 10, 1145862. https://doi.org/10.3389/fnut.2023.1145862

Chen HG, Zhang YHP. (2015) New biorefineries and sustainable agriculture: Increased food, biofuels, and ecosystem security. *Renew. Sust. Energy Rev.* 47, 117–132.

Chen R, Qin Z, Han J, Wang M, Taheripour F, Tyner W, O'Connor D, Duffield J. (2018) Life cycle energy and greenhouse gas emission effects of biodiesel in the United States with induced land use change impacts. *Bioresour. Technol.* 251, 249–258.

Chen WH, Lin BJ, Lin YY, Chu YS, Ubando AT, Show PL, Ong HC, Chang JS, Ho SH, Culaba AB, Pétrissans A. (2021) Progress in biomass torrefaction: Principles, applications and challenges. *Prog. Energy Combustion Sci.* 82, 100887.

Cherubini F, Strømman AH. (2011) Life cycle assessment of bioenergy systems: State of the art and future challenges. *Bioresource Tech.* 102(2), 437–451.

Ciuta S, Tsiamis D, Castaldi MJ. (2017) *Gasification of Waste Materials: Technologies for Generating Energy, Gas, and Chemicals from Municipal Solid Waste, Biomass, Nonrecycled Plastics, Sludges, and Wet Solid Wastes*. Academic Press.

Correa DF, Beyer HL, Fargione JE, Hill JD, Possingham HP, Thomas-Hall SR, Schenk PM. (2019) Towards the implementation of sustainable biofuel production systems. *Renew. Sust. Energy Rev.* 107, 250–263.

Das P, Jha CK, Saxena S, Ghosh RK. (2024) Can biofuels help achieve sustainable development goals in India? A systematic review. *Renew. Sust. Energy Rev.* 192, 114246.

Davis R, Tao L, Tan ECD, Biddy MJ, Beckham GT, Scarlata C, Jacobson J, Cafferty K, Ross J, Lukas J, Knorr D. (2013) *Process Design and Economics for the Conversion of Lignocellulosic Biomass to Hydrocarbons: Dilute-acid and Enzymatic Deconstruction of Biomass to Sugars and Biological Conversion of Sugars to Hydrocarbons* (No. NREL/TP-5100-60223). National Renewable Energy Lab. (NREL): Golden, CO, United States.

De Bhowmick G, Sarmah AK, Sen R. (2018) Lignocellulosic biorefinery as a model for sustainable development of biofuels and value added products. *Bioresour. Technol.* 247, 1144–1154.

De Fraiture C, Giordano M, Liao Y. (2008) Biofuels and implications for agricultural water use: Blue impacts of green energy. *Water Policy* 10(S1), 67–81.

de Jong E, Higson A, Walsh P, Wellisch M. (2012) Bio-based chemicals value added products from biorefineries. *IEA Bioenergy, Task42 Biorefinery* 34, 1–33.

De Vrije T, Bakker RR, Budde MA, Lai MH, Mars AE, Claassen PA. (2009) Efficient hydrogen production from the lignocellulosic energy crop Miscanthus by the extreme thermophilic bacteria Caldicellulosiruptor saccharolyticus and Thermotoga neapolitana. *Biotechnol. Biofuels* 2, 1–15.

Demirbas A. (2001) Relationships between lignin contents and heating values of biomass. *Energy Convers. Manag.* 42(2), 183–188.

Demirbas T, Demirbas AH. (2010) Bioenergy, green energy. biomass and biofuels. *Energy Sources, Part A: Recovery, Utilization, and Environmental Effects* 32(12), 1067–1075.

Deng ZN, Huang HR, Huang CM, Verma KK, Luo HB, Jiang SL, Li YJ, Cao HQ, Wu KC, Xu L. (2023) Short-term limited water irrigation influences the photosynthetic pigments, enzymatic and non-enzymatic activities in *Saccharum spontaneum* L. at vegetative stage. *Sugar Tech.* 25, 110–118. https://doi.org/10.1007/s12355-022-01175-7

Deshmukh SS, Jinturkar AM, Gawande JS. (2008) Comparative experimental study of single basin and stepped type solar stil. *Energy Edu. Sci. Technol.* 20, 79.

Dhyani V, Bhaskar T. (2019) Pyrolysis of biomass. In: *Biofuels: Alternative Feedstocks and Conversion Processes for the Production of Liquid and Gaseous Biofuels*. Academic Press, pp. 217–244.

Dirker J, Juggurnath D, Kaya A, Osowade EA, Simpson M, Lecompte S, Noori Rahim Abadi SMA, Voulgaropoulos V, Adelaja AO, Dauhoo MZ, Khoodaruth A. (2019) Thermal energy processes in direct steam generation solar systems: Boiling, condensation and energy storage–A review. *Front. Energy Res.* 6, 147.

Dudley B. (2018) BP statistical review of world energy 2018. *Energy Economic, Centre for Energy Economics Research and Policy.* British Petroleum, Available via https://www.bp. com/en/global/corporate/energy-economics/statistical-review-of-world-energy/electricity. html, 5.

Edrisi SA, Abhilash PC. (2016) Exploring marginal and degraded lands for biomass and bioenergy production: An Indian scenario. *Renew. Sust. Energy Rev.* 54, 1537–1551.

Eisentraut A, Brown A. (2012) Technology roadmap: Bioenergy for heat and power. *Technol. Roadmaps* 2, 1–41.

Farooqui A, Tripathi G, Moheet K, Dubey P, Ahmad S, Husain A, Shamim A, Mahfooz S. (2021) Algal biomass: Potential renewable feedstock for bioenergy production. In: *Bioenergy Research: Integrative Solution for Existing Roadblock*, pp. 85–113.

Faruque MO, Razzak SA, Hossain MM. (2020) Application of heterogeneous catalysts for biodiesel production from microalgal oil—a review. *Catalysts* 10(9), 1025.

Fiore M, Magi V, Viggiano A. (2020) Internal combustion engines powered by syngas: A review. *Appl. Energy* 276, 115415.

Fischer G, Hizsnyik E, Prieler S, Shah M, Van Velthuizen HT. (2009) *Biofuels and Food Security*. The OPEC Fund for International Development (OFID): Vienna, Austria.

Flach B, Lieberz S, Bolla S. (2019) *EU-28 Biofuels Annual EU Biofuels Annual 2019*. USDA Foreign Agricultural Service.

Fullerton DG, Bruce N, Gordon SB. (2008) Indoor air pollution from biomass fuel smoke is a major health concern in the developing world. *Trans. R. Soc. Trop. Med. Hyg.* 102(9), 843–851.

Gada K, Peter AP, Chai WS, Show PL. (2022) Industry perspectives of microalgae 4.0. In: *Microalgae for Environmental Biotechnology.* CRC Press, pp. 363–395.

Gayathri R, Mahboob S, Govindarajan M, Al-Ghanim KA, Ahmed Z, Al-Mulhm N, Vodovnik M, Vijayalakshmi S. (2021) A review on biological carbon sequestration: A sustainable solution for a cleaner air environment, less pollution and lower health risks. *J. King Saud Univ.-Sci.* 33(2), 101282.

Gielen D, Boshell F, Saygin D, Bazilian MD, Wagner N, Gorini R. (2019) The role of renewable energy in the global energy transformation. *Energy Strategy Rev.* 24, 38–50.

Głąb T, Palmowska J, Zaleski T, Gondek K. (2016) Effect of biochar application on soil hydrological properties and physical quality of sandy soil. *Geoderma* 281, 11–20.

González CAD, Sandoval LP. (2020) Sustainability aspects of biomass gasification systems for small power generation. *Renew. Sust. Energy Rev.* 134, 110180.

González-García P. (2018) Activated carbon from lignocellulosics precursors: A review of the synthesis methods, characterization techniques and applications. *Renew. Sust. Energy Rev.* 82, 1393–1414.

Green EM. (2011) Fermentative production of butanol—The industrial perspective. *Curr. Opin. Biotechnol.* 22(3), 337–343.

Guo D-J, Li D-P, Chen Z-D, Yang B, Verma KK, Singh RK, Singh P, Khan Q, Sharma A, Qin Y, Zhang B-Q, Song X-P, Li Y-R. (2023) Effect of endophytic diazotroph *Enterobacter roggenkampii* ED5 on nitrogen-metabolism-related microecology in the sugarcane rhizosphere at different nitrogen levels. *Front. Microbiol.* 14, 1132016. https://doi.org/10.3389/fmicb.2023.1132016

Guo D-J, Li D-P, Singh RK, Singh P, Sharma A, Verma KK, Qin Y, Khan Q, Lu Z, Malviya MK, Song X-P, Xing Y-X and Li Y-R. (2021) Differential protein expression analysis of two sugarcane varieties in response to diazotrophic plant growth-promoting endophyte enterobacter roggenkampii ED5. *Front. Plant Sci.* 12, 727741. https://doi.org/10.3389/fpls.2021.727741

Guo DJ, Li DP, Singh RK, Singh P, Verma KK, Sharma A, Qin Y, Khan Q, Song XP, Malviya MK, Xing YX, Li YR. (2022) Comparative transcriptome analysis of two sugarcane varieties in response to diazitrophic plant growth promoting endophyte *Enterobacter roggenkampii* ED5. *J. Plant Interact.* 17(1), 75–84.

Guo M, Song W, Buhain J. (2015) Bioenergy and biofuels: History, status, and perspective. *Renew. Sust. Energy Rev.* 42, 712–725.

Hanssen SV, Daioglou V, Steinmann ZJ, Frank S, Popp A, Brunelle T, Lauri P, Hasegawa T, Huijbregts MA, Van Vuuren DP. (2020) Biomass residues as twenty-first century bioenergy feedstock—a comparison of eight integrated assessment models. *Clim. Change* 163, 1569–1586.

Haque A, Mondal D, Khan I, Usmani MA, Bhat AH, Gazal U. (2017) Fabrication of composites reinforced with lignocellulosic materials from agricultural biomass. In: *Lignocellulosic Fibre and Biomass-Based Composite Materials.* Woodhead Publishing, pp. 179–191.

Hassan H, Lim JK, Hameed BH. (2016) Recent progress on biomass co-pyrolysis conversion into high-quality bio-oil. *Bioresour. Technol.* 221, 645–655.

Hassan MH, Kalam MA. (2013) An overview of biofuel as a renewable energy source: Development and challenges. *Procedia Eng.* 56, 39–53.

Havlík P, Schneider UA, Schmid E, Böttcher H, Fritz S, Skalský R, Aoki K, De Cara S, Kindermann G, Kraxner F, Leduc S. (2011) Global land-use implications of first and second generation biofuel targets. *Energy Policy* 39(10), 5690–5702.

Hayes DJ. (2009) An examination of biorefining processes, catalysts and challenges. *Catalysis Today* 145(1–2), 138–151.

Hossain N, Zaini JH, Mahlia TMI. (2017) A review of bioethanol production from plant-based waste biomass by yeast fermentation. *Int. J. Technol.* 8(1). https://doi.org/10.14716/ijtech.v8i1.3948

Huang CM, Wu KC, Verma KK, Luo HB, Deng ZN, Shi GY, Cao HQ, Wu XJ, Ye LP, Wei YW, Xu L. (2024) Isolation and expression analysis of cellulose synthase 3 (*Ces3*) genes from sugarcane (*Saccharum officinarum* L.). *Chem. Biol. Technol. Agric.* 11, 71. https://doi.org/10.1186/s40538-024-00597-7

Huang HR, Gao YJ, Malviya MK, Verma KK, Solanki MK, Huang YX, Li X, Deng Y, Yan J, Tang SY, Wang LW, Xu L. (2022) Genetic diversity analysis of sugarcane (*Saccharum* spp. Hybrids) among high-sucrose clones of GT series and commonly used parents by using microsatellite markers in Guangxi, China. *Sugar Tech.* 24, 397–407. https://doi.org/10.1007/s12355-021-01033-y

Hussain A, Saxena A, Irum Farooqui A, Ashfaque M. (2024) Environment of Lignocellulosic Waste to Biofuel. In: *Agricultural Biomass Nanocatalysts for Green Energy Applications,* Springer: Singapore, pp. 19–43.

Huzir NM, Aziz MMA, Ismail SB, Abdullah B, Mahmood NAN, Umor NA, Muhammad SAFAS. (2018) Agro-industrial waste to biobutanol production: Eco-friendly biofuels for next generation. *Renew. Sust. Energy Rev.* 94, 476–485.

Irum AM, Dubey P, Tripathi G, Khujamshukurov NA, Farooqui A, Mishra V. (2024) Lignocellulosic waste to biofuel-paddy straw to bioethanol: Advancement in technology. In: *Paddy Straw Waste for Biorefinery Applications.* Springer Nature: Singapore, pp. 87–107.

Jagadesh M, Dash M, Kumari A, Singh SK, Verma KK, Kumar P, Bhatt R, Sharma SK. (2024) Revealing the hidden world of soil microbes: Metagenomic insights into plant, bacteria, and fungi interactions for sustainable agriculture and ecosystem restoration. *Microbiol. Res.* 285, 127764. https://doi.org/10.1016/j.micres.2024.127764

Jahirul MI, Rasul MG, Chowdhury AA, Ashwath N. (2012) Biofuels production through biomass pyrolysis—a technological review. *Energies* 5(12), 4952–5001.

Jayachandran M, Gatla RK, Rao KP, Rao GS, Mohammed S, Milyani AH, Azhari AA, Kalaiarasy C, Geetha S. (2022) Challenges in achieving sustainable development goal 7: Affordable and clean energy in light of nascent technologies. *Sust. Energy Technol. Assess.* 53, 102692.

Jeffery S, Abalos D, Prodana M, Bastos AC, Van Groenigen JW, Hungate BA, Verheijen F. (2017) Biochar boosts tropical but not temperate crop yields. *Environ. Res. Lett.* 12(5), 053001.

Jeswani HK, Chilvers A, Azapagic A. (2020) Environmental sustainability of biofuels: A review. *Proc. Royal Soc. A* 476(2243), 20200351.

Jørgensen U. (2011) Benefits versus risks of growing biofuel crops: The case of Miscanthus. *Curr. Opin. Environ. Sust.* 3(1–2), 24–30.

Joshi A, Verma KK, Rajput VD, Arora J. (2023) Metabolic engineering of microorganisms in advancing biofuel production. In: *Advances in Lignocellulosic Biofuel Production Systems.* pp. 181–202. https://doi.org/10.1016/B978-0-323-91192-4.00016-X

Joshi A, Verma KK, Rajput VD, Minkina T, Arora J. (2022) Recent advances in metabolic engineering of microorganisms for advancing lignocellulose-derived biofuels. *Bioengineered* 13(4), 8135–8163. https://doi.org/10.1080/21655979.2022.2051856.

Kalak T. (2023) Potential use of industrial biomass waste as a sustainable energy source in the future. *Energies* 16(4), 1783.

Kang KE, Park DH, Jeong GT. (2013) Effects of inorganic salts on pretreatment of Miscanthus straw. *Bioresour. Technol.* 132, 160–165.

Kannegiesser M. (2008) *Value Chain Management in the Chemical Industry: Global Value Chain Planning of Commodities*. Physica-Verlag.

Karan H, Funk C, Grabert M, Oey M, Hankamer B. (2019) Green bioplastics as part of a circular bioeconomy. *Trends Plant Sci.* 24(3), 237–249.

Karmakar A, Daftari T, Sivagami K, Chandan MR, Shaik AH, Kiran B, Chakraborty S. (2023) A comprehensive insight into Waste to Energy conversion strategies in India and its associated air pollution hazard. *Environ. Technol. Innov.* 29, 103017.

Kasozi GN, Zimmerman AR, Nkedi-Kizza P, Gao B. (2010) Catechol and humic acid sorption onto a range of laboratory-produced black carbons (biochars). *Environ. Sci. Technol.* 44(16), 6189–6195.

Kaur M, Sharma HK, Kumar N. (2022) Processing of oilseeds. In: *Agro-Processing and Food Engineering: Operational and Application Aspects.* Springer Nature: Singapore, pp. 483–533.

Khan S, Naushad M, Iqbal J, Bathula C, Ala'a H. (2022) Challenges and perspectives on innovative technologies for biofuel production and sustainable environmental management. *Fuel* 325, 124845.

Koçar G, Civaş N. (2013) An overview of biofuels from energy crops: Current status and future prospects. *Renew. Sust. Energy Rev.* 28, 900–916.

Koh LP, Ghazoul J. (2008) Biofuels, biodiversity, and people: Understanding the conflicts and finding opportunities. *Biol. Conserv.* 141(10), 2450–2460.

Krishna D, Sachan HK, Verma KK. (2022) Abiotic stresses in plants: An overview. In: *Understanding Abiotic Stresses*. Nova Science Publishers, Inc.: New York.

Kszos LA, Downing ME, Wright LL, Cushman JH, McLaughlin SB, Tolbert VR, Tuskan GA, Walsh ME. (2000) Bioenergy feedstock development program status report. *ORNL/TM-2000/292, Oak Ridge National Laboratory, Oak Ridge, TN, USA.*

Kumar G, Dharmaraja J, Arvindnarayan S, Shoban S, Bakonyi P, Saratale GD, Nemestóthy N, Bélafi–Bakó K, Yoon JJ, Kim SH. (2019) A comprehensive review on thermochemical, biological, biochemical and hybrid conversion methods of bio-derived lignocellulosic molecules into renewable fuels. *Fuel* 251, 352–367.

Kumari A, Dash M, Singh SK, Jagadesh M, Mathpal B, Mishra PK, Pandey SK, Verma KK. (2023a) Soil microbes: A natural solution for mitigating the impact of climate change. *Environ. Monit. Assess.* 195, 1436. https://doi.org/10.1007/s10661-023-11988-y

Kumari A, Lakshmi GA, Krishna GK, Patni B, Prakash S, Bhattacharyya M, Singh SK, Verma KK. (2022) Climate change and its impact on crops: A comprehensive investigation for sustainable agriculture. *Agronomy* 12, 3008. https://doi.org/10.3390/agronomy12123008

Kumari A, Singh SK, Mathpal B, Verma KK, Garg VK, Bhattacharyya M, Bhatt R. (2023b) The biosynthesis, mechanism of action, and physiological functions of

melatonin in horticultural plants: A review. *Horticulturae* 9, 913. https://doi.org/10.3390/horticulturae9080913
Kumari D, Singh R. (2018) Pretreatment of lignocellulosic wastes for biofuel production: A critical review. *Renew. Sust. Energy Rev*. 90, 877–891.
Kuthe NV, Ingle PB, Gore VG. (2021) Biomass briquettes as an alternative energy source compare to wood charcoal in boilers. *Int. J. Sci. Res. Mech. Mater. Eng* 5(4), 16–40.
Lachos-Perez D, Torres-Mayanga PC, Abaide ER, Zabot GL, De Castilhos F. (2022) Hydrothermal carbonization and Liquefaction: Differences, progress, challenges, and opportunities. *Bioresour. Technol*. 343, 126084.
Laghari M, Naidu R, Xiao B, Hu Z, Mirjat MS, Hu M, Kandhro MN, Chen Z, Guo D, Jogi Q, Abudi ZN. (2016) Recent developments in biochar as an effective tool for agricultural soil management: A review. *J. Sci. Food Agricult*. 96(15), 4840–4849.
Lama-Muñoz A, del Mar Contreras M, Espínola F, Moya M, Romero I, Castro E. (2020) Characterization of the lignocellulosic and sugars composition of different olive leaves cultivars. *Food Chem*. 329, 127153.
Lee SY, Sankaran R, Chew KW, Tan CH, Krishnamoorthy R, Chu DT, Show PL. (2019) Waste to bioenergy: A review on the recent conversion technologies. *BMC Energy* 1(1), 1–22.
Lemmer A, Merkle W, Baer K, Graf F. (2017) Effects of high-pressure anaerobic digestion up to 30 bar on pH-value, production kinetics and specific methane yield. *Energy* 138, 659–667.
Li A-M, Liao F, Wang M, Chen Z-L, Qin C-X, Huang R-Q, Verma KK, Li Y-R, Que Y-X, Pan Y-Q, et al. (2023) Transcriptomic and proteomic landscape of sugarcane response to biotic and abiotic stressors. *Int. J. Mol. Sci*. 24, 8913. https://doi.org/10.3390/ijms24108913
Li H, Foston MB, Kumar R, Samuel R, Gao X, Hu F, Ragauskas AJ, Wyman CE. (2012) Chemical composition and characterization of cellulose for Agave as a fast-growing, drought-tolerant biofuels feedstock. *RSC Adv*. 2(11), 4951–4958.
Littlejohns J, Rehmann L, Murdy R, Oo A, Neill S. (2018) Current state and future prospects for liquid biofuels in Canada. *Biofuel Res. J*. 5(1), 759–779.
López-Bellido L, Wery J, López-Bellido RJ. (2014) Energy crops: Prospects in the context of sustainable agriculture. *Eur. J. Agron*. 60, 1–12.
Lopez-Hidalgo AM, Alvarado-Cuevas ZD, De Leon-Rodriguez A. (2018) Biohydrogen production from mixtures of agro-industrial wastes: Chemometric analysis, optimization and scaling up. *Energy* 159, 32–41.
Lund H. (2007) Renewable energy strategies for sustainable development. *Energy* 32(6), 912–919.
Lynd LR, Sow M, Chimphango AF, Cortez LA, Brito Cruz CH, Elmissiry M, Laser M, Mayaki IA, Moraes MA, Nogueira LA, Wolfaardt GM. (2015) Bioenergy and African transformation. *Biotechnol. Biofuels* 8, 1–18.
Macke Y. (2017) Growing interest for ethanol brightens prospects (GAIN Report Number: CH17048).
Major J, Lehmann J, Rondon M, Goodale C. (2010) Fate of soil-applied black carbon: Downward migration, leaching and soil respiration. *Global Change Biol*. 16(4), 1366–1379.

Malińska K, Zabochnicka-Świątek M, Dach J. (2014) Effects of biochar amendment on ammonia emission during composting of sewage sludge. *Ecol. Eng.* 71, 474–478.

Malinska K. (2015) Legal and quality aspects of requirements defined for biochar. *Inzynieriai Ochrona Srodowiska* 18, 359–371.

Malviya MK, Li C-N, Lakshmanan P, Solanki MK, Wang Z, Solanki AC, Nong Q, Verma KK, Singh RK, Singh P, Sharma A, Guo D-J, Dessoky ES, Song X-P, Li Y-R. (2022) High-throughput sequencing-based analysis of rhizosphere and diazotrophic bacterial diversity among wild progenitor and closely related species of sugarcane (*Saccharum* spp. inter-specific hybrids). *Front. Plant Sci.* 13, 829337. https://doi.org/10.3389/fpls.2022.829337.

Malviya MK, Solanki MK, Li CN, Wang Z, Zeng Y, Verma KK, Singh RK, Singh P, Huang HR, Yang LT, Song XP, Li YR. (2021) Sugarcane-legume intercropping can enrich the soil microbiome and plant growth. *Front. Sustain. Food Syst.* 5, 606595. https://doi.org/10.3389/fsufs.2021.606595.

Masse D, Gilbert Y, Savoie P, Bélanger G, Parent G, Babineau D. (2010) Methane yield from switchgrass harvested at different stages of development in Eastern Canada. *Bioresour. Technol.* 101(24), 9536–9541.

Masum MFH, Dwivedi P, Anderson WF. (2020) Estimating unit production cost, carbon intensity, and carbon abatement cost of electricity generation from bioenergy feedstocks in Georgia, United States. *Renew. Sust. Energy Rev.* 117, 109514.

Mata TM, Martins AA, Caetano NS. (2010) Microalgae for biodiesel production and other applications: A review. *Renew. Sust. Energy Rev.* 14(1), 217–232.

McDowall W, Anandarajah G, Dodds PE, Tomei J. (2012) Implications of sustainability constraints on UK bioenergy development: Assessing optimistic and precautionary approaches with UK MARKAL. *Energy Policy* 47, 424–436.

Mehmood MA, Ibrahim M, Rashid U, Nawaz M, Ali S, Hussain A, Gull M. (2017) Biomass production for bioenergy using marginal lands. *Sustain. Prod. Consum.* 9, 3–21.

Mishra S, Upadhyay RK. (2021) Review on biomass gasification: Gasifiers, gasifying mediums, and operational parameters. *Mat. Sci. Energy Technol.* 4, 329–340.

Mohammed NI, Kabbashi N, Alade A. (2018) Significance of agricultural residues in sustainable biofuel development. *Agricult. Waste Residues*, 71–88.

Mohan D, Sarswat A, Ok YS, Pittman Jr CU. (2014) Organic and inorganic contaminants removal from water with biochar, a renewable, low cost and sustainable adsorbent–a critical review. *Bioresour. Technol.* 160, 191–202.

Mueller SA, Anderson JE, Wallington TJ. (2011) Impact of biofuel production and other supply and demand factors on food price increases in 2008. *Biomass Bioenergy* 35(5), 1623–1632.

Muh E, Tabet F, Amara S. (2021) Biomass conversion to fuels and value-added chemicals: A comprehensive review of the thermochemical processes. *Curr. Altern. Energy* 4(1), 3–25.

Mutopo P, Beyene A, Haaland H, Boamah F, Widengård M, Skarstein R. (2011) *Biofuels, Land Grabbing and Food Security in Africa.* Bloomsbury Publishing.

Nanda S, Mohammad J, Reddy SN, Kozinski JA, Dalai AK. (2014) Pathways of lignocellulosic biomass conversion to renewable fuels. *Biomass Convers. Biorefinery* 4, 157–191.

Nayono SE. (2010) *Anaerobic Digestion of Organic Solid Waste for Energy Production*. KIT scientific Publishing, vol. 46.
Niculescu R, Clenci A, Iorga-Siman V. (2019) Review on the use of diesel–biodiesel–alcohol blends in compression ignition engines. *Energies* 12(7), 1194.
Nogueira LAH, Capaz RS. (2013) Biofuels in Brazil: Evolution, achievements and perspectives on food security. *Glob. Food Secur.* 2(2), 117–125.
Norfarhana AS, Ilyas RA, Ngadi N, Othman MHD, Misenan MSM, Norrrahim MNF. (2023) Revolutionizing lignocellulosic biomass: A review of harnessing the power of ionic liquids for sustainable utilization and extraction. *Int. J. Biol. Macromol.* 256, 128256.
Obi OF, Pecenka R, Clifford MJ. (2022) A review of biomass briquette binders and quality parameters. *Energies* 15(7), 2426.
Oladunni AA, Ishola MM. (2023) Novel and cost-effective feedstock for sustainable bioethanol production. In: *Bioethanol: A Green Energy Substitute for Fossil Fuels*. Springer International Publishing: Cham, pp. 21–45.
Oliveira MA, Verma KK, Bhatt R, Oliveira TBA. (2022) Impact of green and organic fertilizers on soil fertility and sugarcane productivity. In: Verma KK. et al. (Eds.), *Agro-industrial Perspectives on Sugarcane Production Under Environmental Stress*. https://doi.org/10.1007/978-981-19-3955-6_11.
Omer AM. (2008) Green energies and the environment. *Renew. Sustain. Energy Rev.* 12(7), 1789–1821.
Owusu PA, Asumadu-Sarkodie S. (2016) A review of renewable energy sources, sustainability issues and climate change mitigation. *Cogent Eng.* 3(1), 1167990.
Pan H, Xia Q, Wang Y, Shen Z, Huang H, Ge Z, Li X, He J, Wang X, Li L, Wang Y. (2022) Recent advances in biodiesel production using functional carbon materials as acid/base catalysts. *Fuel Processing Technol.* 237, 107421.
Panahi HKS, Dehhaghi M, Aghbashlo M, Karimi K, Tabatabaei M. (2019) Shifting fuel feedstock from oil wells to sea: Iran outlook and potential for biofuel production from brown macroalgae (ochrophyta; phaeophyceae). *Renew. Sust. Energy Rev.* 112, 626–642.
Park HJ, Dong JI, Jeon JK, Park YK, Yoo KS, Kim SS, Kim J, Kim S. (2008) Effects of the operating parameters on the production of bio-oil in the fast pyrolysis of Japanese larch. *Chem. Eng. J.* 143(1–3), 124–132.
Penín L, López M, Santos V, Alonso JL, Parajó JC. (2020) Technologies for Eucalyptus wood processing in the scope of biorefineries: A comprehensive review. *Bioresour. Technol.* 311, 123528.
Peskett L, Slater R, Stevens C, Dufey A. (2007) Biofuels, agriculture and poverty reduction. https://odi.cdn.ngo/media/documents/100_bF4ehiq.pdf
Popp J, Lakner Z, Harangi-Rákos M, Fari M. (2014) The effect of bioenergy expansion: Food, energy, and environment. *Renew. Sust. Energy Rev.* 32, 559–578.
Prasad S, Ingle AP. (2019) Impacts of sustainable biofuels production from biomass. In: *Sustainable Bioenergy*. Elsevier, pp. 327–346
Qin Z, Zhuang Q, Cai X, He Y, Huang Y, Jiang D, Lin E, Liu Y, Tang Y, Wang MQ. (2018) Biomass and biofuels in China: Toward bioenergy resource potentials and their impacts on the environment. *Renew. Sust. Energy Rev.* 82, 2387–2400.
Ragauskas AJ., et al. (2014) The path forward for biofuels and biomaterials. *Science* 311(5760), 484–489.

Rangaswamy S, Kumar GS, Kuppusamy C. (2021) Membrane technology for vegetable oil processing—Current status and future prospects. *Comprehensive Rev. Food Sci. Food Safety* 20(5), 5015–5042.
Rial RC. (2024) Biofuels versus climate change: Exploring potentials and challenges in the energy transition. *Renew. Sust. Energy Rev.* 196, 114369.
Rondon MA, Molina D, Hurtado M, Ramirez J, Lehmann J, Major J, Amezquita E. (2006) Enhancing the productivity of crops and grasses while reducing greenhouse gas emissions through bio-char amendments to unfertile tropical soils. In: *18th World Congress of Soil Science.* International Union of Soil Sciences: Philadelphia, PA, pp. 9–15.
Saleem M. (2022) Possibility of utilizing agriculture biomass as a renewable and sustainable future energy source. *Heliyon* 8(2).
Salvi BL. (2022) Transesterification methods. In: *Production of Biodiesel from Non-Edible Sources.* Elsevier, pp. 117–151.
Sanchez PDC, Aspe MMT, Sindol KN. (2022) An overview on the production of bio-briquettes from agricultural wastes: Methods, processes, and quality. *J. Agricult. Food Eng.* 1, 0036.
Sarkar N, Chakraborty D, Dutta R, Agrahari P, Bharathi SD, Singh AA, Jacob S. (2021) A comprehensive review on oilseed cakes and their potential as a feedstock for integrated biorefinery. *J. Adv. Biotechnol. Exp. Therapeutics* 4(3), 376.
Sarkar N, Ghosh SK, Bannerjee S, Aikat K. (2012) Bioethanol production from agricultural wastes: An overview. *Renew. Energy* 37(1), 19–27.
Sarker TR, Nanda S, Dalai AK, Meda V. (2021) A review of torrefaction technology for upgrading lignocellulosic biomass to solid biofuels. *BioEnergy Res.* 14, 645–669.
Saxena RC, Adhikari DK, Goyal HB. (2009) Biomass-based energy fuel through biochemical routes: A review. *Renew. Sust. Energy Rev.* 13(1), 167–178.
Schobert H. (2013) *Chemistry of Fossil Fuels and Biofuels.* Cambridge University Press.
Searchinger TD. (2010) Biofuels and the need for additional carbon. *Environ. Res. Lett.* 5(2), 024007.
Sebothoma, D. (2021). Thermodynamic evaluation of the gasification of municipal solid waste. Faculty of Engineering and the Built Environment, Department of Mechanical Engineering. http://hdl.handle.net/11427/36063
Shah F, Wu W. (2019) Soil and crop management strategies to ensure higher crop productivity within sustainable environments. *Sustainability* 11(5), 1485.
Shahbaz M, AlNouss A, Ghiat I, Mckay G, Mackey H, Elkhalifa S, Al-Ansari T. (2021) A comprehensive review of biomass based thermochemical conversion technologies integrated with CO_2 capture and utilisation within BECCS networks. *Resour. Conserv. Recycl.* 173, 105734.
Shahbeik H, Panahi HKS, Dehhaghi M, Guillemin GJ, Fallahi A, Hosseinzadeh-Bandbafha H, Amiri H, Rehan M, Raikwar D, Latine H, Pandalone B. (2024) Biomass to biofuels using hydrothermal liquefaction: A comprehensive review. *Renew. Sust. Energy Rev.* 189, 113976.
Shamsi M, Obaid AA, Farokhi S, Bayat A. (2022) A novel process simulation model for hydrogen production via reforming of biomass gasification tar. *Int. J. Hydrogen Energy* 47(2), 772–781.
Sharma A, Singh RK, Singh P, Vaishnav A, Guo DJ, Verma KK, Li DP, Song XP, Malviya MK, Khan N, Lakshmanan P, Li YR. (2021a) Insights into the bacterial and

nitric oxide-induced salt tolerance in sugarcane and their growth-promoting abilities. *Microorganisms* 9, 2203. https://doi.org/10.3390/microorganisms9112203.

Sharma A, Song X-P , Singh RK, Vaishnav A, Gupta S, Singh P, Guo D-J, Verma KK, Li Y-R. (2022) Impact of carbendazim on cellular growth, defence system and plant growth promoting traits of *Priestia megaterium* ANCB-12 isolated from sugarcane rhizosphere. *Front. Microbiol.* 13, 1005942. https://doi.org/10.3389/fmicb.2022.1005942

Sharma B, Larroche C, Dussap CG. (2020) Comprehensive assessment of 2G bioethanol production. *Bioresour. Technol.* 313, 123630.

Sharma HK, Xu C, Qin W. (2019) Biological pretreatment of lignocellulosic biomass for biofuels and bioproducts: An overview. *Waste Biomass Valorization* 10, 235–251.

Sharma N, Jaiswal KK, Kumar V, Vlaskin MS, Nanda M, Rautela I, Tomar MS, Ahmad W. (2021b) Effect of catalyst and temperature on the quality and productivity of HTL bio-oil from microalgae: A review. *Renew. Energy* 174, 810–822.

Sheikh MMI, Kim CH, Park HH, Nam HG, Lee GS, Jo HS, Lee JY, Kim JW. (2015) A synergistic effect of pretreatment on cell wall structural changes in barley straw (*Hordeum vulgare* L.) for efficient bioethanol production. *J. Sci. Food Agricult.* 95(4), 843–850.

Shen Y. (2014) Carbon dioxide bio-fixation and wastewater treatment via algae photochemical synthesis for biofuels production. *RSC Adv.* 4(91), 49672–49722.

Show KY, Lee DJ, Zhang ZP. (2011) Production of biohydrogen: Current perspectives and future prospects. *Biofuels* 467–479.

Sieminski A. (2014) International energy outlook. *Energy Information Administration (EIA)* 18, 2.

Sikkema R, Proskurina S, Banja M, Vakkilainen E. (2021) How can solid biomass contribute to the EU's renewable energy targets in 2020, 2030 and what are the GHG drivers and safeguards in energy-and forestry sectors? *Renew. Energy* 165, 758–772.

Sims RE, Hastings A, Schlamadinger B, Taylor G, Smith P. (2006) Energy crops: Current status and future prospects. *Global Change Biol.* 12(11), 2054–2076.

Sims RE, Mabee W, Saddler JN, Taylor M. (2010) An overview of second generation biofuel technologies. *Bioresour. Technol.* 101(6), 1570–1580.

Sims RE. (2004) Biomass, bioenergy and biomaterials: Future prospects. In: *Biomass and Agriculture—Sustainability Markets and Policies OECD,* Paris, pp. 37–61.

Singh P, Singh RK, Li HB, Guo DJ, Sharma A, Verma KK, Solanki MK, Upadhyay SK, Lakshmanan P, Yang LT, Li YR. (2023) Nitrogen fixation and phytohormone stimulation of sugarcane plant through plant growth promoting diazotrophic *Pseudomonas. Biotechnol. Genetic Eng. Rev.* 40(1), 15–35. https://doi.org/10.1080/02648725.2023.2177814

Singh P, Singh RK, Song QQ, Li HB, Guo DJ, Malviya MK, Verma KK, Song XP, Lakshmanan P, Yang LT, Li YR. (2021) Comparative analysis of protein and differential responses of defense-related gene and enzyme activity reveals the long-term molecular responses of sugarcane inoculated with *Sporisorium scitamineum. J. Plant Interact.* 16(1), 12–29.

Singh RK, Singh P, Li HB, Song QQ, Guo DJ, Solanki MK, Verma KK, Malviya MK, Song XP, Lakshmanan P, Yang LT, Li YR. (2020) Diversity of nitrogen-fixing rhizobacteria associated with sugarcane: A comprehensive study of plant-microbe interactions for growth enhancement in *Saccharum* spp. *BMC Plant Biol.* 20, 220. https://doi.org/10.1186/s12870-020-02400-9.

Smith JL, Collins HP, Bailey VL. (2010) The effect of young biochar on soil respiration. *Soil Biol. Biochem.* 42(12), 2345–2347.

Smith P, Davis SJ, Creutzig F, Fuss S, Minx J, Gabrielle B, Kato E, Jackson RB, Cowie A, Kriegler E, Van Vuuren DP. (2016) Biophysical and economic limits to negative CO_2 emissions. *Nat. Clim. Change* 6(1), 42–50.

Solanki MK, Verma KK, Dastogeer KMG, Mora-Poblete F, Mundra S. (2023) Microbial resilience in plant nutrient management towards sustainable farming. *Front. Microbiol.* 14, 1280811. https://doi.org/10.3389/fmicb.2023.1280811

Solarte-Toro JC, Chacón-Pérez Y, Cardona-Alzate CA. (2018) Evaluation of biogas and syngas as energy vectors for heat and power generation using lignocellulosic biomass as raw material. *Electron. J. Biotechnol.* 33, 52–62.

Song X-P, Mo F, Verma KK, Wei J, Zhang X, Yang L, Li YR. (2019) Effect of sugarcane smut (*Ustilago scitaminea* Syd.) on ultrastructure and biochemical indices of sugarcane. *Biomed. J. Sci. Tech. Res.* 17(1), 002950. https://doi.org/10.26717/BJSTR.2019.17.002950.

Song XP, Mo F, Yan M, Zhang X, Zhang B, Huang X, Huang D, Pan Y, Verma KK, Li YR. (2022) Effect of smut infection on the photosynthetic physiological characteristics and related defense enzymes of sugarcane. *Life* 12, 1201. https://doi.org/10.3390/life12081201

Song XP, Verma KK, Tian DD, Zhang XQ, Liang YJ, Huang X, Li CN, Li YR. (2021) Exploration of silicon functions to integrate with biotic stress tolerance and crop improvement. *Biol. Res.* 54, 19. https://doi.org/10.1186/s40659-021-00344-4

Spokas KA, Baker JM, Reicosky DC. (2010) Ethylene: Potential key for biochar amendment impacts. *Plant Soil* 333, 443–452.

Stewart JR, Toma YO, Fernandez FG, Nishiwaki AYA, Yamada T, Bollero G. (2009) The ecology and agronomy of Miscanthus sinensis, a species important to bioenergy crop development, in its native range in Japan: A review. *GCB Bioenergy* 1(2), 126–153.

Strezov V, Anawar HM, Strezov V. (2019) *Renewable Energy Systems from Biomass.* CRC Press: Boca Raton, FL.

Tian D, Qin L, Verma KK, Wei L, Li J, Li B, Zhou W, He Z, Wei D, Huang S, Long S, Huang Q, Li C, Wei S. (2023) Transcriptomic and metabolomic differences between banana varieties which are resistant or susceptible to *Fusarium wilt. PeerJ* 11, e16549. https://doi.org/10.7717/peerj.16549

Tripathi G, Dubey P, Yadav P, Salijonova S, Farooqui A. (2023) Advancement in biodiesel production methodologies using different feedstock. In: *Wastewater Resource Recovery and Biological Methods.* Springer International Publishing: Cham, pp. 323–341.

Tripathi G, Sharma D, Mishra P, Mishra V, Irum Dubey P, Ahmad S, Mishra V, Farooqui A. (2024) Fabrication of activated charcoal from paddy waste for bioethanol production. In *Paddy Straw Waste for Biorefinery Applications.* Springer Nature: Singapore, pp. 127–149.

Usmani Z, Sharma M, Gupta P, Karpichev Y, Gathergood N, Bhat R, Gupta VK. (2020) Ionic liquid based pretreatment of lignocellulosic biomass for enhanced bioconversion. *Bioresour. Technol.* 304, 123003.

Ussiri DA, Lal R. (2014) Miscanthus agronomy and bioenergy feedstock potential on minesoils. *Biofuels* 5(6), 741–770.

Van Sluisveld MA, Hof AF, Carrara S, Geels FW, Nilsson M, Rogge K, Turnheim B, Van Vuuren DP. (2020) Aligning integrated assessment modelling with socio-technical transition insights: An application to low-carbon energy scenario analysis in Europe. *Technol. Forecast. Soc. Change* 151, 119177.

Verma KK, Joshi A, Song X-P, Liang Q, Xu L, Huang H-r, Wu K-C, Seth CS, Arora J and Li Y-R. (2024a) Regulatory mechanisms of plant rhizobacteria on plants to the adaptation of adverse agroclimatic variables. *Front. Plant Sci.* 15, 1377793. https://doi.org/10.3389/fpls.2024.1377793

Verma KK, Joshi A, Song X-P, Singh S, Kumari A, Arora J, Singh SK, Solanki MK, Seth CS, Li Y-R. (2024b) Synergistic interactions of nanoparticles and plant growth promoting rhizobacteria enhancing soil-plant systems: A multigenerational perspective. *Front. Plant Sci.* 15, 1376214. https://doi.org/10.3389/fpls.2024.1376214

Verma KK, Li DM, Singh M, Rajput VD, Malviya MK, Minkina T, Singh RK, Singh P, Song XP, Li YR. (2020a) Interactive role of silicon and plant-rhizobacteria mitigating abiotic stresses: A new approach for sustainable agriculture and climate change. *Plants* 9, 1055. https://doi.org/10.3390/plants9091055.

Verma KK, Rajput VD, Singh M, Verma CL, Minkina T. (2022a) Influence of waterlogging on plant physiology and molecular biology. In: *Understanding Abiotic Stresses*. Nova Science Publishers, Inc.: New York.

Verma KK, Siddiqui N, Singh M. (2022b) Understanding the plant physiological and molecular adaptation strategies to soil salinity. In: *Understanding Abiotic Stresses*. Nova Science Publishers, Inc.: New York. https://doi.org/10.52305/UBUJ4024

Verma KK, Singh P, Song X-P, Malviya MK, Singh RK, Chen G-L, Solomon S, Li YR. (2020b) Mitigating climate change for sugarcane improvement: Role of silicon in alleviating abiotic stresses. *Sugar Tech.* 22(5), 741–749. https://doi.org/10.1007/s12355-020-00831-0.

Verma KK, Singh RK, Song QQ, Singh P, Zhang B-Q, Song X-P, Chen G-L, Li YR. (2019a) Silicon alleviates drought stress of sugarcane plants by improving antioxidant responses. *Biomed. J. Sci. Tech. Res.* 002957. https://doi.org/10.26717/BJSTR.2019.17.002957

Verma KK, Song X, Degu HD, Guo D, Joshi A, Huang H, Xu L, Singh M, Huang D, Rajput VD, Li YR. (2023a) Recent advances in nitrogen and nano-nitrogen fertilizers for sustainable crop production: A mini-review. *Chem. Biol. Technolog. Agricult.* 10, 1–14. https://doi.org/10.1186/s40538-023-00488-3

Verma KK, Song X, Verma CL, Huang H, Singh M, Xu L, Wu J, Khan ZH, Zhang B, Li Y. (2023b) Mathematical modeling of climate and fluoride effects on sugarcane photosynthesis with silicon nanoparticles. *Plant Physiol. Biochem.* 204, 108089. https://doi.org/10.1016/j.plaphy.2023.108089

Verma KK, Song XP, Budeguer F, Nikpay A, Enrique R, Singh M, Zhang B, Wu, JM, Li YR. (2022c) Genetic engineering: An efficient approach to mitigating biotic and abiotic stresses in sugarcane cultivation. *Plant Signal. Behav.* 17(1), e2108253.

Verma KK, Song XP, Joshi A, Rajput VD, Singh M, Sharma A, Singh RK, Li DM, Arora J, Minkina T, Li YR. (2022d) Nanofertilizer possibilities for healthy soil, water and food in future: An overview. Front. *Plant Sci.* 13, 865048. https://doi.org/10.3389/fpls.2022.865048

Verma KK, Song XP, Joshi A, Tian DD, Rajput VD, Singh M, Arora J, Minkina TM, Li YR. (2022e) Recent trends in nano-fertilizer for advancing sustainable agriculture under the era of climate change to ensure food security for future generations. *Nanomaterials* 12, 173. https://doi.org/10.3390/nano12010173.

Verma KK, Song XP, Li DM, Singh M, Wu JM, Singh RK, Sharma A, Zhang BQ, LI YR. (2022f) Silicon and soil microorganisms improves rhizospheric soil health with bacterial community, plant growth, performance and yield. *Plant Signal. Behav.* 17(1), e2104004.

Verma KK, Song XP, Lin B, Guo DJ, Singh M, Rajput VD, Singh RK, Singh P, Sharma A, Malviya MK, Chen GL, Li YR. (2022g) Silicon induced drought tolerance in crop plants: Physiological adaptation strategies. *Silicon* 14(6), 2473–2487. https://doi.org/10.1007/s12633-021-01071-x

Verma KK, Song XP, Rajput VD, Boldyreva V, Zhang BQ, Minkina T, Li YR. (2022h) Morpho-physiological, biochemical and ultrastructural modifications on sugarcane to prolonged water deficit. Verma KK. et al. (Eds.), *Agro-industrial Perspectives on Sugarcane Production Under Environmental Stress*. https://doi.org/10.1007/978-981-19-3955-6_11.

Verma KK, Song XP, Rajput VD, Solomon S, Li YR, Rao GP. (2022i) *Agro-industrial Perspectives on Sugarcane Production Under Environmental Stress*. Springer Nature: Singapore, https://doi.org/10.1007/978-981-19-3955-6

Verma KK, Song XP, Singh M, Tian DD, Rajput VD, Minkina T, Li YR. (2023c) Association of silicon and soil microorganisms induces stress mitigation, increasing plant productivity. de Mello Prado R. (Ed.), *Benefits of Silicon in the Nutrition of Plants*. pp. 299–328. https://doi.org/10.1007/978-3-031-26673-7_17

Verma KK, Song XP, Tian DD, Guo DJ, Chen ZL, Zhong CS, Nikpay A, Singh M, Rajput VD, Singh RK, Minkina T, Li YR. (2021a) Influence of silicon on biocontrol strategies to manage biotic stress for crop protection, performance and improvement. *Plants* 10, 2163. https://doi.org/10.3390/plants10102163

Verma KK, Song XP, Tian DD, Singh M, Verma CL, Rajput VD, Singh RK, Sharma A, Singh P, Malviya MK, Li YR. (2021b) Investigation of defensive role of silicon during drought stress induced by irrigation capacity in sugarcane: Physiological and biochemical characteristics. *ACS Omega* 6, 19811–19821.

Verma KK, Song XP, Verma CL, Chen ZL, Rajput VD, Wu KC, Liao F, Chen GL, Li YR. (2021c) Functional relationship between photosynthetic leaf gas exchange in response to silicon application and water stress mitigation in sugarcane. *Biol. Res*. 54, 15. https://doi.org/10.1186/s40659-021-00338-2

Verma KK, Song XP, Verma CL, Malviya MK, Guo DJ, Rajput VD, Sharma A, Wei KJ, Chen GL, Solomon S, Li YR. (2021d) Predication of photosynthetic leaf gas exchange of sugarcane (*Saccharum* spp.) leaves in response to leaf positions to foliar spray of potassium salt of active phosphorus under limited water irrigation. *ACS Omega* 6, 2396–2409.

Verma KK, Song X-P, Xu L, Huang H-R, Liang Q, Seth CS, Li Y-R. (2023d) Nano-microplastic and agro-ecosystems: A mini-review. *Front. Plant Sci*. 14, 1283852. https://doi.org/10.3389/fpls.2023.1283852

Verma KK, Song XP, Yadav G, Degu HD, Parvaiz A, Singh M, Huang HR, Mustafa G, Xu L, Li YR. (2022j) Impact of agroclimatic variables on proteo-genomics in sugarcane (*Saccharum* spp.) plant productivity. *ACS Omega* 7, 22997–23008.

Verma KK, Song XP, Zeng Y, Guo DJ, Sing M, Rajput VD, Malviya MK, Wei KJ, Sharma A, Li DP, Chen GL, Li YR. (2021e) Foliar application of silicon boosts growth, photosynthetic leaf gas exchange, antioxidative response and resistance to limited water irrigation in sugarcane (*Saccharum officinarum* L.). *Plant Physiol. Biochem.* 166, 582–592.

Verma KK, Song XP,Zeng Y, Li DM, Guo DJ, Rajput VD, Chen GL, Barakhov A, Minkina TM, Li YR. (2020c) Characteristics and correlation of leaf stomata and its relationship with photosynthesis on *Saccharum* spp. under different irrigation and silicon application. *ACS Omega* 5, 24145–24153.

Verma KK, Vatsal S, Singh M, Rajput VD, Minkina T. (2022k) Coping with water stress: Effects and adaptive mechanisms to plants. In: *Understanding Abiotic Stresses*. Nova Science Publishers, Inc.: New York.

Verma KK, Wu KC, Verma CL, Li DM, Malviya MK, Singh RK, Singh P, Chen GL, Song XP, Li YR. (2020d) Developing mathematical model for diurnal dynamics on photosynthetic responses in sugarcane responsive to different irrigation and silicon application. *PeerJ* 8, e10154. https://doi.org/10.7717/peerj.10154.

Verma KK, Wu K-C, Singh P, Malviya MK, Singh RK, Song X-P, Li YR. (2019b) The protective role of silicon in sugarcane under water stress: Photosynthesis and antioxidant enzymes. *Biomed. J. Sci. Tech. Res.* 002685. https://doi.org/0.26717/BJSTR.2019.15.002685

Von Braun J. (2008) Rising food prices: What should be done? Steigende Nahrungsmittelpreise: Was sollte getan werden? La hausse des prix alimentaires: Que doit-on faire? *EuroChoices* 7(2), 30–35.

Wang T, Park AHA, Shi Y, Gadikota G. (2019) Carbon dioxide capture and utilization—Closing the carbon cycle. *Energy Fuels* 33(3), 1693–1693.

Wang D, Portis Jr AR, Moose SP, Long SP. (2008) Cool C_4 photosynthesis: Pyruvate Pi dikinase expression and activity corresponds to the exceptional cold tolerance of carbon assimilation in Miscanthus× giganteus. *Plant Physiol.* 148(1), 557–567.

Wang X, Yang Z, Liu X, Huang G, Xiao W, Han L. (2020) The composition characteristics of different crop straw types and their multivariate analysis and comparison. *Waste Manag.* 110, 87–97.

Wijayanta AT, Goto M, Kamiya N. (2015) Great potency of seaweed waste biomass from the carrageenan industry for bioethanol production by peracetic acid–ionic liquid pretreatment. *Biomass Bioenergy* 81, 63–69.

Won W, Maravelias CT. (2017) Thermal fractionation and catalytic upgrading of lignocellulosic biomass to biofuels: Process synthesis and analysis. *Renew. Energy* 114, 357–366.

World Bioenergy Association. (2013) *Biogas-an Important Renewable Energy Source*. World Bioenergy Association: Stockholm, Sweden.

Wu K-C, Huang C-M, Verma KK, Deng Z-N, Huang H-R, Pang T, Cao H-Q, Luo H-B, Jiang S-L, Xu L. (2022) Transcriptomic responses of *Saccharum spontaneum* roots in response to polyethylene glycol–6000 stimulated drought stress. *Front. Plant Sci.* 13, 992755. https://doi.org/10.3389/fpls.2022.992755

Xu L, Deng Z-N, Wu K-C, Malviya MK, Solanki MK, Verma KK, Pang T, Li Y-J, Liu X-Y, Kashyap BK, Dessoky ES, Wang W-Z and Huang H-R. (2022) Transcriptome analysis reveals a gene expression pattern that contributes to sugarcane bud propagation induced by Indole-3-Butyric acid. *Front. Plant Sci.* 13, 852886. https://doi.org/10.3389/fpls.2022.852886

Yamakawa CK, Qin F, Mussatto SI. (2018) Advances and opportunities in biomass conversion technologies and biorefineries for the development of a bio-based economy. *Biomass Bioenergy* 119, 54–60.

Yang J, Xu H, Jiang J, Zhang N, Xie J, Zhao J, Bu Q, Wei M. (2020) Itaconic acid production from undetoxified enzymatic hydrolysate of bamboo residues using Aspergillus terreus. *Bioresour. Technol.* 307, 123208.

Yohanes AS. (2021) *Development of Small-scale Systems for Power Generation and Hydrogen Production by Biomass Thermochemical Conversion.* Ph.D. Dissertation, Graduate School of Science and Technology, Hirosaki University.

Yoo CG, Meng X, Pu Y, Ragauskas AJ. (2020) The critical role of lignin in lignocellulosic biomass conversion and recent pretreatment strategies: A comprehensive review. *Bioresour. Technol.* 301, 122784.

Yun YM, Lee MK, Im SW, Marone A, Trably E, Shin SR, Kim MG, Cho SK, Kim DH. (2018) Biohydrogen production from food waste: Current status, limitations, and future perspectives. *Bioresour. Technol.* 248, 79–87.

Zakaria R, Aziz HA, Wang LK, Hung YT. (2021) Combustion and incineration. *Solid Waste Eng. Manag.* 1, 345–397.

Zhang Z, Lohr L, Escalante C, Wetzstein M. (2010) Food versus fuel: What do prices tell us? *Energy Policy* 38(1), 445–451.

Zhang H, Wang Q, Liu Y, Cui J, Ma X, Gu M, Xia M. (2015) Coupling effects of water availability and pH on switchgrass and the optimization of these variables for switchgrass productivity determined by response surface methodology. *Biomass Bioenergy* 83, 393–402.

Zheng F-J, Lin B, Yang Y-X, Fang X-C, Verma KK, Chen G-L. (2023) Efficacy and functionality of sugarcane original vinegar on mice. *Front. Microbiol.* 14, 1224666. https://doi.org/10.3389/fmicb.2023.1224666

Zhuang Q, Qin Z, Chen M. (2013) Biofuel, land and water: Maize, switchgrass or Miscanthus? *Environ. Res. Lett.* 8(1), 015020.

Zhuang X, Liu J, Wang C, Zhang Q, Ma L. (2022) A review on the stepwise processes of hydrothermal liquefaction (HTL): Recovery of nitrogen sources and upgrading of biocrude. *Fuel* 313, 122671.

Zimmerman AR, Gao B, Ahn MY. (2011) Positive and negative carbon mineralization priming effects among a variety of biochar-amended soils. *Soil Biol. Biochem.* 43(6), 1169–1179.

Zohuri B. (2023) Navigating the global energy landscape balancing growth, demand, and sustainability. *J. Mater. Sci. Apl. Eng.* 2(4).

Zou Y, Xu X, Li L, Yang F, Zhang S. (2018) Enhancing methane production from U. lactuca using combined anaerobically digested sludge (ADS) and rumen fluid pre-treatment and the effect on the solubilization of microbial community structures. *Bioresour. Technol.* 254, 83–90.

Zuazo VHD, Pleguezuelo CRR, Flanagan D, Tejero IG, Fernández JLM. (2011) Sustainable land use and agricultural soil. In: *Alternative Farming Systems, Biotechnology, Drought Stress and Ecological Fertilisation*, pp. 107–192.

CHAPTER 10

Impact of Irrigation Management on Sugarcane in Response to Adverse Environmental Variables

MUSA OPEYEMI AHMED,[1,2*] ADEPOJU MORONKOLA IDRIS,[1,3*] FAITH EWELUEGIM ENAHORO-OFAGBE,[1,3] and KINGSLEY O. IWUOZOR[1,4]

[1]*Nigeria Sugar Institute, Ilorin, Kwara State, Nigeria*

[2]*Department of Agricultural and Environmental Engineering, University of Ibadan, Ibadan, Oyo State, Nigeria*

[3]*Department of Agronomy, University of Ilorin, Ilorin, Kwara State, Nigeria*

[4]*Department of Pure and Industrial Chemistry, Nnamdi Azikiwe University, Akwa, Anambra State, Nigeria*

*Corresponding author

ABSTRACT

Sugarcane is a C_4 crop and requires the right conditions to thrive. However, climate change has been reported in the literature to affect most agricultural systems through biotic and abiotic stresses, and the reoccurrence of abiotic stress is expected to decrease the yield of sugarcane over the coming decades by 50%. Therefore, to rectify the issues relating to stress conditions, there is a need to understand proper irrigation management. This chapter carefully reviews literature that has been published to study how sugarcane reacts to various stress conditions and how irrigation can be managed to improve its yield. It was observed that some of the stress

Sugarcane Production: Environmental Impacts and Socioeconomic Issues.
Krishan K. Verma, Xiu-Peng Song, Lin Xu, Rajan Bhatt, and Yang-Rui Li (Eds.)

DOI: 10.1201/9781779641137-10

conditions drastically affected sugarcane at the early and grand growth phases, thereby causing a reduction in sugarcane growth, quality, and yield. The mechanism through which proper irrigation management remediates this problem of unfavorable stress conditions and its impact on sugarcane yield was critically discussed.

10.1 INTRODUCTION

Sugarcane, grown primarily for producing ethanol and sugar in tropical and subtropical regions, is important to the global economy (Dlamini, 2021; Verma et al., 2023a, 2023b). Over the past 10 years, yields from this crop have increased noticeably, accounting for 80% of the sugar produced globally (Azevedo et al., 2011; Verma et al., 2019a, 2019b; 2022c). As a result, the effect of sugarcane on the economy is undeniable, and its usage in various ways cannot be viewed as prospective but rather as an actuality. Sugarcane belongs to the Poaceae family and is mainly grown within a latitude of 36.7°N–31°S of the equator (Kumar et al., 2021). It is growing erect up to 5 or 6 m long and producing multiple stems. Sugarcane can store up to 50% of its dry weight in sucrose in the plant stalks (Kumar et al., 2021; Bhatt et al., 2023a, 2023b; Verma et al., 2023d). Apart from being a crucial crop for the manufacturing of sugar and alcohol, its biomass is also used for bio-electricity generation, paper, and fertilizers (Afiomah and Iwuozor, 2020; Dlamini, 2021; Iwuozor et al., 2022; Rabelo et al., 2020; Chen et al., 2020; Zheng et al., 2023).

Sugarcane has a lengthy growth cycle, which presents significant environmental challenges that impact its development and growth (Misra et al., 2020; Verma et al., 2020a, 2020b; Chen et al., 2023). The decline in sugarcane yield is attributed to the stress that the cane undergoes. Sugarcane is subjected to biotic and abiotic stresses. Biotic stress results from harm to plants caused by other living things, such as nematodes, weeds, insect pests, disease-causing organisms, and allelopathic chemicals (Rana et al., 2016; Joshi et al., 2022; Tian et al., 2023). On the other hand, abiotic stress occurs naturally due to changing climatic conditions, thereby causing a reduction in plant yield. Amidst the aforementioned type of stress, abiotic stress is the most concerned in the light of irrigation (Krishna et al., 2022; Kumari et al., 2022, 2023a, 2023b).

Abiotic stress has emerged as a persistent issue due to the sudden changes in climatic conditions (Jun et al., 2023; Joshi et al., 2023;

Huang et al., 2024). Abiotic stressors directly or indirectly influence sugarcane by changing the crop's metabolism, growth, and development (Misra et al., 2020; Jagadesh et al., 2024; Verma et al., 2024a, 2024b). As a result of climate change, abiotic stresses on sugarcane affect the health of the soil (Guo et al., 2021, 2022, 2023), the growth and development of the crop, its synthesis and accumulation of sugar, chemical composition, the availability of seed cane, as well as aggravating other abiotic/biotic stresses that increase losses (Shrivastava et al., 2017; Verma et al., 2020c, 2020d, 2020e; Singh et al., 2020, 2021). Abiotic stress includes drought, salinity, waterlogging, and temperature stress (Rana et al., 2016; Chaudhary et al., 2023). Various researchers have reported the impact of these abiotic stresses on sugarcane (Misra et al., 2016; Misra et al., 2020; Meena et al., 2020; Shrivastava et al., 2017; Deng et al., 2023). For example, Misra et al. (2020) reported that drought has posed a severe threat to the entire Indian nation. Roughly 2.97 lakh acres of land used for sugarcane production are particularly vulnerable to drought, which can hinder the crop at different stages of growth. Waterlogging reported by Gomathi et al. (2015) in sugarcane fields has caused piping in stalks and roots, development of adventitious roots, high rate of tiller mortality, reduced growth and development, as well as reduction in stalk weight and 5–45% loss of cane yield. The problem of salinity has been reported by Meena et al. (2020) to have been one of the major global problems affecting the production of crops, with 33% of irrigated land and 20% of cultivated land suffering from salt accumulation globally. Hence, proper irrigation management must be carried out to mitigate the effects of these abiotic stresses on sugarcane.

Management of irrigation is a critical topic that needs to be extensively discussed. Using various methods and technologies to properly manage water resources is one practical way to address the overwhelming issue of sugarcane under stress. Many studies have looked into the potential for effective irrigation management techniques (Cahn and Johnson, 2017; Saccon, 2018; Li et al., 2023; Singh et al., 2023). However, most of these studies heavily focus on drought conditions (Pereira et al., 2002; Singels et al., 2019). Therefore, to the author's knowledge, no research has studied the impact of irrigation management on sugarcane under different stress conditions. In light of the issues above, this chapter focuses on the irrigation management concepts applied to reducing the effects of sugarcane under different stress scenarios. When the information provided in the article is considered, it will improve knowledge of irrigation management approaches used to control the different stresses sugarcane experiences to increase cane output.

10.2 IRRIGATION MANAGEMENT

A set of technical choices comprising the features of the crop, climate, water, soil, and irrigation system is referred to as irrigation management (de Oliveira et al., 2015; Oliveira et al., 2022; Huang et al., 2022). A producer can achieve higher productivity levels, save water and energy, and help preserve the environment by properly managing irrigation in conjunction with other farming methods. A well-run management strategy includes identifying the ideal time to begin irrigation and calculating the amount of water the crop will need, using information about the timing of water application or the irrigation equipment's displacement rate.

10.2.1 IMPORTANT PARAMETERS IN THE MANAGEMENT OF IRRIGATION

The soil provides natural water storage for plants. The water that plants can access from the soil is present between the field capacity (FC) and the permanent wilting (PW) point of the soil.

10.2.1.1 SATURATED HUMIDITY

Soil is said to be saturated when all the pores are filled with water. Continued application of water at this stage will result in downward movement and air replacement. Plant growth is impeded due to a lack of aeration (Sharma et al., 2022; Solanki et al., 2023).

10.2.1.2 FIELD CAPACITY (FC)

This is the moisture content of the soil after the drainage of gravitational water has ceased. Hence, the water stored in the soil macro pores has drained out by gravity action and has been replaced with air. This level of wetness encourages plants to take up greater nutrients and water uptake.

It usually occurs in the field two days after adequate rainfall or irrigation (Verma et al., 2021a, 2021b, 2021c; 2022a). The retention curve approach is typically used in laboratories to measure FC. According to this method, depending on the texture, structure, and amount of organic matter in the soil, the value of the FC moisture is represented by the balanced moisture with tension ranging from 6 to 33 kPa.

10.2.1.3 LIMIT OF STRESS (LS)

This is the theoretical point where water is less readily available for the plant. At this stage, the plant's root requires more effort to extract water since water from soil micro-pores is not equally easy to access. A balanced moisture with a tension of two bars is required at this stage.

10.2.1.4 PERMANENT WILTING POINT (PW)

This is the moisture content of the soil that is unavailable for plant use. The soil moisture at this stage is tightly held to the soil particle and requires a suction pressure greater than 15 bar. Ultimately, the plant remains wilted and will die if no water is added. The PW point is determined in the laboratory using the pressure plate chamber at a suction pressure of 15 bar.

10.2.1.5 SOIL DENSITY

The relationship between a dry soil sample's mass and volume are called soil density. An Uhland sampler, whose cylinder is implanted in the soil at a medium depth of the soil layer explored by the roots of the plants, can be used to make this assessment. The sample is ready after the cylinder has been removed, and it is then placed in an oven to dry for 24 hr at a temperature of roughly 105°C to calculate its mass. The volume is calculated using the diameter and height measurements of the soil sample. The density is then calculated using the following equation:

$$Ds = \frac{Ms}{Vs}$$

where D_s: soil density, g/cm^3; M_s: dry soil mass, g; and V_s: soil sample volume, cm^3.

10.2.1.6 SOIL POROSITY

It is the fraction of the bulk volume of soil containing air and water. In other words, it is the percentage of pores (micro and macro pores) present in the soil. Clay has been observed to have the highest porosity, 0.60, while sandy soil has the lowest, 0.28.

10.2.2 IRRIGATION MANAGEMENT STRATEGIES

10.2.2.1 IRRIGATION SCHEDULING

Irrigation scheduling is coordinating water application to crops to match their water usage. When done correctly, the right amount of water is applied at the right frequency to produce the best possible yield. Irrigation scheduling involves planning to ensure adequate water is given to plants to help them withstand other external stressors, making it an essential part of management (Singels et al., 1998). Irrigation scheduling considers factors such as the growth stage of the crops, the weather conditions, and the soil's water storage capacity. By carefully planning and executing irrigation schedules, farmers can ensure that their crops receive the water they need to grow and thrive. This can help to maximize yields and minimize the negative effects of water stress on the crops.

In irrigation scheduling, the inputs considered include rainfall and irrigation. Outputs to consider include the water used by crops through processes like evapotranspiration (ET) and water loss through deep percolation. To manage the impact of water stress on plants, irrigation scheduling will then involve estimating the capacity of soil to hold water, estimating the water needs of crops to determine how much water to apply and when, and replacing soil moisture used up with irrigation water. The water balance method is commonly used to determine when irrigation is necessary. This method balances the output (water loss) with the input (water supply). When the soil moisture reaches a predetermined target allowable for depletion, called the management allowable depletion level, irrigation becomes necessary. The irrigation applied will be the amount of water needed to bring the soil back to FC. Factors that determine irrigation water requirement are reference evapotranspiration (ET_o), crop evapotranspiration, ET_c (also known as crop water requirement), and effective rainfall (P_e).

The tensiometer is also used to schedule irrigation. It measures the soil moisture potential (tension or suction). The tensiometer consists of a hollow tube joined to a porous ceramic tip at the base and a vacuum gauge and removable cap at the top. The mode of operation simply involves moving out of water from within the tensiometer through the ceramic tip when the soil is dry. The water loss creates a vacuum in the tensiometer, which is then recorded by the vacuum gauge as a suction reading. For irrigation scheduling, the tensiometer must be calibrated with the FC and the PW point of the soil. Irrigation begins when the tensiometer reads a predetermined level close to the value of the limit of stress. After irrigation or rainfall, water moves back through the ceramic tip, and the vacuum in the tensiometer is reduced.

10.2.2.2 FULL IRRIGATION MANAGEMENT

Full irrigation management is the irrigation management carried out to prevent water stress impact during the entire growing season of a plant (Irmak and Rudnick, 2014). Scheduling irrigation to avoid water stress necessitates thorough knowledge of the irrigation system (such as system capacity and efficiency) as well as regular and precise information on crop status and field characteristics, including current and anticipated crop water demands and plant soil water availability. The volume of irrigation application must be sufficient to meet crop water needs while still being moderate enough to minimize percolation and surface runoff of water and nutrients beneath the crop root zone.

10.2.2.3 DEFICIT AND LIMITED IRRIGATION MANAGEMENT

As previously indicated, there are instances when applying less water than the crop needs is the only alternative when full irrigation control is not an option. So, to improve or optimize yield per unit of applied water, researchers have created deficit and/or limited irrigation management systems. Limited irrigation distributes all available water in fixed amounts throughout the growing season, regardless of the crop development stage. In contrast, deficit irrigation withholds water at crop growth phases less vulnerable to water stress than others. Due to this, the primary distinction between deficit and limited irrigation approaches is that the former takes crop growth stages into account while the latter does not (Irmak and Rudnick, 2014; Verma et al., 2021d, 2021e).

Deficit irrigation is a technique that aims to maximize the efficiency of irrigation to produce higher yields with less water. It is typically used in situations when there is a water allotment or severe water shortage. This is done by carefully regulating the amount of water provided to crops, particularly during critical growth stages, to optimize the use of irrigation resources. Farmers can improve their yields and make the most of their irrigation water by using this method (Verma et al., 2022b, 2022c; Wu et al., 2022).

As regards limited irrigation, water must be available at the time of irrigation when using the limited irrigation management method, which involves administering less water than the crop ETa. Limited irrigation management is used for a variety of reasons. These include water allocations, delays in irrigation scheduling, and operational savings from pumping water. Insufficient irrigation system capacity to meet agricultural water demands is also among them. Different levels of water stress might emerge from limited irrigation, depending on the amount and distribution of precipitation (Verma et al., 2022d, 2022e; Xu et al., 2022). Due to the significant diversity in precipitation timing and volume from one place to another, it is crucial to establish locally specialized deficit or limited irrigation solutions for the effectiveness of these approaches (Irmak and Rudnick, 2014).

10.3 IMPACT OF STRESS ON SUGARCANE: DROUGHT, WATERLOGGING, SALINITY, AND HEAT STRESS

Sugarcane is the most important sugar crop highly sensitive to stress (Song et al., 2022; Malviya et al., 2022). Generally, stresses are unfavorable condition that poses a threat to crop productivity due to their ability to reduce metabolic activities, inhibit physiological processes, and cause a decline in growth and development, as well as the yield of sugarcane during its active growing phase of sugarcane (Shrivastava et al., 2017; Singh et al., 2019; Misra et al., 2020; Anas et al., 2021a, 2021b; Huang et al., 2024). Biotic and abiotic stresses are major categories of stress that influence sugarcane productivity. Abiotic stresses such as drought, waterlogging, salinity, and heat are major abiotic stresses that contribute adversely to sugarcane growth, development, sucrose accumulation, recovery, and cane yield (Shrivastava et al., 2017; Song et al., 2021; Sharma et al., 2021). However, water resources management has proven effective in alleviating the detrimental effect of water stress in sugarcane (Song et al., 2019; Verma et al., 2022f, 2022g, 2022h).

10.3.1 IMPACT OF IRRIGATION CONTROL ON SUGARCANE DURING A DROUGHT

Irrigation, which means supplementary of water during drought, a situation caused by water deficit, has always been the sustainable approach to solving the problem of drought in the sugarcane-producing regions of the world (Santos et al., 2019; Verma et al., 2022i, 2023b). Drought is a significant abiotic stress that affects sugarcane growth and development, affecting over 30% of arable land use in sugarcane-cultivated regions. Drought is mostly established when transpiration exceeds rainfall during the sugarcane active growing period. However, irrigation is employed to supplement the deficit in rainfall to ensure balance (Misra et al., 2020; Verma et al., 2020b). According to Barbosa et al. (2015) and Inman-Bamber and Smith (2005), water is essential for proper physiological processes for sugarcane growth, while its deficit will result in low photosynthesis due to impairment in biochemical processes and reduced leaf gas exchange, and consequently result in a reduction in growth (Malviya et al., 2021; Verma et al., 2022j, 2022k; Xu et al., 2022).

Inman-Bamber and Smith (2005) stated that irrigation management is crucial for efficient water use in a manner that will not be detrimental to crops, as sugarcane morphology responds differently to irrigation. Hence, water deficit at an early stage of sugarcane growth does not significantly affect the crop yield and sucrose level. For example, five (5) months of water deficit at the elongation stage of sugarcane, when the leaf area index (LAI) is less than two (LAI < 2), will reduce the leaf area significantly from 1.8 to 0.9 but will not significantly reduce yield and sucrose accumulation due to ability to recover faster during and low ET at early stage of growth when irrigated (Inman-Bamber and Smith, 2005; Wiedenfeld, 2000).

However, 19% reduction in total biomass and 27% in sucrose accumulation has been attributed to water stress as reported by (Inman-Bamber et al., 2008), which implies that water stress can be both beneficial and detrimental depending on the grower's objectives and the stage of the crop. Although sugarcane is regarded as high biomass crop which requires large amount of water for optimum production, the reduction in cane yield and sugar content has been attributed to the grand growth stage when the crop canopy is greater than 70% as a result of increased ET and insufficient water application of supplementary water at this stage will balance the water loss by ET for crop optimum production (Wiedenfeld, 2000; Inman-Bamber and Smith, 2005; Inman-Bamber et al., 2008).

Shorter internode, reduction in LAI, reduced stem girth, reduction in photosynthesis, and consequently yield reduction have been attributed to 12–80% moisture depletion during the sugarcane growth stage (Table 10.1). Further increase in water stress caused by drought will decrease yield by reducing the stalk elongation rate (Thomas et al., 1978; Lowveld, 1996; Wiedenfeld, 2000). Conversely, the effect of water stress is primarily related to the level/degree of stress in relation to ET demand rather than the crop growth stages to which the stress occurred. Therefore, understanding the ET during crop growth will ensure efficient irrigation management (Wiedenfeld, 2000).

de Almeida Silva et al. (2011) suggested sufficient irrigation during crop emergence, tillering, and canopy establishment. However, the consumptive water use by the sugarcane crop is more effective at grand growth stage than the former. Moreover, effective irrigation management for limited water conditions is beneficial at an early stage of sugarcane growth. Hence, early water application prior to grand growth, full canopy cover is necessary to balance water loss by ET. Full irrigation management practice can be used in areas susceptible to drought stress. Therefore, sufficient irrigation during crop emergence, tillering, and canopy establishment will help reduce the impact of drought stress on sugarcane. This causes stalk elongation, increased LAI, increased stomata conductance, and increased yield.

10.3.2 IMPACT OF IRRIGATION CONTROL ON SUGARCANE DURING A WATERLOGGING STRESS

Waterlogging is the major water stress that seriously impacts sugarcane productivity, especially in areas with a high-water table. Globally, climate change has caused frequent and severe waterlogging conditions in

TABLE 10.1 Drought Level Effect on Physiological Parameters of Sugarcane

Stress Condition	Crop Stage at Which Stress is Applied	Observations	References
60 days of stress	Early growth stage (60–120 DAT)	Total biomass, shoot, and root dry mass	Dinh et al. (2017)
60 days of stress	Tilling up to the grand growth phase (60–120 DAP)	Stalk diameter, number of internodes, intermodal length, cane height, leaf length and width, fresh and dry root weight	Misra et al. (2020)
Water stress	Formative growth phase (60–150 DAP)	internode length, internode number, cane length, and cane diameter	Hemaprabha et al. (2013)
20 days of stress	Tilling phase	Number of leaves, plant height, and stalk diameter	Vu et al. (2019)
20% stress	Tilling phase (88–120 DAP)	Stalk height, stalk diameter, leaf area, and tiller mass	Endres et al. (2016)
Water stress	Formative growth phase	Stalk weight, stalk number, stalk diameter, and stalk height	Silva et al. (2008)

sugarcane-growing regions of the world (Meena et al., 2014). Waterlogging occurs when the soil water table increases, so the root zone becomes saturated with water. This is a condition that results in a plant metabolic shift from aerobic conditions to anaerobic conditions due to significant depletion of oxygen in the root zone, thereby hampering the growth and development of sugarcane (Gilbert et al., 2007, 2008; Sasidharan et al., 2017; Singh et al., 2019). Waterlogging constitutes about an 18–64% reduction in cane yield depending on genotype, stress duration, growth stage, and management practice in terms of drainage (Singh et al., 2019).

According to Gomathi et al. (2015) and Singh et al. (2019), sugarcane is mainly susceptible to waterlogged conditions during the early stage of crop development, especially in the first 3–4months and considerably tolerant at 5–9 months, depending on the water level, waterlogged duration, or submerged condition, also varietal dependent. Moreover, the severity of this stress on the morphology and physiological processes of sugarcane depends on the variety, time, and level of waterlogging in the field (Hasan et al., 2003). Waterlogging conditions result in yield reduction, tiller abortion, reduced stalk weight, stalk breakage, decreased photosynthesis, imbalance in nutrition, decreased root development, and reduced leaf area. Furthermore, researchers have attributed poor germination and poor crop establishment to waterlogged conditions that persist up to 3 days after planting due to reduced dissolved oxygen around the cane set and significantly affecting bud sprouting. Therefore, the construction of good field drains is essential to ameliorate the effect of waterlogging during this developmental phase. During irrigation, we must consider the height of the water table as it is inversely proportional to the irrigation interval, which means the higher the water table, the less frequent irrigation and vice versa.

However, deficit irrigation management can be practiced to mitigate the impact of waterlogging on sugarcane. This is done by carefully regulating the amount of water provided to crops, particularly during critical growth stages, in order to optimize the use of irrigation resources. The water must be applied with a high degree of uniformity to prevent excessive deep percolation in some parts of the field, which could cause waterlogging. Limited irrigation practice will give the soil proper aeration for the sugarcane to thrive. Hence, the impact will result in yield increase, proper photosynthesis, stalk elongation, and increased tillering ability. Irrigation systems appropriate to this management approach include drip irrigation, linear move sprinkler systems, center-pivot sprinkler systems, and low-energy precise application irrigation machines. Furthermore, the withdrawal of water and the construction of infield drains have been the age-old management approaches to waterlogging (Gomathi et al., 2015; Jain et al., 2017; Singh et al., 2019).

10.3.3 IMPACT OF IRRIGATION CONTROL ON SUGARCANE DURING A HEAT STRESS

Heat stress is another form that impacts sugarcane productivity in sugarcane-producing regions worldwide. Heat stress is simply a rise in temperature beyond a favorable threshold level for plant growth, which may result in 2.5–10% yield decline, inhibiting plant physiological processes and poor morphological performance (Hasanuzzaman et al., 2013; Zhao and Li, 2015). According to Hasanuzzaman et al. (2013), the formative stage and grand growth stages of sugarcane were found to be more sensitive to high temperatures. The report has shown a decrease in chlorophyll fluorescence F_V/F_m, leaf RWC (relative water content), nitrogen reductase NR activities, proline accumulation, and other metabolic activities with genotype under heat stress. In contrast, the heat stress process constitutes little or no effect on the noninduced stress genotype. In this report, heat stress increases the total soluble solids of sugarcane during these two growth phases (Figure 10.1). Sugarcane yield and its components were reported to have been affected by heat stress with a significant decrease in tillering, elongation, and stalk characteristics at temperatures greater than 35°C, consequently resulting in yield decrease across sugarcane-producing regions (Mall et al., 2016). At different sugarcane developmental phases, temperatures above or below the optimum temperature threshold of 20–40°C may slow the growth and development of sugarcane and drastically reduce production (Figure 10.1).

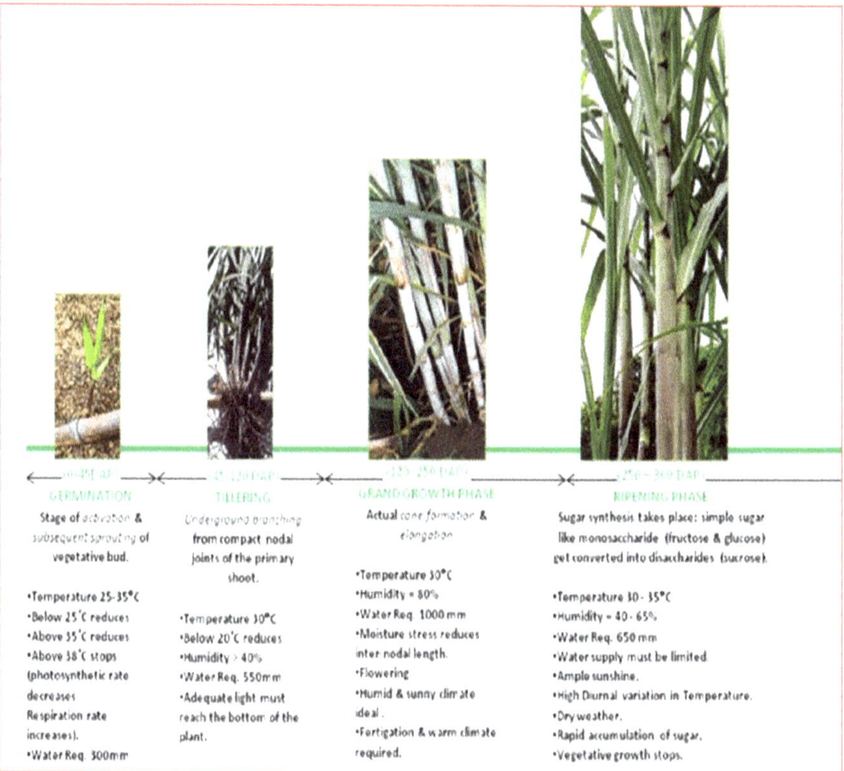

FIGURE 10.1 Relative effect of heat stress at different developmental phases of sugarcane.

Source: Adapted from Mall et al. (2016).

Similarly, increased photorespiration due to high temperature may reduce sugar accumulation (Gawander, 2007; Teixeira et al., 2013; Rao et al., 2015; Jaiphong et al., 2016; Mall et al., 2016). However, effective irrigation or rainfall prior to or during heat stress could help reduce the effect on sugarcane productivity, as the soil moisture content reduces at heat stress due to increased evaporation (Hunsigi, 2012; Santos et al., 2015). Furthermore, full irrigation management practice is usually advised to be practiced to mitigate the effect of heat stress on the sugarcane. This practice will help reduce the rising temperature of both soil and plant, which results from increasing heat stress. By so doing, the leaf RWC will increase morphological processes and yield increase.

10.3.4 IMPACT OF IRRIGATION CONTROL ON SUGARCANE DURING A SALINITY STRESS

Abiotic stress potentially results in soil salinity due to the concentration of soluble salt (e.g., sodium) on the soil surface, which leads to an increase in the electrical conductivity (EC) of the soil (Sharma and UPPAL, 1999; Wiedenfeld, 2008). Salinity has been reported to have caused a 25–50% reduction in sugarcane germination at 45 DAP, a situation caused by inadequate water around the buds or the direct effect of sodium salt on germination. The number of leaves, LAI, and tillering are reduced by 25–50%, 35–75%, and 20–60%, respectively. Hence, this significant effect on the sugarcane growth component will impact the photosynthetic activities of the plant, ultimately resulting in yield loss (Sharma and UPPAL, 1999; Vasantha et al., 2010).

Low transpiration rate, stomata closure, and reduced sucrose accumulation have been reportedly caused by salinity. Moreover, research revealed that a decline in sucrose percentage, yield, and its components (stalk length, stalk height, girth, and internode length) is the result of salinity during the active growth stages of sugarcane. Also, cane quality, such as sucrose, purity, pol, and CCS (commercial cane sugar), has been significantly affected by salinity (Sharma and UPPAL, 1999; Vasantha et al., 2010; Wiedenfeld, 2008). To prevent the adverse effects of salinity, hence, a proper irrigation management plan must be carried out. Irrigation management recommends the use of subsurface drainage to prevent accumulation of salt, periodic rainfall, introduction of leaching factors to wash away soil from the root zone, and the use of the electrostatic precipitation method when irrigation water quality analysis EC and TDS exceed 3.0 dS/m and 2000 mg/l, respectively (Enciso and Wiedenfeld,

2005; Wiedenfeld, 2008; de Oliveira et al., 2015; Culas and Baig, 2020; Wiedenfeld, 2000).

10.4 CONCLUSION

According to the analysis presented above, abiotic stressors are the main obstacles to increasing sugarcane crop production globally. Hence, to address the problem of the nation's sugar security, it is vital to reduce the negative effects of these pressures on the sugarcane crops' productivity and production. This chapter presented the impact of different abiotic stresses on sugarcane, such as drought, waterlogging, heat, and salinity. The adverse effects of the stress above include reduced LAI, decreased photosynthesis rate, reduced cane elongation, poor germination, and reduced yield and quality. Since sugarcane is critically vulnerable to water stress conditions, the amount of irrigation water, when to irrigate it, and how to irrigate it are the questions that must be considered carefully. The selection of proper irrigation management is vital to maximizing sugarcane quality, yield, and water use efficiency while minimizing environmental impacts. It was concluded that using irrigation management practices such as irrigation scheduling and deficit or limited irrigation will help reduce the adverse effects of these stresses and increase sugarcane quality and yield.

KEYWORDS

- **abiotic stress**
- **irrigation management**
- **growth and development**
- **productivity**
- **drought**
- **waterlogging**
- **salinity**
- **heat stress**
- **sugarcane**

REFERENCES

Afiomah CS, Iwuozor KO. (2020) Nutritional and phytochemical properties of beta vulgaris linnaeus (Chenopodiaceae)—A review. *Nigerian J. Pharm. Appl. Sci. Res.* 9(4), 38–44.

Anas M, Verma KK, Haq IU, Naeem M, Li Q, Liao F, Xiang YX, Li YR. (2021a) Characterization of exotic, native and wild-type genotypes of sugarcane (*Saccharum* spp. Hybrids) for internal nitrogen use efficiency under different nitrogen levels. *Sugar Tech.* 23, 1258–1267. https://doi.org/10.1007/s12355-021-01014-1

Anas M, Verma KK, Riaz M, Qiang L, Liao F, Liu Y, Li YR. (2021b) Physio-morphological and biochemical mechanism of nitrogen use efficiency in sugarcane (*Saccharum* spp.) genotypes under different growth stages and nitrogen levels. *J. Plant Interact.* 16, 332–343.

Azevedo RA, Carvalho RF, Cia MC, Gratão PL. (2011) Sugarcane under pressure: An overview of biochemical and physiological studies of abiotic stress. *Trop. Plant Biol.* 4(1), 42–51.

Barbosa A, Guidorizi K, Catuchi T, Marques T, Ribeiro R, Souza G. (2015) Biomass and bioenergy partitioning of sugarcane plants under water deficit. *Acta Physiol. Plant.* 37(8), 1–8.

Bhatt R, Verma KK, Jamwal NS. (2023a) Can paddy-compost eliminate N fertilization for climate-smart sugarcane? *Sugar Tech.* 25, 1036–1046. https://doi.org/10.1007/s12355-023-01270-3

Bhatt R, Verma KK, Kumar R, Sanghera GS. (2023b) Foliar application of potassium salt of active phosphorus (PSAP) mitigates insect pests and improves yield along with sugarcane quality in response to agroclimatic conditions of Punjab. *Sugar Tech.* 25(3), 660–669.

Cahn MD, Johnson LF. (2017) New approaches to irrigation scheduling of vegetables. *Horticulture* 3(2), 28.

Chaudhary P, Xu M, Ahamad L, Chaudhary A, Kumar G, Adeleke BS, Verma KK, Hu D-M, Siric I, Kumar P, Popescu SM, Fayssal SA. (2023) Application of synthetic consortia for improvement of soil fertility, pollution remediation, and agricultural productivity: A review. *Agronomy* 13, 643. https://doi.org/10.3390/agronomy13030643.

Chen GL, Zheng FJ, Lin B, Lao SB, He J, Huang Z, Zeng Y, Sun J, Verma KK. (2020) Phenolic and volatile compounds in the production of sugarcane vinegar. *ACS Omega* 5, 30587–30595.

Chen G-L, Zheng F-J, Lin B, Yang Y-X, Fang X-C, Verma KK, Yang L-F. (2023) Vinegar: A potential source of healthy and functional food with special reference to sugarcane vinegar. *Front. Nutr.* 10, 1145862. https://doi.org/10.3389/fnut.2023.1145862

Culas RJ, Baig IA. (2020) Impacts of irrigation water user allocations on water quality and crop productivity: The LCC irrigation system in Pakistan. *Irrigation Drainage* 69(1), 38–51.

de Almeida Silva M, Jifon JL, Sharma V, da Silva JA, Caputo MM, Damaj MB, Ferro MI. (2011) Use of physiological parameters in screening drought tolerance in sugarcane genotypes. *Sugar Tech.* 13(3), 191–197.

de Oliveira RA, Ramos MM, de Aquino LA. (2015) *Irrigation Management Sugarcane*. Elsevier, pp. 161–183.

Deng ZN, Huang HR, Huang CM, Verma KK, Luo HB, Jiang SL, Li YJ, Cao HQ, Wu KC, Xu L. (2023) Short-term limited water irrigation influences the photosynthetic pigments, enzymatic and non-enzymatic activities in *Saccharum spontaneum* L. at vegetative stage. *Sugar Tech.* 25, 110–118. https://doi.org/10.1007/s12355-022-01175-7

Dinh TH, Watanabe K, Takaragawa H, Nakabaru M, Kawamitsu Y. (2017) Photosynthetic response and nitrogen use efficiency of sugarcane under drought stress conditions with different nitrogen application levels. *Plant Prod. Sci.* 20(4), 412–422.

Dlamini PJ. (2021) Drought stress tolerance mechanisms and breeding effort in sugarcane: A review of progress and constraints in South Africa. *Plant Stress* 2, 100027.

Enciso J, Wiedenfeld B. (2005) Irrigation guidelines based on historical weather data in the Lower Rio Grande Valley of Texas. *Agric. Water Manag.* 76(1), 1–7.

Endres L, Oliveira NG, Ferreira VM, Silva JV, Barbosa GV, Maia Junior SO. (2016) Morphological and physiological response of sugarcane under abiotic stress to neonicotinoid insecticides. *Theoretical Exp. Plant Physiol.* 28(4), 347–355.

Gawander J. (2007) Impact of climate change on sugar-cane production in Fiji. *World Meteorol. Organization Bull.* 56(1), 34–39.

Gilbert R, Rainbolt C, Morris D, McCray J. (2008) Sugarcane growth and yield responses to a 3-month summer flood. *Agric. Water Manag.* 95(3), 283–291.

Gilbert RA, Rainbolt CR, Morris DR, Bennett AC. (2007) Morphological responses of sugarcane to long-term flooding. *Agron. J.* 99(6), 1622–1628.

Gomathi R, Gururaja Rao P, Chandran K, Selvi A. (2015) Adaptive responses of sugarcane to waterlogging stress: An overview. *Sugar Tech.* 17(4), 325–338.

Guo D-J, Li D-P, Chen Z-D, Yang B, Verma KK, Singh RK, Singh P, Khan Q, Sharma A, Qin Y, Zhang B-Q, Song X-P, Li Y-R. (2023) Effect of endophytic diazotroph *Enterobacter roggenkampii* ED5 on nitrogen-metabolism-related microecology in the sugarcane rhizosphere at different nitrogen levels. *Front. Microbiol.* 14, 1132016. https://doi.org/10.3389/fmicb.2023.1132016

Guo D-J, Li D-P, Singh RK, Singh P, Sharma A, Verma KK, Qin Y, Khan Q, Lu Z, Malviya MK, Song X-P, Xing Y-X and Li Y-R. (2021) Differential protein expression analysis of two sugarcane varieties in response to diazotrophic plant growth-promoting endophyte enterobacter roggenkampii ED5. *Front. Plant Sci.* 12, 727741. https://doi.org/10.3389/fpls.2021.727741

Guo DJ, Li DP, Singh RK, Singh P, Verma KK, Sharma A, Qin Y, Khan Q, Song XP, Malviya MK, Xing YX, Li YR. (2022) Comparative transcriptome analysis of two sugarcane varieties in response to diazitrophic plant growth promoting endophyte *Enterobacter roggenkampii* ED5. *J. Plant Interact.* 17(1), 75–84.

Hasan MF, Alam M, Jabber M, Begum M, Miah M. (2003) Effects of water-logging on juice quality and yield of sugarcane. *Pak. J. Biol. Sci.* 6(13), 1151–1155.

Hasanuzzaman M, Nahar K, Alam MM, Roychowdhury R, Fujita M. (2013) Physiological, biochemical, and molecular mechanisms of heat stress tolerance in plants. *Int. J. Mol. Sci.* 14(5), 9643–9684.

Hemaprabha G, Swapna S, Lavanya DL, Sajitha B, Venkataramana S. (2013) Evaluation of drought tolerance potential of elite genotypes and progenies of sugarcane (*Saccharum* sp. hybrids). *Sugar Tech.* 15(1), 9–16.

Huang CM, Wu KC, Verma KK, Luo HB, Deng ZN, Shi GY, Cao HQ, Wu XJ, Ye LP, Wei YW, Xu L. (2024) Isolation and expression analysis of cellulose synthase 3 (*Ces3*) genes from sugarcane (*Saccharum officinarum* L.). *Chem. Biol. Technol. Agric.* 11, 71. https://doi.org/10.1186/s40538-024-00597-7

Huang HR, Gao YJ, Malviya MK, Verma KK, Solanki MK, Huang YX, Li X, Deng Y, Yan J, Tang SY, Wang LW, Xu L. (2022) Genetic diversity analysis of sugarcane (*Saccharum* spp. Hybrids) among high-sucrose clones of GT series and commonly used parents by using microsatellite markers in Guangxi, China. *Sugar Tech.* 24, 397–407. https://doi.org/10.1007/s12355-021-01033-y

Hunsigi G. (2012) *Production of Sugarcane: Theory and Practice*. Springer, vol. 21.

Inman-Bamber N, Bonnett G, Spillman M, Hewitt M, Jackson J. (2008) Increasing sucrose accumulation in sugarcane by manipulating leaf extension and photosynthesis with irrigation. *Australian J. Agric. Res.* 59(1), 13–26.

Inman-Bamber N, Smith DM. (2005) Water relations in sugarcane and response to water deficits. *Field Crops Res.* 92(2–3), 185–202.

Irmak S, Rudnick DR. (2014) *Corn Irrigation Management under Water-limiting Conditions*. Unviersity of Nebraska-Lincoln: Extension.

Iwuozor KO, Anyanwu VU, Olaniyi BO, Mbamalu PS, Adeniyi AG. (2022) Adulteration of sugar: A growing global menace. *Sugar Tech.* 24, 914–919. https://doi.org/10.1007/s12355-022-01122-6

Jagadesh M, Dash M, Kumari A, Singh SK, Verma KK, Kumar P, Bhatt R, Sharma SK. (2024) Revealing the hidden world of soil microbes: Metagenomic insights into plant, bacteria, and fungi interactions for sustainable agriculture and ecosystem restoration. *Microbiol. Res.* 285, 127764. https://doi.org/10.1016/j.micres.2024.127764

Jain R, Singh A, Singh S, Singh SP, Srivastava VK, Chandra A, Solomon S. (2017) Physio-biochemical characterization of sugarcane genotypes for waterlogging tolerance. *World J. Agric. Sci.* 13, 90–97.

Jaiphong T, Tominaga J, Watanabe K, Nakabaru M, Takaragawa H, Suwa R, Kawamitsu Y. (2016) Effects of duration and combination of drought and flood conditions on sugarcane leaf photosynthesis, growth and sugar content. *Plant Prod. Sci.* 19(3), 427–437.

Joshi A, Verma KK, Rajput VD, Arora J. (2023) Metabolic engineering of microorganisms in advancing biofuel production. In: *Advances in Lignocellulosic Biofuel Production Systems*. pp. 181–202. https://doi.org/10.1016/B978-0-323-91192-4.00016-X

Joshi A, Verma KK, Rajput VD, Minkina T, Arora J. (2022) Recent advances in metabolic engineering of microorganisms for advancing lignocellulose-derived biofuels. *Bioengineered* 13(4), 8135–8163. https://doi.org/10.1080/21655979.2022.2051856.

Jun Z, Franca UA, Iwuozor KO. (2023) Relationship between agricultural production, energy consumption, and climate change in Nigeria. In: *Climate Change Impacts on Nigeria: Environment and Sustainable Development*. Springer: Singapore, pp. 563–580.

Krishna D, Sachan HK, Verma KK. (2022) Abiotic stresses in plants: An overview. In: *Understanding Abiotic Stresses*. Nova Science Publishers, Inc.: New York.

Kumar D, Malik N, Sengar RS. (2021) Physio-biochemical insights into sugarcane genotypes under water stress. *Biol. Rhythm Res.* 52(1), 92–115.

Kumari A, Dash M, Singh SK, Jagadesh M, Mathpal B, Mishra PK, Pandey SK, Verma KK. (2023a) Soil microbes: A natural solution for mitigating the impact of climate change. *Environ. Monit. Assess.* 195, 1436. https://doi.org/10.1007/s10661-023-11988-y

Kumari A, Lakshmi GA, Krishna GK, Patni B, Prakash S, Bhattacharyya M, Singh SK, Verma KK. (2022) Climate change and its impact on crops: A comprehensive investigation for sustainable agriculture. *Agronomy* 12, 3008. https://doi.org/10.3390/agronomy12123008

Kumari A, Singh SK, Mathpal B, Verma KK, Garg VK, Bhattacharyya M, Bhatt R. (2023b) The biosynthesis, mechanism of action, and physiological functions of melatonin in horticultural plants: A review. *Horticulturae* 9, 913. https://doi.org/10.3390/horticulturae9080913

Li A-M, Liao F, Wang M, Chen Z-L, Qin C-X, Huang R-Q, Verma KK, Li Y-R, Que Y-X, Pan Y-Q, et al. (2023) Transcriptomic and proteomic landscape of sugarcane response to biotic and abiotic stressors. *Int. J. Mol. Sci.* 24, 8913. https://doi.org/10.3390/ijms24108913

Lowveld E. (1996) Effect of irrigation regime on cane and sugar yields of variety nco376in the south-east lowveld of zimbabwe. *Proc. S. Afr. Sug. Technolass* 70.

Mall R, Sonkar G, Bhatt D, Sharma N, Baxla A, Singh K. (2016) Managing impact of extreme weather events in sugarcane in different agro-climatic zones of Uttar Pradesh. *Mausam* 67(1), 233–250.

Malviya MK, Li C-N, Lakshmanan P, Solanki MK, Wang Z, Solanki AC, Nong Q, Verma KK, Singh RK, Singh P, Sharma A, Guo D-J, Dessoky ES, Song X-P, Li Y-R. (2022) High-throughput sequencing-based analysis of rhizosphere and diazotrophic bacterial diversity among wild progenitor and closely related species of sugarcane (*Saccharum* spp. inter-specific hybrids). *Front. Plant Sci.* 13, 829337. https://doi.org/*10.3389/fpls.2022.829337.*

Malviya MK, Solanki MK, Li CN, Wang Z, Zeng Y, Verma KK, Singh RK, Singh P, Huang HR, Yang LT, Song XP, Li YR. (2021) Sugarcane-legume intercropping can enrich the soil microbiome and plant growth. *Front. Sustain. Food Syst.* 5, 606595. https://doi.org/10.3389/fsufs.2021.606595.

Meena MR, Kumar R, Chinnaswamy A, Karuppaiyan R, Kulshreshtha N, Ram B. (2020) Current breeding and genomic approaches to enhance the cane and sugar productivity under abiotic stress conditions. *3 Biotech* 10(10), 1–18.

Meena V, Dotaniya M, Coumar V, Rajendiran S, Kundu S, Subba Rao A. (2014) A case for silicon fertilization to improve crop yields in tropical soils. *Proc. Nat. Acad. Sci. India, Sec. B: Biol. Sci.* 84(3), 505–518.

Misra V, Solomon S, Ansari MI. (2016) Impact of drought on post-harvest quality of sugarcane crop. *Adv. Life Sci.* 20(5), 9496–9505.

Misra V, Solomon S, Mall A, Prajapati C, Hashem A, AbduAllah EF, Ansari MI. (2020) Morphological assessment of water stressed sugarcane: A comparison of waterlogged and drought affected crop. *Saudi J. Biol. Sci.* 27(5), 1228–1236.

Oliveira MA, Verma KK, Bhatt R, Oliveira TBA. (2022) Impact of green and organic fertilizers on soil fertility and sugarcane productivity. In: Verma KK. et al. (Eds.), *Agro-industrial Perspectives on Sugarcane Production Under Environmental Stress*. https://doi.org/10.1007/978-981-19-3955-6_11.

Pereira LS, Oweis T, Zairi A. (2002) Irrigation management under water scarcity. *Agric. Water Manag.* 57(3), 175–206.

Rabelo SC, de Paiva LBB, Pin TC, Pinto LFR, Tovar LP, Nakasu PYS. (2020) Chemical and energy potential of sugarcane, In: *Sugarcane Biorefinery, Technology and Perspectives*, Elsevier: Netherlands, pp. 141–163.

Rana D, Dass A, Rajanna G, Kaur R. (2016) Biotic and abiotic stress management in pulses. *Ind. J. Agron.* 61, S238-S248.

Rao BB, Chowdary PS, Sandeep V, Pramod V, Rao V. (2015) Spatial analysis of the sensitivity of wheat yields to temperature in India. *Agric. Forest Meteorol.* 200, 192–202.

Saccon P. (2018) Water for agriculture, irrigation management. *Appl. Soil Ecol.* 123, 793–796.

Santos F, Borém A, Caldas C. (2015) *Sugarcane: Agricultural Production, Bioenergy and Ethanol*. Academic Press.

Santos LC, Coelho RD, Barbosa FS, Leal DP, Júnior EFF, Barros TH,. Ribeiro NL. (2019) Influence of deficit irrigation on accumulation and partitioning of sugarcane biomass under drip irrigation in commercial varieties. *Agric. Water Manag.* 221, 322–333.

Sasidharan R, Bailey-Serres J, Ashikari M, Atwell BJ, Colmer TD, Fagerstedt K, Hill RD. (2017) Community recommendations on terminology and procedures used in flooding and low oxygen stress research. *New Phytol.* 214(4), 1403–1407.

Sharma A, Singh RK, Singh P, Vaishnav A, Guo DJ, Verma KK, Li DP, Song XP, Malviya MK, Khan N, Lakshmanan P, Li YR. (2021) Insights into the bacterial and nitric oxide-induced salt tolerance in sugarcane and their growth-promoting abilities. *Microorganisms* 9, 2203. https://doi.org/10.3390/microorganisms9112203.

Sharma A, Song X-P , Singh RK, Vaishnav A, Gupta S, Singh P, Guo D-J, Verma KK, Li Y-R. (2022) Impact of carbendazim on cellular growth, defence system and plant growth promoting traits of *Priestia megaterium* ANCB-12 isolated from sugarcane rhizosphere. *Front. Microbiol.* 13, 1005942. doi:10.3389/fmicb.2022.1005942

Sharma S, Uppal KSK. (1999) Influence of salt stress on growth and quality on sugarcane. *Ind. Sugar* 22, 1–4.

Shrivastava A, Pathak A, Misra V, Srivastava S, Swapna M, Shukla S. (2017) Sugarcane crop: Its tolerance towards abiotic stresses. In: *Abiotic Stress Management for Resilient Agriculture*. Springer, pp. 375–397.

Silva MdA, Silva JAGd, Enciso J, Sharma V, Jifon J. (2008) Yield components as indicators of drought tolerance of sugarcane. *Scientia Agricola* 65, 620–627.

Singels A, Kennedy A, Bezuidenhout C. (1998) *Irricane: A Simple Computerised Irrigation Scheduling Method for Sugarcane.* Paper presented at the Proc S Afr Sug Technol Ass.

Singels A, Paraskevopoulos A, Mashabela M. (2019) Farm level decision support for sugarcane irrigation management during drought. *Agric. Water Manag.* 222, 274–285.

Singh P, Singh RK, Li HB, Guo DJ, Sharma A, Verma KK, Solanki MK, Upadhyay SK, Lakshmanan P, Yang LT, Li YR. (2023) Nitrogen fixation and phytohormone stimulation of sugarcane plant through plant growth promoting diazotrophic *Pseudomonas*. *Biotechnol. Genetic Eng. Reviews* 1–21. https://doi.org/10.1080/02648725.2023.2177814

Singh P, Singh RK, Song QQ, Li HB, Guo DJ, Malviya MK, Verma KK, Song XP, Lakshmanan P, Yang LT, Li YR. (2021) Comparative analysis of protein and differential responses of defense-related gene and enzyme activity reveals the long-term molecular responses of sugarcane inoculated with *Sporisorium scitamineum*. *J. Plant Interact.* 16(1), 12–29.

Singh RK, Singh P, Li HB, Song QQ, Guo DJ, Solanki MK, Verma KK, Malviya MK, Song XP, Lakshmanan P, Yang LT, Li YR. (2020) Diversity of nitrogen-fixing rhizobacteria associated with sugarcane: A comprehensive study of plant-microbe interactions for growth enhancement in *Saccharum* spp. *BMC Plant Biol.* 20, 220. https://doi.org/10.1186/s12870-020-02400-9.

Singh S, Singh S, Pathak A, Pandey N. (2019) Assessment of waterlogging induced physiobiochemical changes in sugarcane varieties and its association with waterlogging tolerance. *J. Environ. Biol.* 40(3), 384–392.

Solanki MK, Verma KK, Dastogeer KMG, Mora-Poblete F, Mundra S. (2023) Microbial resilience in plant nutrient management towards sustainable farming. *Front. Microbiol.* 14, 1280811. https://doi.org/10.3389/fmicb.2023.1280811

Song X-P, Mo F, Verma KK, Wei J, Zhang X, Yang L, Li YR. (2019) Effect of sugarcane smut (*Ustilago scitaminea* Syd.) on ultrastructure and biochemical indices of sugarcane. *Biomed J. Sci. Tech. Res.* 17(1), 002950. https://doi.org/10.26717/BJSTR.2019.17.002950.

Song XP, Mo F, Yan M, Zhang X, Zhang B, Huang X, Huang D, Pan Y, Verma KK, Li YR. (2022) Effect of smut infection on the photosynthetic physiological characteristics and related defense enzymes of sugarcane. *Life* 12, 1201. https://doi.org/10.3390/life12081201

Song XP, Verma KK, Tian DD, Zhang XQ, Liang YJ, Huang X, Li CN, Li YR. (2021) Exploration of silicon functions to integrate with biotic stress tolerance and crop improvement. *Biol. Res.* 54, 19. https://doi.org/10.1186/s40659-021-00344-4

Teixeira EI, Fischer G, Van Velthuizen H, Walter C, Ewert F. (2013) Global hot-spots of heat stress on agricultural crops due to climate change. *Agric. Forest Meteorol.* 170, 206–215.

Thomas J, Salinas F, Namken, L. (1978) Growth and yield of sugarcane as affected by row spacing and irrigation regime. *Paper Presented at the Proceedings—New Series.* American Society of Sugar Cane Technologists (USA).

Tian D, Qin L, Verma KK, Wei L, Li J, Li B, Zhou W, He Z, Wei D, Huang S, Long S, Huang Q, Li C, Wei S. (2023) Transcriptomic and metabolomic differences between banana varieties which are resistant or susceptible to *Fusarium wilt. PeerJ* 11, e16549. https://doi.org/10.7717/peerj.16549

Vasantha S, Venkataramana S, Gururaja Rao P, Gomathi R. (2010) Long term salinity effect on growth, photosynthesis and osmotic characteristics in sugarcane. *Sugar Tech.* 12(1), 5–8.

Verma KK, Joshi A, Song X-P, Liang Q, Xu L, Huang H-r, Wu K-C, Seth CS, Arora J and Li Y-R. (2024a) Regulatory mechanisms of plant rhizobacteria on plants to the adaptation of adverse agroclimatic variables. *Front. Plant Sci.* 15, 1377793. https://doi.org/10.3389/fpls.2024.1377793

Verma KK, Joshi A, Song X-P, Singh S, Kumari A, Arora J, Singh SK, Solanki MK, Seth CS, Li Y-R. (2024b) Synergistic interactions of nanoparticles and plant growth promoting rhizobacteria enhancing soil-plant systems: A multigenerational perspective. *Front. Plant Sci.* 15, 1376214. https://doi.org/10.3389/fpls.2024.1376214

Verma KK, Li DM, Singh M, Rajput VD, Malviya MK, Minkina T, Singh RK, Singh P, Song XP, Li YR. (2020a) Interactive role of silicon and plant-rhizobacteria mitigating abiotic stresses: A new approach for sustainable agriculture and climate change. *Plants* 9, 1055. https://doi.org/10.3390/plants9091055.

Verma KK, Rajput VD, Singh M, Verma CL, Minkina T. (2022a) Influence of waterlogging on plant physiology and molecular biology. In: *Understanding Abiotic Stresses*. Nova Science Publishers, Inc.: New York.

Verma KK, Siddiqui N, Singh M. (2022b) Understanding the plant physiological and molecular adaptation strategies to soil salinity. In: *Understanding Abiotic Stresses*. Nova Science Publishers, Inc.: New York. https://doi.org/10.52305/UBUJ4024

Verma KK, Singh P, Song X-P, Malviya MK, Singh RK, Chen G-L, Solomon S, Li YR. (2020b) Mitigating climate change for sugarcane improvement: Role of silicon in alleviating abiotic stresses. *Sugar Tech.* 22(5), 741–749. https://doi.org/10.1007/s12355-020-00831-0.

Verma KK, Singh RK, Song QQ, Singh P, Zhang B-Q, Song X-P, Chen G-L, Li YR. (2019a) Silicon alleviates drought stress of sugarcane plants by improving antioxidant responses. *Biomed J. Sci. Tech. Res.* 002957. https://doi.org/10.26717/BJSTR.2019.17.002957

Verma KK, Song X, Degu HD, Guo D, Joshi A, Huang H, Xu L, Singh M, Huang D, Rajput VD, Li YR. (2023a) Recent advances in nitrogen and nano-nitrogen fertilizers for sustainable crop production: A mini-review. *Chem. Biol. Technol. Agric.* 10, 1–14. https://doi.org/10.1186/s40538-023-00488-3

Verma KK, Song X, Verma CL, Huang H, Singh M, Xu L, Wu J, Khan ZH, Zhang B, Li Y. (2023b) Mathematical modeling of climate and fluoride effects on sugarcane photosynthesis with silicon nanoparticles. *Plant Physiol. Biochem.* 204, 108089. https://doi.org/10.1016/j.plaphy.2023.108089

Verma KK, Song XP, Budeguer F, Nikpay A, Enrique R, Singh M, Zhang B, Wu, JM, Li YR. (2022c) Genetic engineering: An efficient approach to mitigating biotic and abiotic stresses in sugarcane cultivation. *Plant Signal. Behav.* 17(1), e2108253.

Verma KK, Song XP, Joshi A, Rajput VD, Singh M, Sharma A, Singh RK, Li DM, Arora J, Minkina T, Li YR. (2022d) Nanofertilizer possibilities for healthy soil, water and food in future: An overview. *Front. Plant Sci.* 13, 865048. *https://doi.org/*10.3389/fpls.2022.865048

Verma KK, Song XP, Joshi A, Tian DD, Rajput VD, Singh M, Arora J, Minkina TM, Li YR. (2022e) Recent trends in nano-fertilizer for advancing sustainable agriculture under the era of climate change to ensure food security for future generations. *Nanomaterials* 12, 173. *https://doi.org/10.3390/nano12010173.*

Verma KK, Song XP, Li DM, Singh M, Wu JM, Singh RK, Sharma A, Zhang BQ, LI YR. (2022f) Silicon and soil microorganisms improves rhizospheric soil health with bacterial community, plant growth, performance and yield. *Plant Signal. Behav.* 17(1), e2104004.

Verma KK, Song XP, Lin B, Guo DJ, Singh M, Rajput VD, Singh RK, Singh P, Sharma A, Malviya MK, Chen GL, Li YR. (2022g) Silicon induced drought tolerance in crop plants: Physiological adaptation strategies. *Silicon* 14(6), 2473–2487. https://doi.org/10.1007/s12633-021-01071-x

Verma KK, Song XP, Rajput VD, Boldyreva V, Zhang BQ, Minkina T, Li YR. (2022h) Morpho-physiological, biochemical and ultrastructural modifications on sugarcane to prolonged water deficit. In: Verma KK. et al. (Eds.), *Agro-industrial Perspectives on Sugarcane Production Under Environmental Stress*. https://doi.org/10.1007/978-981-19-3955-6_11.

Verma KK, Song XP, Rajput VD, Solomon S, Li YR, Rao GP. (2022i) *Agro-industrial Perspectives on Sugarcane Production Under Environmental Stress*. Springer Nature: Singapore, https://doi.org/10.1007/978-981-19-3955-6

Verma KK, Song XP, Tian DD, Guo DJ, Chen ZL, Zhong CS, Nikpay A, Singh M, Rajput VD, Singh RK, Minkina T, Li YR. (2021a) Influence of silicon on biocontrol strategies to manage biotic stress for crop protection, performance and improvement. *Plants* 10, 2163. https://doi.org/10.3390/plants10102163

Verma KK, Song XP, Tian DD, Singh M, Verma CL, Rajput VD, Singh RK, Sharma A, Singh P, Malviya MK, Li YR. (2021b) Investigation of defensive role of silicon during drought stress induced by irrigation capacity in sugarcane: Physiological and biochemical characteristics. *ACS Omega* 6, 19811–19821.

Verma KK, Song XP, Verma CL, Chen ZL, Rajput VD, Wu KC, Liao F, Chen GL, Li YR. (2021c) Functional relationship between photosynthetic leaf gas exchange in response to silicon application and water stress mitigation in sugarcane. *Biol. Res.* 54, 15. https://doi.org/10.1186/s40659-021-00338-2

Verma KK, Song XP, Verma CL, Malviya MK, Guo DJ, Rajput VD, Sharma A, Wei KJ, Chen GL, Solomon S, Li YR. (2021d) Predication of photosynthetic leaf gas exchange of sugarcane (*Saccharum* spp.) leaves in response to leaf positions to foliar spray of potassium salt of active phosphorus under limited water irrigation. *ACS Omega* 6, 2396–2409.

Verma KK, Song X-P, Xu L, Huang H-R, Liang Q, Seth CS, Li Y-R. (2023d) Nano-microplastic and agro-ecosystems: A mini-review. *Front. Plant Sci.* 14, 1283852. https://doi.org/10.3389/fpls.2023.1283852

Verma KK, Song XP, Yadav G, Degu HD, Parvaiz A, Singh M, Huang HR, Mustafa G, Xu L, Li YR. (2022j) Impact of agroclimatic variables on proteo-genomics in sugarcane (*Saccharum* spp.) plant productivity. *ACS Omega* 7, 22997–23008.

Verma KK, Song XP, Zeng Y, Guo DJ, Sing M, Rajput VD, Malviya MK, Wei KJ, Sharma A, Li DP, Chen GL, Li YR. (2021e) Foliar application of silicon boosts growth, photosynthetic leaf gas exchange, antioxidative response and resistance to limited water irrigation in sugarcane (*Saccharum officinarum* L.). *Plant Physiol. Biochem.* 166, 582–592.

Verma KK, Song XP,Zeng Y, Li DM, Guo DJ, Rajput VD, Chen GL, Barakhov A, Minkina TM, Li YR. (2020c) Characteristics and correlation of leaf stomata and its relationship with photosynthesis on *Saccharum* spp. under different irrigation and silicon application. *ACS Omega* 5, 24145–24153.

Verma KK, Vatsal S, Singh M, Rajput VD, Minkina T. (2022k) Coping with water stress: Effects and adaptive mechanisms to plants. In: *Understanding Abiotic Stresses*. Nova Science Publishers, Inc.: New York.

Verma KK, Wu KC, Verma CL, Li DM, Malviya MK, Singh RK, Singh P, Chen GL, Song XP, Li YR. (2020d) Developing mathematical model for diurnal dynamics on photosynthetic responses in sugarcane responsive to different irrigation and silicon application. *PeerJ* 8, e10154. https://doi.org/10.7717/peerj.10154.

Verma KK, Liu X-H, Wu K-C, Singh RK, Song QQ, Malviya MK, Song X-P, Singh P, Verma CL, Li YR. (2020e) The impact of silicon on photosynthetic and biochemical responses of sugarcane under different soil moisture levels. *Silicon* 12, 1355–1367.

Verma KK, Wu K-C, Singh P, Malviya MK, Singh RK, Song X-P, Li YR. (2019b) The protective role of silicon in sugarcane under water stress: Photosynthesis and antioxidant enzymes. *Biomed J. Sci. Tech. Res.* 002685. https://doi.org/0.26717/BJSTR.2019.15.002685

Vu NT, Vu DC, Pham TT. (2019) Growth and physiological responses of sugarcane to drought stress at an early growth stage. *Tap chi Khoa hoc Nong nghiep Viet Nam/Vietnam J. Agric. Sci.* 2(4), 451–460.

Wiedenfeld RP. (2000) Water stress during different sugarcane growth periods on yield and response to N fertilization. *Agric. Water Manag.* 43(2), 173–182.

Wiedenfeld B. (2008) Effects of irrigation water salinity and electrostatic water treatment for sugarcane production. *Agric. Water Manag.* 95(1), 85–88.

Wu K-C, Huang C-M, Verma KK, Deng Z-N, Huang H-R, Pang T, Cao H-Q, Luo H-B, Jiang S-L, Xu L. (2022) Transcriptomic responses of *Saccharum spontaneum* roots in response to polyethylene glycol–6000 stimulated drought stress. *Front. Plant Sci.* 13, 992755. https://doi.org/10.3389/fpls.2022.992755

Xu L, Deng Z-N, Wu K-C, Malviya MK, Solanki MK, Verma KK, Pang T, Li Y-J, Liu X-Y, Kashyap BK, Dessoky ES, Wang W-Z and Huang H-R. (2022) Transcriptome analysis reveals a gene expression pattern that contributes to sugarcane bud propagation induced by indole-3-butyric acid. *Front. Plant Sci.* 13, 852886. https://doi.org/10.3389/fpls.2022.852886

Zhao D, Li Y-R. (2015) Climate change and sugarcane production: Potential impact and mitigation strategies. *Int. J. Agron.* 2015, 547386.

Zheng F-J, Lin B, Yang Y-X, Fang X-C, Verma KK, Chen G-L. (2023) Efficacy and functionality of sugarcane original vinegar on mice. *Front. Microbiol.* 14, 1224666. https://doi.org/10.3389/fmicb.2023.1224666

CHAPTER 11

Climate-Smart Technologies in Sugarcane

MANDAPELLI SHARATH CHANDRA,[1*] RAJAN BHATT,[2]
RAYAPATI KARTHIK,[3] PRAVEEN V. KADAM,[4] CH. PRAGATHI KUMARI,[1]
L. PEACE RAISING,[5] and NALLAGATLA VINOD KUMAR[6]

[1]*All India Coordinated Research Project on Integrated Farming System, Professor Jayashankar Telangana State Agricultural University, Hyderabad, Telangana, India*

[2]*Krishi Vigyan Kendra, Punjab Agricultural University, Amritsar, Punjab, India*

[3]*College of Agriculture, Professor Jayashankar Telangana State Agricultural University, Hyderabad, Telangana, India*

[4]*Agronomy, Keladi Shivappa Nayaka University of Agricultural and Horticultural Sciences, Shivamogga, Karnataka, India*

[5]*Department of Agricultural Sciences, Sharda University, Greater Noida, Uttar Pradesh, India*

[6]*Malla Reddy University, Maisammaguda, Hyderabad, Telangana, India*

*Corresponding author

ABSTRACT

Significant transformation is required in agriculture within developing nations to address the interconnected challenges of ensuring food security while also mitigating climate change. The physiological and metabolic processes of crops are negatively impacted by global warming; the

Sugarcane Production: Environmental Impacts and Socioeconomic Issues.
Krishan K. Verma, Xiu-Peng Song, Lin Xu, Rajan Bhatt, and Yang-Rui Li (Eds.)

DOI: 10.1201/9781779641137-11

sugarcane crop is not an exception. Climate change and other sustainability-related concerns also have an impact on it. Numerous climate-smart technologies (CSTs) have been suggested by academics as solutions to these problems in order to increase sugarcane production and guarantee its long-term sustainability. CSTs in sugarcane farming refer to a variety of strategies designed to increase yield, adaptability, and sustainability while reducing the effects of climate change. It can be a smart idea to use CST interventions such as optimizing different crop production methods/techniques, planting techniques, nutrient management techniques, insect-pest and infection control alternatives, such as cultural and physical methods, the use of bioagents, biopesticides, botanicals, light/pheromone, along with combo traps, and so forth. Sustainable soil quality and cane production are further improved by preserving its appropriateness during drought and floods, recycling biomass waste in an environmentally beneficial manner, and using it safely. This chapter provides a thorough overview of CSTs, which are the best ways to lessen the carbon, water, fertilizer, along with pesticide footprints associated with sugarcane production. This will benefit cane farmers' livelihoods by enhancing cane quality, yield, and overall growth.

11.1 INTRODUCTION

The equatorial belt, spanning tropical and subtropical regions, has experienced the effects of rising temperatures and climate change, resulting in altered weather patterns characterized by more frequent and intense droughts, storms, and heat waves (Verma et al., 2019a, 2019b; Huang et al., 2024). These climate extremes have significantly impacted both livestock and crops in the region. Sugarcane, classified as a C_4 plant, exhibits a higher tolerance to a wide range of climatic variations compared to C_3 plants (Bhatt 2020, 2023a, 2023b; Verma et al., 2022a, 2022b; Chen et al., 2023; Jagadesh et al., 2024). However, despite this inherent resilience, changing climatic conditions have led to reductions in cane production and productivity globally (Kumari et al., 2022, 2023a, 2023b; Solanki et al., 2023; Verma et al., 2023a, 2023b). As sugarcane is a long-duration crop, it experiences the impact of all four seasons, including extreme temperatures, and is vulnerable to various insect pests, diseases, and weed infestations in the region (Guo et al., 2021, 2022; Oliveira et al., 2022). Climate-smart technologies

(CSTs) in sugarcane cultivation encompass a range of approaches aimed at enhancing productivity, resilience, and sustainability while mitigating the impact of climate change. Implementing appropriate CSTs not only reduces greenhouse gas emissions but also enhances the adaptation capacity of agricultural systems to climate change. These technologies contribute to increased cane yields and improved food security (Verma et al., 2021a, 2021b; Li et al., 2023; Deng et al., 2023).

Generally, sugarcane is cultivated primarily for sugar production to meet demand. However, it also yields several other products, including bagasse, molasses, syrup, and jiggery (Xu et al., 2022, 2005; Wu et al., 2022; Verma et al., 2023c, Zheng et al., 2023). Utilizing sugarcane trash as mulch offers multiple benefits, including the retention of soil moisture, regulation of soil temperature, facilitation of nutrient uptake by sugarcane plants, and enhancement of yield and quality parameters (Huang et al., 2022; Malviya et al., 2021, 2022; Gao et al., 2023; Singh et al., 2023). Additionally, various CSTs, as previously discussed, play a crucial role in reducing carbon emissions, water usage, fertilizer usage, energy usage, and pesticide usage, thereby improving input efficiency. It is important to note that these CSTs may vary in their applicability depending on site-specific conditions and situations and may not be universally applicable (Verma et al., 2020a, 2020b; Singh et al., 2020, 2021).

Further advancements in sugarcane production could involve the recovery of sugarcane straw for bioenergy purposes, such as bioelectricity generation and second-generation ethanol production. Implementing best management practices throughout the sugarcane production chain, including the complete cessation of preharvest burning, ensuring accurate straw recovery rates, and enhancing sugarcane quality, can contribute to sustainable and environmentally friendly sugarcane cultivation practices.

11.2 EFFECT OF CLIMATE CHANGE ON SUGARCANE

Climate change is expected to primarily have adverse effects on agricultural production, food security, and economic development, especially in developing countries (Hannah et al., 2005; Chen et al., 2020; Chaudhary et al., 2023; Joshi et al., 2023; Verma et al., 2024a, 2024b). Furthermore, sugarcane is heavily affected by the long-term impacts of climate change, as well as local weather patterns and seasonal variations. Climate conditions are crucial for plant growth and development, potentially causing crop damage. Additionally, climate change can significantly affect microorganisms, directly or indirectly influencing crop growth and yield (Anas et al., 2021a, 2021b; Sharma et al., 2021, 2022). Higher levels of CO_2 and temperatures will affect both plant growth and water balance (Kimball et al., 2002). Increased atmospheric CO_2 concentrations mainly enhance CO_2 diffusion into leaves, leading to higher photosynthetic rates in C_3 plants under various radiation intensities (Lawlor et al., 1991).

Temperature plays a crucial role in the development of sugarcane, influencing various stages from inception to maturity. It directly impacts plant growth, photosynthesis, and other biochemical processes, including bud development. The efficiency of photosynthesis in sugarcane exhibits a linear relationship with temperature, ranging from 8°C to a peak of 34°C. However, cool nights and early morning temperatures, reaching 14°C in winter and 20°C in summer, significantly impede photosynthesis the following day. Leaf growth is restricted at temperatures below 14–19°C. Consequently, cool night temperatures combined with sunny days can slow growth rates and carbon consumption, even though photosynthesis may continue (Gawander 2007; Verma et al., 2020c, 2020d; Joshi et al., 2022)). Gbetibouo and Hassan (2005) used a Ricardian model to evaluate the impact of climate change on South Africa's field crops, including sugarcane. Their study indicates that field crop production in South Africa is highly vulnerable to even minor temperature changes due to the narrow tolerance range for increased temperatures, as opposed to variations in precipitation. Climate change is expected to significantly impact global sugarcane productivity, highlighting the need for climate-smart practices. The adoption of CSTs in sugarcane agriculture can enhance sugarcane output and sugar recovery, while also boosting farmers' income and economic sustainability (Bhatt et al., 2023). Sugarcane growers are adapting to the challenges posed by climate change by implementing various strategies discussed below to maintain and enhance yields.

11.3 DROUGHT-TOLERANT VARIETIES

Developing and adopting sugarcane varieties that are resilient to drought conditions can help mitigate the impact of water scarcity, which is exacerbated by climate change. Drought-tolerant sugarcane varieties are increasingly seen as essential components of climate change resilience strategies in sugarcane-growing regions (Verma et al., 2020d, 2020e; 2021c; 2022c, 2022d; Dlamini 2021). Drought-tolerant varieties of sugarcane are crucial for regions where water scarcity is a challenge or where irrigation resources are limited (Verma et al., 2022e, 2022f; 2023a).

Researchers shall focus on climate-resilient varieties across multiple locations without compromise in productivity. One of the major constraints in achieving the production potential of sugarcane is the frequent occurrence of severe drought periods. As a high water-demanding tropical crop, sugarcane boasts the highest productivity per drop of water consumed. However, water deficit stress during various stages of its growth can result in a 30–70% loss in productivity (Verma et al., 2021d, 2021e; 2022g, 2022h). Additionally, sucrose accumulation and recovery can be reduced by up to 5%, leading to significant losses in sugar productivity. The efficiency of the sugar industry largely depends on the availability of improved sugarcane varieties with better juice quality and tolerance to biotic and abiotic stresses (Krishna et al., 2022; Verma et al., 2022i, 2022j, 2022k). Meeting the increased demand for sugarcane production can be achieved by identifying location-specific varieties that enhance production per unit area and time (Appunu et al., 2017). Meeting the growing demand for sugarcane production entails identifying location-specific varieties tailored for enhanced yield per unit area and time. Thus, the development of varieties surpassing existing ones becomes crucial for boosting sugarcane production (Verma et al., 2023c, Huang et al., 2024).

Newly developed sugarcane varieties with purported drought tolerance undergo extensive field trials and evaluation processes. These trials assess the performance of the varieties under various levels of water stress and compare them to standard commercial cultivars (Figure 11.1). Traits such as yield, sucrose content, stalk quality, and overall plant vigor are evaluated to determine the suitability of the varieties for commercial cultivation. Drought-tolerant sugarcane varieties exhibit different stress tolerance and avoidance mechanisms. For instance, varieties N41 and N26 perform well under mild drought conditions. Variety N12 minimizes its leaf surface area by rolling its leaves, while N19 conserves water by closing its leaf stomata and ceasing growth until the soil moisture content becomes optimal again (Snyman et al., 2004; Eksteen et al., 2014).

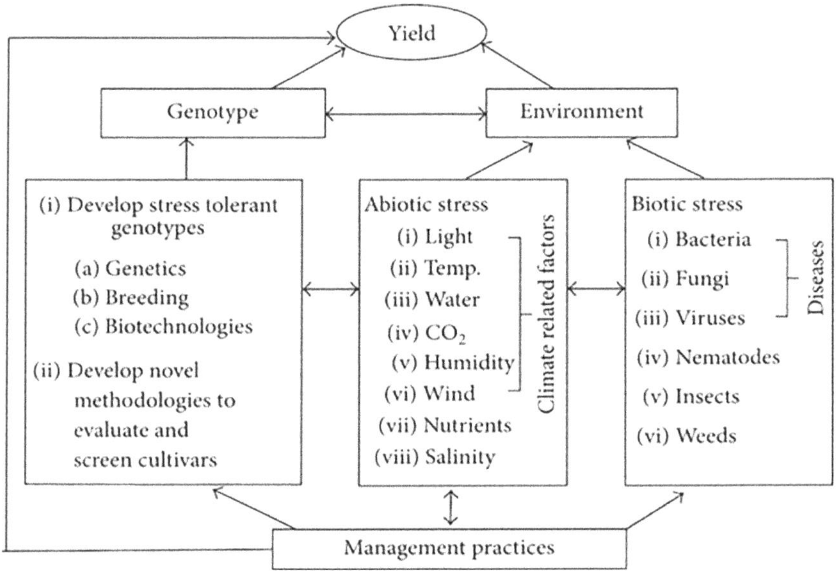

FIGURE 11.1 A flowchart to demonstrate major factors (genotype, environment, and management practices) influencing sugarcane yield as well as their interactions.

Source: Reprinted from Zhao and Li (2015). https://creativecommons.org/licenses/by/3.0/

Despite the potential benefits of drought-tolerant sugarcane varieties, their adoption may face challenges. Farmers may be hesitant to adopt new varieties due to concerns about unfamiliar agronomic practices, market acceptance of new cultivars, and the initial investment required for transitioning to new varieties. Extension services and agricultural outreach programs play a crucial role in educating farmers about the advantages of adopting drought-tolerant varieties and providing support for their implementation. Research into drought-tolerant sugarcane varieties continues, with ongoing efforts to further enhance their performance under water-limited conditions. Advances in genomics, molecular breeding, and biotechnology hold promise for accelerating the development of improved varieties with enhanced drought resilience, ultimately contributing to the sustainability and profitability of sugarcane farming in drought-prone regions.

11.4 PRECISION IRRIGATION

Although sugarcane is an irrigated crop, it is commonly grown in moisture-stressed climates in most states. In subtropical and tropical India, the irrigation water requirements for sugarcane range from approximately 1600 to 2300 mm and 2000 to 3500 mm per year, respectively (Shukla et al., 2019). Implementing precision irrigation techniques, such as drip or sprinkler irrigation, helps optimize water usage by delivering water directly to the root zone of sugarcane plants. This reduces water wastage and increases efficiency. Climate-smart irrigation in sugarcane involves the application of innovative irrigation practices that enhance water use efficiency, resilience to climate variability, and sustainability of sugarcane cultivation. Climate-smart irrigation systems incorporate soil moisture sensors to continuously monitor the moisture content in the soil. This data helps farmers optimize irrigation scheduling, ensuring that sugarcane plants receive the right amount of water at the right time, thereby avoiding both water stress and waterlogging. Integrating weather data into irrigation scheduling allows farmers to adjust irrigation practices based on climatic conditions such as temperature, humidity, and rainfall forecasts. This proactive approach helps optimize water use efficiency and adapt to changing climate patterns (Figure 11.1).

Irrigation is given in alternate furrows, skipping one furrow in between, and one row is skipped. This strategy is practical and effective in saving 30–40% of irrigation water without compromising cane productivity. Altering irrigation furrows each time has also reduced water use in the alternate furrow approach, which is suggested in locations where water is scarce. Variable rate irrigation (VRI) systems adjust irrigation application rates across the sugarcane field based on spatial variability in soil properties, topography, and crop water requirements. By tailoring water delivery to specific areas within the field, VRI maximizes water use efficiency and minimizes unnecessary water application. Utilizing advanced software platforms enables farmers to remotely monitor and manage irrigation systems, analyze data, and make informed decisions regarding irrigation scheduling and water management practices. This technology-driven approach improves operational efficiency and resource utilization.

Climate-smart irrigation practices can be combined with water conservation measures such as mulching, cover cropping, and soil conservation techniques. These integrated approaches help preserve soil moisture, reduce water losses, and enhance the resilience of sugarcane cultivation to climate variability and water scarcity. By adopting climate-smart irrigation strategies, sugarcane farmers can mitigate the impacts of climate change, improve water use efficiency, and ensure the long-term sustainability of sugarcane production systems.

11.5 IMPROVED CROP MANAGEMENT PRACTICES

Improved crop practices in sugarcane encompass a range of techniques and strategies aimed at enhancing productivity, sustainability, and resilience to environmental stresses. Introducing practices like conservation tillage, intercropping, and crop rotation can help improve soil health, reduce erosion, and enhance carbon sequestration, contributing to climate change mitigation. Choosing sugarcane varieties with high yields, disease resistance, and tolerance to environmental stresses is essential for maximizing productivity and reducing risks. The optimal technique for surface and subsurface drip irrigation systems is paired row planting, which maximizes resource usage efficiency and increases cane output. The furrow irrigated raised bed (FIRB) method contributes to farmers' food security and economic prosperity (Singh et al., 2002).

Similarly, the ridge and furrow method, along with the wide bed and furrow system, offers advantages such as *in situ* rainwater conservation in furrows, improved aeration, and effective surplus water management in the seedbed and rhizosphere. Adjusting planting density based on soil fertility, climate conditions, and variety characteristics can enhance light interception, airflow, and overall crop performance. Implementing precision land preparation techniques, such as laser leveling and minimum tillage, ensures uniform soil conditions and proper drainage, facilitating optimal root development and nutrient uptake. Applying fertilizers judiciously based on soil nutrient status, crop requirements, and growth stages helps prevent nutrient deficiencies or excesses, promoting balanced plant growth and higher yields. Introducing crop rotation with legumes or other nonsugarcane crops helps break pest and disease cycles, improve soil health, and diversify income sources. Utilizing organic mulches or cover crops helps conserve soil moisture, suppress weeds, enhance soil structure, and promote beneficial soil microbial activity. By integrating these improved crop practices, sugarcane growers can enhance productivity, profitability, and sustainability while mitigating environmental impacts and adapting to changing climatic conditions.

11.6 CLIMATE-RESILIENT CROPPING SYSTEMS

Climate-resilient cropping systems in sugarcane involve the implementation of practices and techniques that enhance the adaptability and resilience of sugarcane cultivation to climate variability and change. Integrating sugarcane with other crops in agroforestry or mixed cropping systems can enhance resilience to climate variability, diversify income sources, and improve ecosystem services. Intercropping with one row of summer moong or summer mash between two rows of sugarcane has been recommended to improve cane farmers' livelihoods, yielding an extra 375–500 kg ha^{-1} of summer moong or summer mash (PAU 2022). The Punjab Agricultural University in Ludhiana, Punjab, India, has recommended various intercrops with sugarcane to enhance the livelihoods of cane farmers (Table 11.1). Autumn-planted sugarcane germinates before winter but remains nearly dormant until spring, providing ample time for winter crops to be grown as intercrops. During this period, sugarcane does not require significant growth. Vegetables are increasingly in demand due to their essential nutrients, which are vital in modern diets. Consequently, recent years have seen numerous changes in cropping schemes. Intercropping pulses with sugarcane expand the area under pulse crops, reduces weeds, and improves soil health. Although intercrops have minimal impact on cane yield, they enhance soil fertility (Figure 11.1). Additionally, mentha and okra (*Abelmoschus esculentus*) are also used as intercrops with sugarcane.

11.7 IMPORTANCE OF SUGARCANE RATOON MANAGEMENT

Ratooning is a method of growing sugarcane by cultivating the 2nd, 3rd, or 4th crop from the stubble left after harvesting the initial seed crop. These ratoon crops mature earlier than the original seed crop but typically receive less attention. Ratooning reduces the costs associated with seedbed preparation, seed, and planting activities, though it generally results in lower yields compared to plant crops due to factors like soil compaction (Verma 2002), poor soil fertility (Hunsigi 2001), and inadequate fertilizer application (Sundara and Tripathi 1999). Ratoon crops require an additional 25% nitrogen compared to plant crops and should be fertilized with 80 kg K_2O ha^{-1} of potassium to prevent lodging and enhance sugar recovery (Bhatt 2021a, 2021b). Effective ratoon management includes harvesting near the end of January, removing water shoots, intercultural operations using a tractor-drawn tiller, gap filling in March, and regular inspections for pests and insects in May and June. Additionally, utilizing

TABLE 11.1 Sugarcane Intercrops in the Autumn (PAU 2022)

Name of Intercrop	Sowing/Planting Time	Seed Rate (kg ha−1)	Spacing (cm)	Fertilizers Recommended ha−1	Harvesting Time	Remarks
Wheat (*Triticum* spp.)	Last week of October–15th November	40	20	62.5 kg N, 30 kg P_2O_5, and 30 kg K_2O	Mid--April	–
Gram (*Cicer arietinum*)	25th October–10th November	30	30	15 kg N (32.5 kg urea) + 20 kg P_2O_5(125 kg SSP)	April	–
Potato (*Solanum tuberosum*)	20th September–15th October	2000	–	87.5 kg N (195 kg urea), 40 kg P_2O_5 (250 kg SSP), and 87.5 kg K_2O (150 kg MOP)	End December	Wheat can be taken after the harvesting of potato at the end of December
Cabbage (*Brassica oleracea*)	Last week of October to November	–	–	62.5 kg N (135 kg urea), 31.3 kg P_2O_5(195 kg SSP), and 31.3 kg K_2O (50 kg MOP)	January and February	Transplant 4–5 weeks, an old seedling
Radish (*Raphanus sativus*)	October	11.25	30	62.5 kg N (135 kg urea) and 30 kg P_2O_5(187.5 kg SSP)	January	–
Peas (*Pisum sativum*)	October	55	30	35 kg N (77.5 kg urea) and 40 kg P_2O_5(1250 kg SSP)	–	–
Tomato (*Solanum lycopersicum*)	Nursery: October–November Transplanting: (November–December)	0.125 (nursery)	At the center between cane rows	75 kg N, 31.3 kg P_2O_5, and 31.3 kg K_2O	End March to mid-May	Protect intercrop from winter
Garlic (*Allium sativum*)	4th week of September–1st week of October	280–312 kg	15	25-ton FYM, 62.5 kg N (135 kg urea), and 30 kg P_2O_5 (187.5 kgSSP)	April	Use 212.5–237.5 kg garlic seed ha^{-1}

sugarcane residues, such as bagasse and straw, for bioenergy production can help reduce greenhouse gas emissions and dependence on fossil fuels, thereby contributing to climate change mitigation.

11.8 CARBON SEQUESTRATION PRACTICES

Implementing practices like reforestation, afforestation, and improved land management can enhance carbon sequestration in sugarcane-growing areas, contributing to climate change mitigation efforts. In the planning of an agroforestry system, selecting the arrangement of trees is crucial for the success of the production system. Tree arrangements should take into account the intended use of the trees, their future growth, and the resource needs of both the trees and the crops to be integrated into the intercropping system (Prasad et al., 2010). Agroforestry systems based on sugarcane, incorporating trees such as *Tectona grandis, Gmelina arborea, Eucalyptus, Casuarina equisetifolia*, and *Leucaena leucocephala*, have been shown to yield higher benefit-to-cost ratios compared to sole sugarcane cultivation. The inclusion of trees in farming systems not only enhances diversity but also provides additional income and employment opportunities to farm households, besides meeting wood requirements (Mali et al., 2017).

11.9 WEATHER FORECASTING AND ADVISORY SERVICES

Weather forecasting plays a crucial role in sugarcane farming, as weather conditions directly influence crop growth, development, and yield. Weather forecasts are essential for determining the optimal timing for planting and harvesting sugarcane. Planting decisions are often based on forecasts of rainfall patterns and temperature trends to ensure adequate soil moisture for germination and early growth. Similarly, harvesting schedules may be adjusted based on forecasts to avoid periods of heavy rainfall, which can impede harvesting operations and degrade cane quality. In regions prone to frost events, weather forecasts are critical for implementing frost protection measures to safeguard sugarcane crops. Forecasts indicating the likelihood of frost allow farmers to take preemptive actions such as applying protective coatings, deploying frost fans, or irrigating fields to mitigate the impact of freezing temperatures on sugarcane plants.

Among the myriad management decisions on farms, optimizing the use of available irrigation water can prove both intricate and demanding. Growers grapple with conflicting objectives of minimizing water usage while maximizing yields and profitability, all while contending with uncertainty regarding future weather and seasonal climate patterns. The unpredictability of climate often prompts conservative management approaches, particularly in irrigation, which may sacrifice productivity to mitigate the risk of losses during unfavorable years (Jones et al., 2000). In regions where irrigation is indispensable for sugarcane cultivation, weather forecasts play a pivotal role in devising irrigation schedules. Predictions regarding precipitation, temperature, and evapotranspiration rates are utilized to gauge crop water requirements and determine the timing of irrigation applications to uphold optimal soil moisture levels (Figure 11.1).

Weather data, including historical weather patterns and forecasted conditions, are often integrated into yield forecasting models for sugarcane. By analyzing weather variables such as temperature, precipitation, solar radiation, and humidity, researchers and agricultural economists can make predictions about future sugarcane yields, which are valuable for planning production, marketing, and resource allocation. With changing climate patterns and increased frequency of extreme weather events, weather forecasting becomes even more critical for sugarcane farmers to adapt to changing conditions. Long-term climate forecasts and projections help farmers anticipate shifts in temperature and precipitation patterns, allowing them to make informed decisions about crop management practices, varietal selection, and investment in irrigation infrastructure.

Advances in technology, such as the integration of weather data into farm management software and mobile applications, enable farmers to access real-time weather information and forecasts directly in the field. This empowers farmers to make timely decisions based on the latest weather updates, improving overall farm efficiency and productivity. Weather forecasting is indispensable for sugarcane farming, guiding decisions related to planting, irrigation, pest and disease management, frost protection, yield forecasting, and climate change adaptation. Access to accurate and timely weather information is essential for maximizing crop productivity and minimizing risks associated with adverse weather conditions. Enhanced seasonal forecasts, coupled with improved communication, comprehension, and capacity to utilize these forecasts in decision-making, will greatly benefit farmers in terms of productivity and profitability. Additionally, it will mitigate their exposure to financial risks and broader environmental risks under variable climatic conditions (An-Vo et al., 2019).

11.10 INTEGRATED PESTS AND DISEASES MANAGEMENT

Unchecked insects and pests pose significant threats, particularly to crops like sugarcane. The rapid proliferation of these pests, influenced by various factors such as climate, rainfall, temperature, and the availability of food sources within the crop itself, can lead to substantial losses. India harbors over 200 insect species, with a handful considered highly destructive. Termites, for instance, are subterranean insects that target the stalks and shoots of sugarcane, exhibiting heightened activity during dry spells, notably from April to June and October, though they remain active throughout the year. Farmers combat termite infestations by applying pesticides to the stalks during planting (Cheavegatti-Gianotto et al., 2011; Song et al., 2019, 2021, 2022; Tian et al., 2023). Remedial measures also include incorporating well-decomposed farmyard manure, clearing stubble and debris from previous crops, and introducing chemicals into the furrows during planting. Another prevalent threat is the internode borer, a significant pest in peninsular India (Gupta 1957), particularly active during or shortly after internode formation and persisting throughout the plant's life cycle (Ananthanarayana and Balasubramanian, 1980). Infestation by the internode borer leads to a reduction in internode length and girth, resulting in a significant decrease in cane yield (David et al., 1979) and a decline in juice quality, including reduced sucrose content (David and Ranganathan 1960). Mealybugs play a crucial role in virus transmission and the growth, of sooty mold fungus due to the large amount of honeydew secreted by the insect (Eid et al., 2011). Deterioration in juice quality and a decrease in sucrose content are the eventual outcomes. Control measures include destroying affected leaves at harvest, selecting proper and healthy seed canes, and spraying insecticides. The spray material should be applied in a manner that allows it to trickle down the internodes into the leaf sheaths underneath to kill the nymphs. Whitefly infestation can be mitigated by avoiding ratoon in low-lying areas, burning all trash from harvested crops, destroying affected parts, and using foliar sprays (Srivastava and Rai 2012).

Red rot disease is influenced by various factors such as temperature, pH, nutrition, and environmental conditions. It is a significant disease not only in the Indian subcontinent but also in other parts of the world.

Management strategies include selecting healthy setts treated with heat therapy, planting resistant sugarcane varieties, and rotating crops with paddy, onion, garlic, linseed, and green manure crops. Precautions should be taken to prevent the spread of the disease through water, and water from infected crop fields should not be allowed to flow into other areas where crops are grown. Set rot disease can cause young cane crops to exhibit patchy, uneven germination over large areas (Raid 1990). This disease is particularly prevalent in low-lying areas and poorly drained soils. It can be controlled by treating the setts with chemical fungicides before planting (Vijaya et al., 2007). Usage of biological control agents, cultural practices, along with chemicals, should be used to control diseases and pests in sugarcane in this climate change era, as dependence on a single method might not be beneficial in the long term. Apart from drought stress, sugarcane diseases also seriously affect cane and juice yield in major sugarcane varieties (Viswanathan et al., 2014).

11.11 CLIMATE-SMART TECHNOLOGIES IN FARMING

The era of climate change is a universal and critical challenge to global food security, and acknowledging the imperative to enhance current agricultural management practices and stewardship of natural resources. FAO (2016) advocates for strengthening the adaptive capacity of rural communities. This involves equipping them to effectively navigate climate changes and increased variability, while also fostering resilience to climatic shocks. The agricultural sector must undergo transformation to meet the escalating demand for food from the expanding global population. This necessitates a shift in how we manage land, water, soil nutrients, and genetic resources, ensuring their more efficient and sustainable utilization to enhance agricultural productivity and resilience (Figure 11.1). Adaptation to climate change and variability is essential within agricultural production systems to safeguard food and livelihood security for farming communities.

11.12 CONCLUSION

Sugarcane crops experience all four seasons and endure environmental extremes like temperature fluctuations, making them vulnerable to insect pests, diseases, and weed infestations. Implementing suitable CSTs not only reduces greenhouse gas emissions but also enhances the ability of agricultural systems to adapt to climate change, leading to higher cane yields and improved food security. Adopting a range of CSTs, from seed selection to harvest practices, aids in mitigating the adverse effects of global warming by reducing carbon, water, fertilizer, and pesticide footprints while simultaneously enhancing soil health, growth, yield, and overall sustainability of sugarcane cultivation. Further, precision irrigation techniques like drip or sprinkler systems optimize water usage in sugarcane, delivering water directly to the root zone for increased efficiency. Integrating climate-smart irrigation with soil moisture sensors and weather data optimizes scheduling, preventing water stress and adapting to climate change. Combining climate-smart irrigation with water conservation measures enhances resilience to climate variability, mitigates climate change impacts, and ensures long-term sugarcane production sustainability.

The FIRB approach boosts sugarcane output while simultaneously improving water and fertilizer input efficiency, especially in the sugarcane–wheat combination. Use of sugarcane trash as mulch conserves soil moisture, regulates temperature, enhances nutrient uptake in canes, and improves yield and quality. Various CSTs help reduce carbon, water, fertilizer, energy, and pesticide footprints while enhancing input use efficiency, although their applicability may vary based on specific conditions. Advancements could include utilizing sugarcane straw for bioenergy, implementing optimal management practices to maximize growth, yield, and quality, thereby addressing soil health and farmers' livelihoods. These research goals aim to increase soil carbon sequestration, reduce anthropogenic CO_2 emissions, and ultimately benefit both soil health and sugarcane farmers. Consequently, policies and programs aimed at promoting CSTs should focus on site-specific technologies that are relevant to the local farmers. Collaboration among sugarcane growers, researchers, government agencies, and industry stakeholders facilitates the exchange of knowledge, resources, and best practices for adapting to climate change challenges.

KEYWORDS

- **sugarcane**
- **climate-smart technologies**
- **resilience**
- **sustainability**
- **crop improvement**

REFERENCES

Ananthanarayana K, Balasubramanian M. (1980) Distribution and intensity of internode borer of sugarcane, Chilo sacchariphagus indicus (K.) during different periods of plant growth. *Proc. Ann. Conven. Deccan Sugar Technol. Assoc.* 30, 81–88.

Anas M, Verma KK, Haq IU, Naeem M, Li Q, Liao F, Xiang YX, Li YR. (2021a) Characterization of exotic, native and wild-type genotypes of sugarcane (*Saccharum* spp. Hybrids) for internal nitrogen use efficiency under different nitrogen levels. *Sugar Tech.* 23, 1258–1267. https://doi.org/10.1007/s12355-021-01014-1

Anas M, Verma KK, Riaz M, Qiang L, Liao F, Liu Y, Li YR. (2021b) Physio-morphological and biochemical mechanism of nitrogen use efficiency in sugarcane (*Saccharum* spp.) genotypes under different growth stages and nitrogen levels. *J. Plant Interact.* 16, 332–343.

An-Vo DA, Mushtaq S, Reardon-Smith K, Kouadio L, Attard S, Cobon D, Stone R. (2019) Value of seasonal forecasting for sugarcane farm irrigation planning. *Eur. J. Agron.* 104, 37–48.

Appunu C, Mohanraj K, Hemaprabha G, Durai AA, Mahadevaiah C, Nair NV, Bakshi Ram BR. (2017) Co 06022—A sugarcane early maturing and drought tolerant variety suitable for Tamil Nadu and Pondicherry. *J. Sugarcane Res.* 7(2), 83–92.

Bhatt R, Majumder D, Tiwari AK, Singh SR, Prasad S, Palanisamy G. (2023) Climate-smart technologies for improving sugarcane sustainability in India—A review. *Sugar Tech.* 25(1), 1–14.

Bhatt R, Singh P, Ali OM, Latef AAHA, Laing AM, Hossain A. (2021b) Yield and quality of ratoon sugarcane are improved by applying potassium under irrigation to potassium deficient soils. *Agronomy* 11(7), 1381. https://doi.org/10.3390/agronomy11071381.

Bhatt R, Verma KK, Jamwal NS. (2023a) Can paddy-compost eliminate N fertilization for climate-smart sugarcane? *Sugar Tech.* 25, 1036–1046. https://doi.org/10.1007/s12355-023-01270-3

Bhatt R, Verma KK, Kumar R, Sanghera GS. (2023b) Foliar application of potassium salt of active phosphorus (PSAP) mitigates insect pests and improves yield along with sugarcane quality in response to agroclimatic conditions of Punjab. *Sugar Tech.* 25(3), 660–669.

Bhatt R. (2020) Resources management for sustainable sugarcane production. In: Kumar S, Meena RS, Jhariya MK. (Eds.), *Resources use efficiency in agriculture*. Singapore: Springer, pp. 647–693. https://doi.org/10.1007/978-981-15-6953-1_18.

Bhatt R. (2021a) Does straw mulch alter soil evaporation, yield, and quality of sugarcane? *Proc. Ind. Nat. Sci. Academy* 87, 524–533. https://doi.org/10.1007/s43538-021-00048-z.

Chaudhary P, Xu M, Ahamad L, Chaudhary A, Kumar G, Adeleke BS, Verma KK, Hu D-M, Siric I, Kumar P, Popescu SM, Fayssal SA. (2023) Application of synthetic consortia for improvement of soil fertility, pollution remediation, and agricultural productivity: A review. *Agronomy* 13, 643. https://doi.org/10.3390/agronomy13030643.

Cheavegatti-Gianotto A, De Abreu HMC, Arruda P, BespalhokFilho JC, Burnquist WL, Creste S, di Ciero L, Ferro JA, de Oliveira Figueira AV, de Sousa Filgueiras T, Grossi-de-Sá MDF. (2011) Sugarcane (*Saccharum officinarum*): A reference study for the regulation of genetically modified cultivars in Brazil. *Trop. Plant Biol.* 4, 62–89.

Chen GL, Zheng FJ, Lin B, Lao SB, He J, Huang Z, Zeng Y, Sun J, Verma KK. (2020) Phenolic and volatile compounds in the production of sugarcane vinegar. *ACS Omega* 5, 30587−30595.

Chen G-L, Zheng F-J, Lin B, Yang Y-X, Fang X-C, Verma KK, Yang L-F. (2023) Vinegar: A potential source of healthy and functional food with special reference to sugarcane vinegar. *Front. Nutr.* 10, 1145862. https://doi.org/10.3389/fnut.2023.1145862

David H, Sithanantham S, Velayutham B. (1979) Some aspects of losses due to internode borer in sugarcane in Tamil Nadu. *Proc. Ann. Conven. Deccan Sugar Technol. Assoc.* 29, 27–40.

David IL, Ranganathan V. (1960) Influence of internode borer on the quality of juice in sugarcane. *Ind. J. Sugarcane Res. Dev.* 4, 209–212.

Deng ZN, Huang HR, Huang CM, Verma KK, Luo HB, Jiang SL, Li YJ, Cao HQ, Wu KC, Xu L. (2023) Short-term limited water irrigation influences the photosynthetic pigments, enzymatic and non-enzymatic activities in *Saccharum spontaneum* L. at vegetative stage. *Sugar Tech.* 25, 110–118. https://doi.org/10.1007/s12355-022-01175-7

Dlamini PJ. (2021) Drought stress tolerance mechanisms and breeding effort in sugarcane: A review of progress and constraints in South Africa. *Plant Stress* 2, 100027.

Eid MA, El-Shabrawy HA, Yakoub RS. (2011) An attempt to study the impact of Pink Mealybug infestation *Saccharicoccus sacchari* Ckll. on chemicals and allelochemicals of some sugarcane cultivars. *Acad. J. Entomol.* 4(1), 23–29.

Eksteen A, Singels A, Ngxaliwe S. (2014) Water relations of two contrasting sugarcane genotypes. *Field Crops Res.* 168, 86–100.

FAO (United Nations Food and Agriculture Organisation) (2016) In: Ruane J, Dargie J D, Daly C (Eds.), *The Proceedings of the FAO International Symposium on the Role of Agriculture and Biotechnologies in Sustainable Food System and Nutrition.* FAO Publications: Rome, pp. 66–79. Ed's.

Gawander J. (2007) Impact of climate change on sugarcane production in Fiji. *WMO Bull.* 56(1), 34–39.

Gbetibouo GA, Hassan RM. (2005) Measuring the economic impact of climate change on major South African field crops: A Ricardian approach. *Glob. Planet Change* 47, 143–152.

Guo D-J, Li D-P, Chen Z-D, Yang B, Verma KK, Singh RK, Singh P, Khan Q, Sharma A, Qin Y, Zhang B-Q, Song X-P, Li Y-R. (2023) Effect of endophytic diazotroph *Enterobacter roggenkampii* ED5 on nitrogen-metabolism-related microecology in the sugarcane rhizosphere at different nitrogen levels. *Front. Microbiol.* 14, 1132016. https://doi.org/10.3389/fmicb.2023.1132016

Guo D-J, Li D-P, Singh RK, Singh P, Sharma A, Verma KK, Qin Y, Khan Q, Lu Z, Malviya MK, Song X-P, Xing Y-X and Li Y-R (2021) Differential protein expression analysis of two sugarcane varieties in response to diazotrophic plant growth-promoting endophyte *Enterobacter roggenkampii* ED5. *Front. Plant Sci.* 12, 727741. https://doi.org/10.3389/fpls.2021.727741

Guo DJ, Li DP, Singh RK, Singh P, Verma KK, Sharma A, Qin Y, Khan Q, Song XP, Malviya MK, Xing YX, Li YR. (2022) Comparative transcriptome analysis of two sugarcane varieties in response to diazitrophic plant growth promoting endophyte *Enterobacter roggenkampii* ED5. *J. Plant Interact.*, 17(1), 75–84.

Gupta BD. (1957) A note on the scientific and common names of sugarcane pests in India. *Ind. J. Sugarcane Res. Dev.* 2, 9–13.

Hannah L, Lovejoy TE, Schneider SH. (2005) Biodiversity and climate change in context. In: Lovejoy TE, Hannah L. (Eds.) *Climate Change and Biodiversity*. Yale University Press, New Haven.

Huang CM, Wu KC, Verma KK, Luo HB, Deng ZN, Shi GY, Cao HQ, Wu XJ, Ye LP, Wei YW, Xu L. (2024) Isolation and expression analysis of cellulose synthase 3 (*Ces3*) genes from sugarcane (*Saccharum officinarum* L.). *Chem. Biol. Technol. Agric.* 11, 71. https://doi.org/10.1186/s40538-024-00597-7

Huang HR, Gao YJ, Malviya MK, Verma KK, Solanki MK, Huang YX, Li X, Deng Y, Yan J, Tang SY, Wang LW, Xu L. (2022) Genetic diversity analysis of sugarcane (*Saccharum* spp. Hybrids) among high-sucrose clones of GT series and commonly used parents by using microsatellite markers in Guangxi, China. *Sugar Tech.* 24, 397–407. https://doi.org/10.1007/s12355-021-01033-y

Hunsigi G. (2001) Ratoon management. In: *Sugarcane in Agriculture and Industry*. Prism Books: Bangalore, pp. 217.

Jagadesh M, Dash M, Kumari A, Singh SK, Verma KK, Kumar P, Bhatt R, Sharma SK. (2024) Revealing the hidden world of soil microbes: Metagenomic insights into plant, bacteria, and fungi interactions for sustainable agriculture and ecosystem restoration. *Microbiol. Res.* 127764. https://doi.org/10.1016/j.micres.2024.127764

Jones JW, Hansen JW, Royce FS, Messina CD. (2000) Potential benefits of climate forecasting to agriculture. *Agric. Ecosyst. Environ.* 82(1), 169–184.

Joshi A, Verma KK, Rajput VD, Arora J. (2023) Metabolic engineering of microorganisms in advancing biofuel production. In: *Advances in Lignocellulosic Biofuel Production Systems*. pp. 181–202. https://doi.org/10.1016/B978-0-323-91192-4.00016-X

Joshi A, Verma KK, Rajput VD, Minkina T, Arora J. (2022) Recent advances in metabolic engineering of microorganisms for advancing lignocellulose-derived biofuels. *Bioengineered* 13(4), 8135–8163. https://doi.org/10.1080/21655979.2022.2051856.

Kimball BA, Kobayashi K, Bindi M. (2002) Responses of agricultural crops to free-air CO_2 enrichment. *Adv. Agronomy* 77, 293–368.

Krishna D, Sachan HK, Verma KK. (2022) Abiotic stresses in plants: An overview. In: *Understanding Abiotic Stresses*. Nova Science Publishers, Inc.: New York.

Kumari A, Dash M, Singh SK, Jagadesh M, Mathpal B, Mishra PK, Pandey SK, Verma KK. (2023a) Soil microbes: A natural solution for mitigating the impact of climate change. *Environ. Monit. Assess.* 195, 1436. https://doi.org/10.1007/s10661-023-11988-y

Kumari A, Lakshmi GA, Krishna GK, Patni B, Prakash S, Bhattacharyya M, Singh SK, Verma KK. (2022) Climate change and its impact on crops: A comprehensive investigation for sustainable agriculture. *Agronomy* 12, 3008. https://doi.org/10.3390/agronomy12123008

Kumari A, Singh SK, Mathpal B, Verma KK, Garg VK, Bhattacharyya M, Bhatt R. (2023b) The biosynthesis, mechanism of action, and physiological functions of melatonin in horticultural plants: A review. *Horticulturae* 9, 913. https://doi.org/10.3390/horticulturae9080913

Lawlor DW, Mitchell RAC. (1991) The effects of increasing CO_2 on crop photosynthesis and productivity-a review of field studies. *Plant Cell Environ.* 14, 807-818.

Li A-M, Liao F, Wang M, Chen Z-L, Qin C-X, Huang R-Q, Verma KK, Li Y-R, Que Y-X, Pan Y-Q, et al. (2023) Transcriptomic and proteomic landscape of sugarcane response to biotic and abiotic stressors. *Int. J. Mol. Sci.* 24, 8913. https://doi.org/10.3390/ijms24108913

Mali SC, Shedage S, Shrivastava PK. (2017) Economic Evaluation of Sugarcane Based Agro forestry Systems. *J. Tree Sci.* 36(1), 34–37.

Malviya MK, Li C-N, Lakshmanan P, Solanki MK, Wang Z, Solanki AC, Nong Q, Verma KK, Singh RK, Singh P, Sharma A, Guo D-J, Dessoky ES, Song X-P, Li Y-R. (2022) High-throughput sequencing-based analysis of rhizosphere and diazotrophic bacterial diversity among wild progenitor and closely related species of sugarcane (*Saccharum* spp. inter-specific hybrids). *Front. Plant Sci.* 13, 829337. *https://doi.org/10.3389/fpls.2022.829337.*

Malviya MK, Solanki MK, Li CN, Wang Z, Zeng Y, Verma KK, Singh RK, Singh P, Huang HR, Yang LT, Song XP, Li YR. (2021) Sugarcane-legume intercropping can enrich the soil microbiome and plant growth. *Front. Sustain. Food Syst.* 5, 606595. https://doi.org/10.3389/fsufs.2021.606595.

Oliveira MA, Verma KK, Bhatt R, Oliveira TBA. (2022) Impact of green and organic fertilizers on soil fertility and sugarcane productivity. In: Verma KK, et al. (Eds.), *Agro-Industrial Perspectives on Sugarcane Production Under Environmental Stress.* https://doi.org/10.1007/978-981-19-3955-6_11.

PAU (Punjab Agricultural University) (2022). *Package of Practices for crops of Punjab-Kharif.* Ludhiana, India: Punjab Agricultural University, pp. 55–66.

Prasad JVNS, Korwar GR, Rao KV, Mandal UK, Rao CAR, Rao GR, Ramakrishna YS, Venkateswarlu B, Rao SN, Kulkarni HD, Rao MR. (2010) Tree row spacing affected agronomic and economic performance of Eucalyptus-based agroforestry in Andhra Pradesh, SouthernIndia. *Agrofor. Syst.* 78, 253–267.

Raid RN. (1990) Fungicidal control of pineapple disease of sugarcane. *J. Amer. Soc. Sugarcane Technol.* 10, 45–50.

Sharma A, Singh RK, Singh P, Vaishnav A, Guo DJ, Verma KK, Li DP, Song XP, Malviya MK, Khan N, Lakshmanan P, Li YR. (2021) Insights into the bacterial and nitric oxide-induced salt tolerance in sugarcane and their growth-promoting abilities. *Microorganisms* 9, 2203. https://doi.org/10.3390/microorganisms9112203.

Sharma A, Song X-P , Singh RK, Vaishnav A, Gupta S, Singh P, Guo D-J, Verma KK, Li Y-R. (2022) Impact of carbendazim on cellular growth, defence system and plant growth promoting traits of *Priestia megaterium* ANCB-12 isolated from sugarcane rhizosphere. *Front. Microbiol.* 13, 1005942. doi:10.3389/fmicb.2022.1005942

Shukla SK, Solomon S, Sharma L, Jaiswal VP, Pathak AD, Singh P. (2019) Green technologies for improving cane sugar productivity and sustaining soil fertility in sugarcane-based cropping system. *Sugar Tech.* 21, 186–196.

Singh P, Singh RK, Li HB, Guo DJ, Sharma A, Verma KK, Solanki MK, Upadhyay SK, Lakshmanan P, Yang LT, Li YR. (2023) Nitrogen fixation and phytohormone stimulation of sugarcane plant through plant growth promoting diazotrophic *Pseudomonas. Biotechnol. Genetic Eng. Reviews* 1–21. https://doi.org/10.1080/02648725.2023.2177814

Singh P, Singh RK, Song QQ, Li HB, Guo DJ, Malviya MK, Verma KK, Song XP, Lakshmanan P, Yang LT, Li YR. (2021) Comparative analysis of protein and differential responses of defense-related gene and enzyme activity reveals the long-term molecular responses of sugarcane inoculated with *Sporisorium scitamineum. J. Plant Interactions* 16(1), 12–29.

Singh RK, Singh P, Li HB, Song QQ, Guo DJ, Solanki MK, Verma KK, Malviya MK, Song XP, Lakshmanan P, Yang LT, Li YR. (2020) Diversity of nitrogen-fixing rhizobacteria associated with sugarcane: A comprehensive study of plant-microbe interactions for growth enhancement in *Saccharum* spp. *BMC Plant Biol.* 20, 220. https://doi.org/10.1186/s12870-020-02400-9.

Singh SA, Yadav M, Malik RK, Singh H. (2002) Furrow irrigated raised bed planting system: Are source conservation technology for increasing wheat productivity in rice-wheat sequence. *Proceedings of the International Workshop*, Hisar, India, 4–6 March 2002, CCSHAU: Hisar.

Snyman SJ, Mhlanga P, Watt MP. (2004) Rapid screening of sugarcane Plantlets for *In vitro* mannitol-induced stress. Sugar Tech. 18, 437–440.

Solanki MK, Verma KK, Dastogeer KMG, Mora-Poblete F, Mundra S. (2023) Microbial resilience in plant nutrient management towards sustainable farming. *Front. Microbiol.* 14, 1280811. https://doi.org/10.3389/fmicb.2023.1280811

Song X-P, Mo F, Verma KK, Wei J, Zhang X, Yang L, Li YR. (2019) Effect of sugarcane smut (*Ustilago scitaminea* Syd.) on ultrastructure and biochemical indices of sugarcane. *Biomed J. Sci. Tech. Res.* 002950. https://doi.org/10.26717/BJSTR.2019.17.002950.

Song XP, Mo F, Yan M, Zhang X, Zhang B, Huang X, Huang D, Pan Y, Verma KK, Li YR. (2022) Effect of smut infection on the photosynthetic physiological characteristics and related defense enzymes of sugarcane. *Life* 12, 1201. https://doi.org/10.3390/life12081201 Song XP, Verma KK, Tian DD, Zhang XQ, Liang YJ, Huang X, Li CN, Li YR. (2021) Exploration of silicon functions to integrate with biotic stress tolerance and crop improvement. *Biol. Res.* 54, 19. https://doi.org/10.1186/s40659-021-00344-4

Srivastava AK, Rai MK. (2012) Sugarcane production: Impact of climate change and its mitigation. *Biodivers. J. Biol. Diversity* 13(4).

Sundara B, Tripathi BK. (1999) Available N changes and N balance under multi ratooning of sugarcane varieties in a tropical vertisol. In: *Proc. XXIII ISSCT Congress*, New Delhi, India, pp. 80–88.

Tian D, Qin L, Verma KK, Wei L, Li J, Li B, Zhou W, He Z, Wei D, Huang S, Long S, Huang Q, Li C, Wei S. (2023) Transcriptomic and metabolomic differences between banana varieties which are resistant or susceptible to *Fusarium wilt. PeerJ.* 11, e16549. https://doi.org/10.7717/peerj.16549

Verma KK, Anas M, Chen Z, Rajput VD, Malviya MK, Verma CL, Singh RK, Singh P, Song XP, Li YR. (2020) Silicon supply improves leaf gas exchange, antioxidant defense system and growth in sugarcane responsive to water limitation. *Plants* 9, 1032. https://doi.org/10.3390/plants9081032.

Verma KK, Joshi A, Song X-P, Liang Q, Xu L, Huang H-r, Wu K-C, Seth CS, Arora J and Li Y-R (2024a) Regulatory mechanisms of plant rhizobacteria on plants to the adaptation of adverse agroclimatic variables. *Front. Plant Sci.* 15, 1377793. https://doi.org/10.3389/fpls.2024.1377793

Verma KK, Joshi A, Song X-P, Singh S, Kumari A, Arora J, Singh SK, Solanki MK, Seth CS, Li Y-R. (2024b) Synergistic interactions of nanoparticles and plant growth promoting rhizobacteria enhancing soil-plant systems: A multigenerational perspective. *Front. Plant Sci.* 15, 1376214. https://doi.org/10.3389/fpls.2024.1376214

Verma KK, Li DM, Singh M, Rajput VD, Malviya MK, Minkina T, Singh RK, Singh P, Song XP, Li YR. (2020a) Interactive role of silicon and plant-rhizobacteria mitigating abiotic stresses: A new approach for sustainable agriculture and climate change. *Plants* 9, 1055. https://doi.org/10.3390/plants9091055.

Verma KK, Rajput VD, Singh M, Verma CL, Minkina T. (2022a) Influence of waterlogging on plant physiology and molecular biology. In: *Understanding Abiotic Stresses Publisher*. Nova Science Publishers, Inc.: New York.

Verma KK, Siddiqui N, Singh M. (2022b) Understanding the plant physiological and molecular adaptation strategies to soil salinity. In: *Understanding Abiotic Stresses*. Nova Science Publishers, Inc.: New York. https://doi.org/10.52305/UBUJ4024

Verma KK, Singh P, Song X-P, Malviya MK, Singh RK, Chen G-L, Solomon S, Li YR. (2020b) Mitigating climate change for sugarcane improvement: Role of silicon in alleviating abiotic stresses. *Sugar Tech.* 22(5), 741–749. https://doi.org/10.1007/s12355-020-00831-0.

Verma KK, Singh RK, Song QQ, Singh P, Zhang B-Q, Song X-P, Chen G-L, Li YR. (2019a) Silicon alleviates drought stress of sugarcane plants by improving antioxidant responses. *Biomed J. Sci. Tech. Res.* 002957. https://doi.org/10.26717/BJSTR.2019.17.002957

Verma KK, Song X, Degu HD, Guo D, Joshi A, Huang H, Xu L, Singh M, Huang D, Rajput VD, Li YR. (2023a) Recent advances in nitrogen and nano-nitrogen fertilizers for sustainable crop production: A mini-review. *Chem. Biol. Technol. Agric.* 10, 1–14. https://doi.org/10.1186/s40538-023-00488-3

Verma KK, Song X, Verma CL, Huang H, Singh M, Xu L, Wu J, Khan ZH, Zhang B, Li Y. (2023b) Mathematical modeling of climate and fluoride effects on sugarcane photosynthesis with silicon nanoparticles. *Plant Physiol. Biochem.* 204, 108089. https://doi.org/10.1016/j.plaphy.2023.108089

Verma KK, Song XP, Budeguer F, Nikpay A, Enrique R, Singh M, Zhang B, Wu, JM, Li YR. (2022c) Genetic engineering: An efficient approach to mitigating biotic and abiotic stresses in sugarcane cultivation. *Plant Signal. Behav.* 17(1), e2108253.

Verma KK, Song XP, Joshi A, Rajput VD, Singh M, Sharma A, Singh RK, Li DM, Arora J, Minkina T, Li YR. (2022d) Nanofertilizer possibilities for healthy soil, water and food in future: An overview. *Front. Plant Sci.* 13, 865048. https://doi.org/10.3389/fpls.2022.865048

Verma KK, Song XP, Joshi A, Tian DD, Rajput VD, Singh M, Arora J, Minkina TM, Li YR. (2022e) Recent trends in nano-fertilizer for advancing sustainable agriculture under the era of climate change to ensure food security for future generations. *Nanomaterials* 12, 173. https://doi.org/10.3390/nano12010173.

Verma KK, Song XP, Li DM, Singh M, Wu JM, Singh RK, Sharma A, Zhang BQ, LI YR. (2022f) Silicon and soil microorganisms improves rhizospheric soil health with bacterial community, plant growth, performance and yield. *Plant Signal. Behav*. 17(1), e2104004.

Verma KK, Song XP, Lin B, Guo DJ, Singh M, Rajput VD, Singh RK, Singh P, Sharma A, Malviya MK, Chen GL, Li YR. (2022g) Silicon induced drought tolerance in crop plants: Physiological adaptation strategies. *Silicon* 14(6), 2473–2487. https://doi.org/10.1007/s12633-021-01071-x

Verma KK, Song XP, Rajput VD, Boldyreva V, Zhang BQ, Minkina T, Li YR. (2022h) Morpho-physiological, biochemical and ultrastructural modifications on sugarcane to prolonged water deficit. In: Verma KK, et al. (Eds.), *Agro-Industrial Perspectives on Sugarcane Production Under Environmental Stress*. https://doi.org/10.1007/978-981-19-3955-6_11.

Verma KK, Song XP, Rajput VD, Solomon S, Li YR, Rao GP. (2022i) *Agro-Industrial Perspectives on Sugarcane Production Under Environmental Stress*. Springer Nature Singapore, https://doi.org/10.1007/978-981-19-3955-6

Verma KK, Song XP, Singh M, Tian DD, Rajput VD, Minkina T, Li YR. (2023c) Association of silicon and soil microorganisms induces stress mitigation, increasing plant productivity. In: de Mello Prado R. (Eds.), *Benefits of Silicon in the Nutrition of Plants*. pp. 299–328. https://doi.org/10.1007/978-3-031-26673-7_17

Verma KK, Song XP, Tian DD, Guo DJ, Chen ZL, Zhong CS, Nikpay A, Singh M, Rajput VD, Singh RK, Minkina T, Li YR/. (2021a) Influence of silicon on biocontrol strategies to manage biotic stress for crop protection, performance and improvement. *Plants* 10, 2163. https://doi.org/10.3390/plants10102163

Verma KK, Song XP, Tian DD, Singh M, Verma CL, Rajput VD, Singh RK, Sharma A, Singh P, Malviya MK, Li YR. (2021b) Investigation of defensive role of silicon during drought stress induced by irrigation capacity in sugarcane: Physiological and biochemical characteristics. *ACS Omega* 6, 19811–19821.

Verma KK, Song XP, Verma CL, Chen ZL, Rajput VD, Wu KC, Liao F, Chen GL, Li YR. (2021c) Functional relationship between photosynthetic leaf gas exchange in response to silicon application and water stress mitigation in sugarcane. *Biol. Res*. 54, 15. https://doi.org/10.1186/s40659-021-00338-2

Verma KK, Song XP, Verma CL, Malviya MK, Guo DJ, Rajput VD, Sharma A, Wei KJ, Chen GL, Solomon S, Li YR. (2021d) Predication of photosynthetic leaf gas exchange of sugarcane (*Saccharum* spp.) leaves in response to leaf positions to foliar spray of potassium salt of active phosphorus under limited water irrigation. *ACS Omega* 6, 2396–2409.

Verma KK, Song X-P, Xu L, Huang H-R, Liang Q, Seth CS, Li Y-R. (2023d) Nano-microplastic and agro-ecosystems: A mini-review. *Front. Plant Sci*. 14, 1283852. https://doi.org/10.3389/fpls.2023.1283852

Verma KK, Song XP, Yadav G, Degu HD, Parvaiz A, Singh M, Huang HR, Mustafa G, Xu L, Li YR. (2022j) Impact of agroclimatic variables on proteo-genomics in sugarcane (*Saccharum* spp.) plant productivity. *ACS Omega* 7, 22997–23008.

Verma KK, Song XP, Zeng Y, Guo DJ, Sing M, Rajput VD, Malviya MK, Wei KJ, Sharma A, Li DP, Chen GL, Li YR. (2021e) Foliar application of silicon boosts growth, photosynthetic leaf gas exchange, antioxidative response and resistance to limited water irrigation in sugarcane (*Saccharum officinarum* L.). *Plant Physiol. Biochem.* 166, 582–592.

Verma KK, Song XP, Zeng Y, Li DM, Guo DJ, Rajput VD, Chen GL, Barakhov A, Minkina TM, Li YR. (2020c) Characteristics and correlation of leaf stomata and its relationship with photosynthesis on *Saccharum* spp. under different irrigation and silicon application. *ACS Omega* 5, 24145–24153.

Verma KK, Vatsal S, Singh M, Rajput VD, Minkina T. (2022k) Coping with water stress: Effects and adaptive mechanisms to plants. In: *Understanding Abiotic Stresses*. Nova Science Publishers, Inc.: New York.

Verma KK, Wu KC, Verma CL, Li DM, Malviya MK, Singh RK, Singh P, Chen GL, Song XP, Li YR. (2020d) Developing mathematical model for diurnal dynamics on photosynthetic responses in sugarcane responsive to different irrigation and silicon application. *PeerJ* 8, e10154. https://doi.org/10.7717/peerj.10154.

Verma KK, Liu X-H, Wu K-C, Singh RK, Song QQ, Malviya MK, Song X-P, Singh P, Verma CL, Li YR. (2020e) The impact of silicon on photosynthetic and biochemical responses of sugarcane under different soil moisture levels. *Silicon* 12, 1355–1367.

Verma KK, Wu K-C, Singh P, Malviya MK, Singh RK, Song X-P, Li YR. (2019b) The protective role of silicon in sugarcane under water stress: Photosynthesis and antioxidant enzymes. *Biomed J. Sci. Tech. Res.* 002685. https://doi.org/0.26717/BJSTR.2019.15.002685

Verma RS (2002) *Sugarcane Ratoon Management*. International Book Distributing Company: Lucknow, India.

Vijaya HK, Kulkarni S, Yashoda R, Hedge YR. (2007) Chemical control of sett rot of sugarcane caused by Ceratocystis paradoxa. *Karnataka J. Agric. Sci.* 20(1), 62–64.

Viswanathan R, Chinnaraja C, Malathi P, Gomathi R, Rakkiyappan P, Neelamathi D, Ravichandran V. (2014) Impact of Sugarcane yellow leaf virus (ScYLV) infection on physiological efficiency and growth parameters of sugarcane under tropical climatic conditions in India. *Acta Physiol. Plant.* 36, 1805–1822.

Wu K-C, Huang C-M, Verma KK, Deng Z-N, Huang H-R, Pang T, Cao H-Q, Luo H-B, Jiang S-L, Xu L. (2022) Transcriptomic responses of *Saccharum spontaneum* roots in response to polyethylene glycol–6000 stimulated drought stress. *Front. Plant Sci.* 13, 992755. https://doi.org/10.3389/fpls.2022.992755

Xu F, Sun RC, Sun JX, Liu CF, He BH, Fan JS. (2005) Determination of cell wall ferulic and p-coumaric acids in sugarcane bagasse. *Analyt. Chem. Acta* 552(1–2), 207–217. https://doi.org/10. 1016/j.aca.2005. 07. 037.

Xu L, Deng Z-N, Wu K-C, Malviya MK, Solanki MK, Verma KK, Pang T, Li Y-J, Liu X-Y, Kashyap BK, Dessoky ES, Wang W-Z and Huang H-R. (2022) Transcriptome analysis reveals a gene expression pattern that contributes to sugarcane bud propagation induced by indole-3-butyric acid. *Front. Plant Sci.* 13, 852886. https://doi.org/10.3389/fpls.2022.852886

Zhao D, Li YR. (2015) Climate change and sugarcane production: Potential impact and mitigation strategies. *Int. J. Agronomy* 1–10.

Zheng F-J, Lin B, Yang Y-X, Fang X-C, Verma KK, Chen G-L. (2023) Efficacy and functionality of sugarcane original vinegar on mice. *Front. Microbiol.* 14, 1224666. https://doi.org/10.3389/fmicb.2023.1224666

CHAPTER 12

Sugarcane Crop Improvement: Current Status, Challenges, and Future Perspectives

KRISHAN K. VERMA,[1] XIU-PENG SONG,[1*] QIANG LIANG,[1] LIN XU,[1] KAMLESH K. MISHRA,[2] and YANG-RUI LI[1*]

[1]*Sugarcane Research Institute, Guangxi Academy of Agricultural Sciences/Key Laboratory of Sugarcane Biotechnology and Genetic Improvement (Guangxi), Ministry of Agriculture and Rural Affairs, Nanning, Guangxi, China*

[2]*Department of Botany, University of Lucknow, Lucknow, Uttar Pradesh, India*

*Corresponding author

ABSTRACT

Sugarcane (*Saccharum* spp. hybrids) is an important cash crop and a renewable bioenergy source worldwide with better plant growth potential in tropical and semitropical areas. China is the world's third-largest sugar consumer and fourth-largest sugar producer, and about 90% of sugar is produced from sugarcane. Sugarcane is used as a raw material in bioethanol production, which is a renewable source of energy. It is a C_4 crop with unique features of sugar and feedstock properties for bioenergy and cogeneration. The maximum level of polyploidy and genomic peculiarities in this crop point toward a complex regulation mechanism for economically vital features such as high sugar concentration, biomass capacity, stress tolerance, etc. Molecular revisions have revealed the understanding of gene interactions, proteins, and metabolites. This chapter discusses action

Sugarcane Production: Environmental Impacts and Socioeconomic Issues.
Krishan K. Verma, Xiu-Peng Song, Lin Xu, Rajan Bhatt, and Yang-Rui Li (Eds.)

DOI: 10.1201/9781779641137-12

mechanisms to upgrade sugarcane plant growth, development, and stress avoidance capacity. It offers advanced strategies for enhancing sugarcane plant performance and production for socioeconomic development.

12.1 INTRODUCTION

Plant development exposed to environmental stresses is required, especially in agricultural crop plants in the dynamic era of climate change and global warming (Krishna et al., 2022; Verma et al., 2024a, 2024b). Plants have established various modifications for surviving in unfavorable atmospheric variables such as biotic and abiotic stress factors through coordinated signaling and response mechanisms. Environmental stresses are deliberated as a global problem of yield reduction (Mahajan and Tuteja 2005; Augustine et al., 2015a; Song et al., 2019, 2021, 2022; Jagadesh et al., 2024). Plant cellular growth and development are vital for maintaining and balancing crop yield under stress conditions. Tolerant plants change their gene expression during stress to balance typical cellular and metabolic activities (Cho and Choi 2009; Verma et al., 2019a, 2019b; Wu et al., 2022; Xu et al., 2022; Huang et al., 2024). Plant biomass accumulation is positively related to the use and production of metabolites during diurnal variation. During the day, the photosynthesis activity in green healthy leaves is the major source of nonstructured carbohydrates that sustains/balances growth, development, and various metabolic activities (Anas et al., 2021a,b; Bhatt et al., 2023a, 2023b; Deng et al., 2023). The significance of sugarcane for the sugar and green-bioenergy sectors, various techniques built on the photosynthetic capability and molecular changes have been suggested to enhance cane productivity (Chen et al., 2020, 2023; Zeng et al., 2023).

Sugarcane accounts for about 80% production of worldwide sugar production (Chen et al., 2020; Verma et al., 2020a, 2023b; Guo et al., 2021, 2022, 2023). Furthermore, it is more effective for bioethanol production, and nearly half of the cane is utilized to produce bioethanol (Ali et al., 2019). Lately, established sugarcane cultivars have a diverse genome, with more polyploidy and aneuploidy. *Saccharum officinarum* genome composition is around 70–80% and 10–20% of *Saccharum spontaneum*, combining *S. officinarum* higher sugar content with *S. spontaneum* hardiness, disease avoidance, and ratooning (Zhang et al., 2018; Garsmeur et al., 2018; Singh et al., 2020, 2021; Malviya et al., 2021). Due to small genetic pools and a complex genome, based on the advanced breeding technologies is more difficult to sugarcane crop betterment (Mustafa et al., 2018; Huang et al., 2022; Li et al., 2023; Verma et al., 2023a, 2023b). A group of researchers associated with the omic strategies has been evaluated in sugarcane crop, that is, transcriptomics and proteomics research concerning enhanced production and sucrose content, and stress avoidance mechanisms (Sharma et al., 2021, 2022; Singh et al., 2023; Tian et al., 2023). The novel information has been attained based on molecular adaptations and mechanisms of sugarcane sensitivity, stress prevention, plant growth, development, and productivity (Barnabas et al., 2015; Kaur et al., 2017; Verma et al., 2020c, 2020d; Solanki et al., 2023).

Recently, various transcriptomic studies have identified a significant number of genes involved in regulating major biological pathways (Augustine et al., 2015b; Malviya et al., 2022; Oliveira et al., 2022). Transcriptomic techniques could classify genes as DNA markers or produce transgenic cane sugar (Li et al., 2016). Moreover, proteomic techniques have shown several differentially expressed proteins and its functions in response to transduction of signal pathways to unfavorable atmospheric conditions (Barnabas et al., 2015). It is vital for sugar and renewable energy sources; various photosynthetic and molecular modification techniques have been projected to enhance sugarcane productivity (Waclawovsky et al., 2010; De Souza et al., 2018; Verma et al., 2020e, 2021a, 2023c, 2023d; Huang et al., 2024). The development of stress tolerance and high-yield sugarcane cultivars is one of the main adaptation approaches to unfavorable environmental conditions (Verma et al., 2021b, 2021c; 2022a, 2022b). Plant breeders can develop a bioinformatics database to design hybridization as within or between cultivars/varieties for the necessity of the breeding programs for stress avoidance, and use of micropropagation techniques, molecular and gene transformation approaches to enhance breeding and selection efficiencies (Inman-Bamber et al., 2012; Zhao and Li 2015; Verma et al., 2021d, 2021e; Li et al., 2023). Recent technical methods and the agricultural challenges described above have resulted in high-emergence development tools to discover and exploit plant genomes for crop betterment. Molecular studies have discovered understanding the interactive role between genes, proteins, and metabolites. This chapter discusses molecular techniques for crop improvement during environmental stresses.

12.2 ROLE OF OMIC APPROACHES ON SUGARCANE PLANT GROWTH, DEVELOPMENT, AND PRODUCTIVITY

Various genes in sugarcane crops under unfavorable conditions have been revealed through molecular analysis (Li et al., 2016; Tripathi et al., 2019). *In silico* approach, that is, arrays of probe hybridization, expressed sequence tags from other related crops have been employed in transcriptome studies to deliver appropriate gene data (Figure 12.1). The Brazilian sugarcane expressed sequence tag (EST) findings are one of the biggest, with about 238,000 ESTs attained from different cDNA libraries (26) created from a broad range of Brazilian plant genotypes by applying different plant tissues (Cardoso-Silva et al., 2014). Conversely, the lack of a comprehensive and accurate sugarcane genome challenges the prediction of transcriptome database and genes application (Xu et al., 2018; Song et al., 2022; Verma et al., 2022c,d,e; Joshi et al., 2022, 2023). Hence, the optimum similarity in the genetic locations between the sugarcane genome and sorghum plants, the reference genome of sorghum is broadly applied as a transcriptomic experiment (Grivet et al., 1994; Wang et al., 2010).

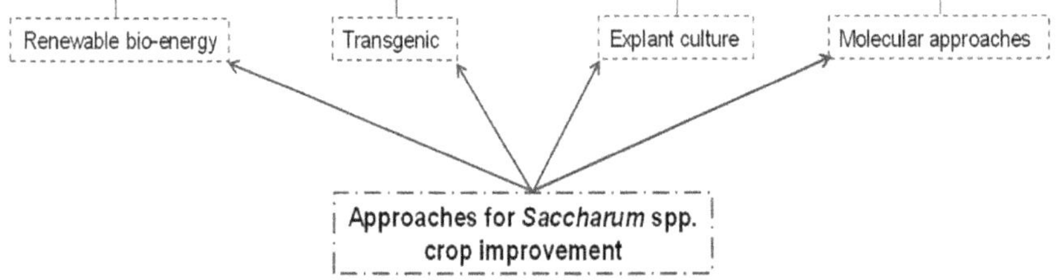

FIGURE 12.1 Different applications for sugarcane crop development/improvement.

Source: Adapted from Mustafa et al. (2018).

Evaluating the transcriptomic effects of stress factors on sugar cane crops may offer insights into protective actions and significant molecular actions to generate stress-tolerant plant genotypes. The effects of transcriptomic factors, that is, WRKY, MYB, bZIP, AP2/DREBP, and zinc finger-like proteins, which act as regulatory agents of metabolic pathways accountable for plant defense mechanisms in unfavorable conditions, have been broadly studied (Song et al., 2011).

Abiotic stresses have detrimental effects on sugarcane plant growth and productivity globally (Verma et al., 2022f, 2022g, 2022h; Kumari et al., 2022, 2023a, 2023b; Chaudhary et al., 2023). The transcriptome profiles of sugarcane drought-tolerant and susceptible cultivars during drought applied the Illumina HiScanSQ system and HiSeq 2500 platforms (Belesini et al., 2017). The various genes, that is, APx, MYB, E3 SUMO-Protein ligase SIZ2, the chief enzyme for flavonoids biosynthesis, that is, coenzyme A ligase, and aquaporin, are induced in stress-avoidance genotypes (Belesini et al., 2017; Guo et al., 2017).

The gene responsible for ABA was also enhanced in drought and temperature stress conditions. da Silva et al. (2017) observed using HT-SuperSAGE approaches to identify various stress avoidance and resistance strategies. The 9831 induced unitag genes present in the roots of sugarcane resistant genotypes were modulated differently in the sensitive genotypes after being stressed by a deficiency of sufficient water irrigation. In the metabolic process of sugarcane plants, various stress-responsive genes play a major role, like attenuation of ethylene stress, root elongation, loss of protein, oxidative detoxification, fatty acid synthesis, transportation of amino acids, carbohydrate metabolism, and pentose phosphate pathway (da Silva et al., 2017; Verma et al., 2022i, 2022j, 2022k). An experiment on *Saccharum narenga* (wild type) subjected to water deficit observed 3389 differentially expressed genes (DEGs) up- and down-regulated. DEGs are also associated with biological factors such as metabolic pathways, blue light response, and single transduction of plant hormones (Liu et al., 2018).

Polyamine oxidase, cytochrome c oxidase, S-adenosylmethionine (SAM) decarboxylase, and thioredoxin encoding genes are deliberated in sugarcane crops during unfavorable conditions and can aid in alleviating stress (Li et al., 2016; Tripathi et al., 2019). The encoding PP2C gene, that is, protein phosphatase, SAM decarboxylase, and two delta-12 oleate desaturases, was induced by ABA and limited water supply. *SodERF3* (ethylene-responsive factor) in sugarcane is induced by abscisic acid and drought situations and can be associated with stress avoidance strategy. Deviations in gene expression can defend adaptation of cells to water stress, and expression of genes exposes the complexities of defense mechanisms (Sreenivasulu et al., 2007; Li et al., 2016). Microarrays are a well-applied application for assessing DEGs due to their high-throughput capability, which simultaneously permits the study of various genes. A total of 15,593 sugarcane expression genes were found, and only 1501 differentially expressed genes during stress conditions (Li et al., 2016). Different demonstrations indicated functional annotation analyses of genes differentially expressed during stress (Zeng et al., 2015).

Differential gene expression profile on the *S. spontaneum* by transcriptomic application during cold-stressed conditions. The analysis of DEGs presents 5840 genes (upregulated 2538 and downregulated 3302) under cold temperature. Several cold-sensitive genes are linked to various metabolic pathways (Selvarajan et al., 2018). The effect of transcriptome profile on sugarcane under less potassium and nitrogen treatment was observed by Zeng et al. (2015) and Yang et al. (2019). Genes linked with freeze resistance stress were found in the profile expression of cold-inducible genes (Azevedo et al., 2011). The 3575 ESTs present in DEGs (165) in sugarcane plants with stress tolerance, signifying that many genes are linked to stress avoidance (Rodrigues et al., 2009). The ESTs may be categorized into different groups based on genotypes: two small ESTs from *S. arundinaceum* and *S. officinarum,* and an extra-large category of ESTs derived from *Saccharum* spp. hybrid, the most advanced sugarcane variety. XDH is a gene that encodes for a putative NAD-dependent dehydrogenase (Xu et al., 1996) linked to oxidative stress in the defensive system subjected to freezing conditions.

The transcriptomic studies on sugarcane plants have been executed with effective and recent omic applications, that is, cDNA microarrays, Roche/454 and Illumina/Solexa, RNA-sequence, qRT-PCR, and microscopy, for evaluating expression profiles of genes from cells/tissues and their influences on cellular and mechanical changes at sugarcane growth period (Kaur et al., 2017; Nishiyama et al., 2014). The mechanical variations and annotation of genes associated with the morphological factors are more important and helpful in sugarcane varieties that improve plant productivity. Transcriptomics studies have been executed to explain the expression of genes and to corroborate expression patterns (Manners et al., 2011). Different proteins linked with the biosynthesis of cellulose and lignin, abscission of leaf, maturity, and photosynthetic CO_2 assimilation rate have been confirmed the growth phases and physiological efficiency (Li et al., 2016; Kasirajan et al., 2018). Recent transcriptomic studies assess the genes associated with the traits of interest, that is, the content of sugar and the tolerance of abiotic stresses for crop betterment.

12.3 SUGARCANE DEVELOPMENT VS PROTEOMICS

In addition to transcriptomic and proteomics techniques, it provides updated insights into complex biological mechanisms (Barnabas et al., 2015; Singh et al., 2018). The functional activities for quantification of proteins and their post-translational derivatives are crucial to studying biological phenomena. While the genome is stagnant, the proteome of an individual dynamically responds to abiotic stresses and metabolic activities via different expression levels and post-translational variations like glycosylation, phosphorylation, methylation, acetylation, etc., further increasing the complexity of the proteome (Barnabas et al., 2015). Isolation and quantification strategies of different genes, like 2D electrophoresis, mass spectrometry, and matrix-assisted laser desorption/ionization-time of flight mass spectrometry, are used to measure the protein levels of differential and comparative expression in sugarcane crop subjected to agroclimatic variables (Barnabas et al. 2015). Also, iTRAQ is one of the main quantitative techniques used in various plant proteomic studies (Velez-Bermudez et al., 2016). Several proteomic research studies have been applied to cane crops using gel-based or gel-free methods in unfavorable stress conditions.

The morpho-physiological state and metabolic development of sugarcane are directly or indirectly associated with various stressors (Barnabas et al., 2015). Abiotic stresses are considered multidimensional as they affect numerous metabolic responses in plants, especially loss of crop production (Khueychai et al., 2015). Metabolic processes are the difference in protein synthesis/degradation that affects stress prevention (Figure 12.1). Proteomics-based demonstrations of stress-tolerant sugarcane functions and pathways have been approved. 2-DE linked to LC-ESI-IT-MS/MS/MS for isolation and detection of certain photosynthetic-related genes and destruction of antioxidative enzymes (Ngamhui et al., 2012). Different genes responded to stress-resistance and sensitive genotypes by 2-DE linked with LC–MS/MS application (Khueychai et al., 2015).

Various proteins, like fructose 1,6-bisphosphate aldolase, germin-like protein, glyceraldehyde 3-phosphate dehydrogenase, and HSPs, were DEGs during normal and soil salinity conditions in *Saccharum* spp. (Murad et al., 2014; Augustine et al., 2015b).

Proteomics techniques have also presented proteome regulation in sugarcane developmental processes and proteome profiling in several plant sections. Sugarcane growth and development processes need to attain full proteomic coverage. Proteomic techniques between putrescine and control situations assessed somatic embryogenesis-related DEGs, such as arabinogalactan proteins, peroxidases, HSPs, glutathione S-transferases, late embryogenesis abundant proteins, and 14-3-3 proteins played key functions in the protection of cells under abiotic stress in plants (Heringer et al., 2017). The cell wall proteins (CWPs) of young and mature sugarcane leaves and stems showed that leaves and young stems had the optimum lipid metabolism rate among all species and enriched 277 sugarcane-labeled CWPs. Transformation of omic approaches is applied to develop genetically modified sugarcane crop varieties (Zhao and Li 2015; Fonseca et al., 2018).

12.4 CONCLUSION

Sugarcane productivity will be directly or indirectly affected by climatic variables. Therefore, agro-scientists and policy decision-makers need to work closely to alleviate the potentially harmful effects of various stresses on crops and enhance sugarcane productivity through multi-disciplinary techniques, that is, consistently developing new sugarcane varieties/cultivars applying breeding and molecular biology. Advanced bioengineering approaches can develop environmentally friendly sugarcane varieties by minimizing the adverse responses. It has been applied to develop sugarcane aspects with tolerance, but some efforts have successfully developed transgenic varieties for commercial release globally. Genetic improvements in transgenic cane lines are not commercially released because of regulatory concerns and uncertainty regarding their development. Recently, Indonesia and Brazil released a transgenic sugarcane cultivar for commercial purposes. Other countries have developed to improve sugarcane programs worldwide for food protection and security in sustainable agriculture in the upcoming years.

KEYWORDS

- **crop improvement**
- **production**
- **stress tolerance**
- **omic approaches**
- **Saccharum spp.**
- **morphological**
- **cellular modifications**

ACKNOWLEDGMENT

The authors would like to thank the Guangxi Academy of Agricultural Sciences, Nanning, Guangxi, China, for providing the necessary facilities for this study.

FUNDING

This research was financially supported by the Guangxi Key R & D Program (Guike AB24010140), Guangxi Natural Science Foundation (2021GXNSFAA220022; 2023GXNSFAA026459; 2023GXNSFAA026460), Guangxi Innovation Teams of Modern Agriculture Technology (nycytxgxcxtd-2021–03), Guangxi Characteristic Crop Experimental Station (GTS2022022), National Key Research and Development Project (2022YFD2301102–07), The National Natural Science Foundation of China (31760415; 32060468), Fund of Guangxi Academy of Agricultural Sciences (2021YT011), Science and Technology Major Project of Guangxi (Guike AA22117002–1), Fundamental Research Fund of Guangxi Academy of Agriculture Sciences (2023YM55), Key Research and Development Program of Nanning (20232060) and Guangxi Key Laboratory of Sugarcane Genetic Improvement Project (21-238-16-K-03-09).

REFERENCES

Ali A, Pan YB, Wang Q-N, Wang J-D, Chen J-L, Gao S-J. (2019) Genetic diversity and population structure analysis of *Saccharum* and *Erianthus* genera using microsatellite (SSR) markers. *Sci. Rep.* 9, 395.

Anas M, Verma KK, Haq IU, Naeem M, Li Q, Liao F, Xiang YX, Li YR. (2021a) Characterization of exotic, native and wild-type genotypes of sugarcane (*Saccharum* spp. hybrids) for internal nitrogen use efficiency under different nitrogen levels. *Sugar Tech.* 23, 1258–1267. https://doi.org/10.1007/s12355-021-01014-1

Anas M, Verma KK, Riaz M, Qiang L, Liao F, Liu Y, Li YR. (2021b) Physio-morphological and biochemical mechanism of nitrogen use efficiency in sugarcane (*Saccharum* spp.) genotypes under different growth stages and nitrogen levels. *J. Plant Interact.* 16, 332–343.

Augustine SM, Cherian AV, Syamaladevi DP, Subramonian N. (2015b) *Erianthus arundinaceus* HSP70 (EaHSP70) Acts as a key regulator in the formation of anisotropic interdigitation in sugarcane (*Saccharum* spp. hybrid) in response to drought stress. *Plant Cell Physiol.* 56, 2368–2380.

Augustine SM, Narayan JA, Syamaladevi DP, Appunu C, Chakravarthi M, Ravichandran V, Tuteja N, Subramonian N. (2015a) Overexpression of EaDREB2 and pyramiding of EaDREB2 with the pea DNA helicase gene (PDH45) enhance drought and salinity tolerance in sugarcane (*Saccharum* spp. hybrid). *Plant Cell Rep.* 34, 247–263.

Azevedo RA, Carvalho RF, Cia MC, Gratao PL. (2011) Sugarcane under pressure: An overview of biochemical and physiological studies of abiotic stress. *Trop. Plant Biol.* 4, 42–51.

Barnabas L, Ramadass A, Amalraj RS, Palaniyandi M, Rasappa V. (2015) Sugarcane Proteomics: An update on current status, challenges, and future prospects. *Proteomics* 15, 1658–1670.

Belesini A, Carvalho F, Telles B, De Castro G, Giachetto P, Vantini J, Carlin S, Cazetta J, Pinheiro D, Ferro M. (2017) De novo transcriptome assembly of sugarcane leaves submitted to prolonged water-deficit stress. *Genet. Mol. Res.* 16, gmr16028845.

Bhatt R, Verma KK, Jamwal NS. (2023a) Can paddy-compost eliminate N fertilization for climate-smart sugarcane? *Sugar Tech.* 25, 1036–1046. https://doi.org/10.1007/s12355-023-01270-3

Bhatt R, Verma KK, Kumar R, Sanghera GS. (2023b) Foliar application of potassium salt of active phosphorus (PSAP) mitigates insect pests and improves yield along with sugarcane quality in response to agroclimatic conditions of Punjab. *Sugar Tech.* 25(3), 660–669.

Cardoso-Silva CB, Costa EA, Mancini MC, Balsalobre TWA, Canesin LEC, Pinto LR, Carneiro MS, Garcia AAF, de Souza AP, Vicentini R. (2014) De novo assembly and transcriptome analysis of contrasting sugarcane varieties. *PLoS ONE* 9, e88462.

Chaudhary P, Xu M, Ahamad L, Chaudhary A, Kumar G, Adeleke BS, Verma KK, Hu D-M, Siric I, Kumar P, Popescu SM, Fayssal SA. (2023) Application of synthetic consortia for improvement of soil fertility, pollution remediation, and agricultural productivity: A review. *Agronomy* 13, 643. https://doi.org/10.3390/agronomy13030643.

Chen G-L, Zheng F-J, Lin B, Lao S-B, He J, Huang Z, Zeng Y, Sun J, Verma KK. (2020) Phenolic and volatile compounds in the production of sugarcane vinegar. *ACS Omega* 5, 30587.

Chen GL, Zheng FJ, Lin B, Lao SB, He J, Huang Z, Zeng Y, Sun J, Verma KK. (2020) Phenolic and volatile compounds in the production of sugarcane vinegar. *ACS Omega* 5, 30587–30595.

Chen G-L, Zheng F-J, Lin B, Yang Y-X, Fang X-C, Verma KK, Yang L-F. (2023) Vinegar: A potential source of healthy and functional food with special reference to sugarcane vinegar. *Front. Nutr.* 10, 1145862. https://doi.org/10.3389/fnut.2023.1145862

Cho EK, Choi YJ. (2009) A nuclear-localized HSP70 confers thermo protective activity and drought-stress tolerance on plants. *Biotechnol. Lett.* 31, 597–606.

da Silva MD, de Oliveira Silva RL, Neto JRCF, Benko-Iseppon AM, Kido EA. (2017) Genotype-dependent regulation of drought-responsive genes in tolerant and sensitive sugarcane cultivars. *Gene* 633, 17–27.

De Souza AP, Grandis A, Arenque-Musa BC, Buckeridge MS. (2018) Diurnal variation in gas exchange and nonstructural carbohydrates throughout sugarcane development. *Funct. Plant Biol.* 45, 865–876.

Deng ZN, Huang HR, Huang CM, Verma KK, Luo HB, Jiang SL, Li YJ, Cao HQ, Wu KC, Xu L. (2023) Short-term limited water irrigation influences the photosynthetic pigments, enzymatic and non-enzymatic activities in *Saccharum spontaneum* L. at vegetative stage. *Sugar Tech.* 25, 110–118. https://doi.org/10.1007/s12355-022-01175-7

Fonseca JG, Calderaneran Pases MJ, de Moraes FE, Cataldi TR, Jamet E, Labate CA. (2018) Cell wall proteome of sugarcane young and mature leaves and stems. *Proteomics* 18, 1700129.

Garsmeur O, Droc G, Antonise R, Grimwood J, Potier B, Aitken K, Jenkins J, Martin G, Charron C, Hervouet C. (2018) A mosaic monoploid reference sequence for the highly complex genome of sugarcane. *Nat. Commun.* 9, 2638.

Grivet L, D'Hont A, Dufour P, Hamon P, Roques D, Glaszmann J-C. (1994) Comparative genome mapping of sugar cane with other species within the Andropogoneae tribe. *Heredity* 73, 500.

Guo D-J, Li D-P, Chen Z-D, Yang B, Verma KK, Singh RK, Singh P, Khan Q, Sharma A, Qin Y, Zhang B-Q, Song X-P, Li Y-R. (2023) Effect of endophytic diazotroph *Enterobacter roggenkampii* ED5 on nitrogen-metabolism-related microecology in the sugarcane rhizosphere at different nitrogen levels. *Front. Microbiol.* 14, 1132016. https://doi.org/10.3389/fmicb.2023.1132016

Guo D-J, Li D-P, Singh RK, Singh P, Sharma A, Verma KK, Qin Y, Khan Q, Lu Z, Malviya MK, Song X-P, Xing Y-X and Li Y-R (2021) Differential protein expression analysis of two sugarcane varieties in response to diazotrophic plant growth-promoting endophyte enterobacter roggenkampii ED5. *Front. Plant Sci.* 12, 727741. https://doi.org/10.3389/fpls.2021.727741

Guo DJ, Li DP, Singh RK, Singh P, Verma KK, Sharma A, Qin Y, Khan Q, Song XP, Malviya MK, Xing YX, Li YR. (2022) Comparative transcriptome analysis of two sugarcane varieties in response to diazitrophic plant growth promoting endophyte *Enterobacter roggenkampii* ED5. *J. Plant Interactions*, 17(1), 75–84.

Guo J, Ling H, Ma J, Chen Y, Su Y, Lin Q, Gao S, Wang H, Que Y, Xu L. (2017) A sugarcane R2R3-MYB transcription factor gene is alternatively spliced during drought stress. *Sci. Rep.* 7, 41922.

Heringer AS, Reis RS, Passamani LZ, de Souza-Filho GA, Santa-Catarina C, Silveira V. (2017) Comparative proteomics analysis of the e_ect of combined red and blue lights on sugarcane somatic embryogenesis. *Acta Physiol. Plant.* 39, 52.

Huang CM, Wu KC, Verma KK, Luo HB, Deng ZN, Shi GY, Cao HQ, Wu XJ, Ye LP, Wei YW, Xu L. (2024) Isolation and expression analysis of cellulose synthase 3 (*Ces3*) genes from sugarcane (*Saccharum officinarum* L.). *Chem. Biol. Technol. Agric.* 11, 71. https://doi.org/10.1186/s40538-024-00597-7

Huang HR, Gao YJ, Malviya MK, Verma KK, Solanki MK, Huang YX, Li X, Deng Y, Yan J, Tang SY, Wang LW, Xu L. (2022) Genetic diversity analysis of sugarcane (*Saccharum* spp. Hybrids) among high-sucrose clones of GT series and commonly used parents by using microsatellite markers in Guangxi, *China. Sugar Tech.* 24, 397–407. https://doi.org/10.1007/s12355-021-01033-y

Inman-Bamber NG, Lakshmanan P, Park S. (2012) Sugarcane for water-limited environments: Theoretical assessment of suitable traits. *Field Crops Res.* 134, 95–104.

Jagadesh M, Dash M, Kumari A, Singh SK, Verma KK, Kumar P, Bhatt R, Sharma SK. (2024) Revealing the hidden world of soil microbes: Metagenomic insights into plant, bacteria, and fungi interactions for sustainable agriculture and ecosystem restoration. *Microbiol. Res.* 285, 127764. https://doi.org/10.1016/j.micres.2024.127764

Joshi A, Verma KK, Rajput VD, Arora J. (2023) Metabolic engineering of microorganisms in advancing biofuel production. In: *Advances in Lignocellulosic Biofuel Production Systems*. pp. 181–202. https://doi.org/10.1016/B978-0-323-91192-4.00016-X

Joshi A, Verma KK, Rajput VD, Minkina T, Arora J. (2022) Recent advances in metabolic engineering of microorganisms for advancing lignocellulose-derived biofuels. *Bioengineered* 13(4), 8135–8163. https://doi.org/10.1080/21655979.2022.2051856.

Kasirajan L, Hoang NV, Furtado A, Botha FC, Henry RJ. (2018) Transcriptome analysis highlights key di_erentially expressed genes involved in cellulose and lignin biosynthesis of sugarcane genotypes varying in fiber content. *Sci. Rep.* 8, 11612.

Kaur L, Dharshini S, Ram B, Appunu C. (2017) Sugarcane Genomics and Transcriptomics. In: *Sugarcane Biotechnology: Challenges and Prospects*. Springer: Cham, Switzerland, pp. 13–32.

Khueychai S, Jangpromma N, Daduang S, Jaisil P, Lomthaisong K, Dhiravisit A, Klaynongsruang S. (2015) Comparative proteomic analysis of leaves, leaf sheaths, and roots of drought-contrasting sugarcane cultivars in response to drought stress. *Acta Physiol. Plant.* 37, 88.

Krishna D, Sachan HK, Verma KK. (2022) Abiotic stresses in plants: An overview. In: *Book: In: Understanding Abiotic Stresses Publisher*. Nova Science Publishers, Inc.: New York.

Kumari A, Dash M, Singh SK, Jagadesh M, Mathpal B, Mishra PK, Pandey SK, Verma KK. (2023a) Soil microbes: A natural solution for mitigating the impact of climate change. *Environ. Monit. Assess.* 195, 1436. https://doi.org/10.1007/s10661-023-11988-y

Kumari A, Lakshmi GA, Krishna GK, Patni B, Prakash S, Bhattacharyya M, Singh SK, Verma KK. (2022) Climate change and its impact on crops: A comprehensive investigation for sustainable agriculture. *Agronomy* 12, 3008. https://doi.org/10.3390/agronomy12123008

Kumari A, Singh SK, Mathpal B, Verma KK, Garg VK, Bhattacharyya M, Bhatt R. (2023b) The biosynthesis, mechanism of action, and physiological functions of melatonin in horticultural plants: A review. *Horticulturae* 9, 913. https://doi.org/10.3390/horticulturae9080913

Li A-M, Liao F, Wang M, Chen Z-L, Qin C-X, Huang R-Q, Verma KK, Li Y-R, Que Y-X, Pan Y-Q, et al. (2023) Transcriptomic and proteomic landscape of sugarcane response to biotic and abiotic stressors. *Int. J. Mol. Sci.* 24, 8913. https://doi.org/10.3390/ijms24108913

Li M, Liang Z, Zeng Y, Jing Y, Wu K, Liang J, He S, Wang G, Mo Z, Tan F. (2016) De novo analysis of transcriptome reveals genes associated with leaf abscission in sugarcane (*Saccharum officinarum* L.). *BMC Genom.* 17, 195.

Liu X, Zhang R, Ou H, Gui Y, Wei J, Zhou H, Tan H, Li Y. (2018) Comprehensive transcriptome analysis reveals genes in response to water deficit in the leaves of *Saccharum narenga* (Nees ex Steud.) hack. *BMC Plant Biol.* 18, 250.

Mahajan S, Tuteja N. (2005) Cold, salinity and drought stresses: An overview. *Arch. Biochem. Biophys.* 444, 139–158.

Malviya MK, Li C-N, Lakshmanan P, Solanki MK, Wang Z, Solanki AC, Nong Q, Verma KK, Singh RK, Singh P, Sharma A, Guo D-J, Dessoky ES, Song X-P, Li Y-R. (2022) High-throughput sequencing-based analysis of rhizosphere and diazotrophic bacterial diversity among wild progenitor and closely related species of sugarcane (*Saccharum* spp. inter-specific hybrids). *Front. Plant Sci.* 13, 829337. *https://doi.org/10.3389/fpls.2022.829337.*

Malviya MK, Solanki MK, Li CN, Wang Z, Zeng Y, Verma KK, Singh RK, Singh P, Huang HR, Yang LT, Song XP, Li YR. (2021) Sugarcane-legume intercropping can enrich the soil microbiome and plant growth. *Front. Sustain. Food Syst.* 5, 606595. https://doi.org/10.3389/fsufs.2021.606595.

Manners JM, Casu RE. (2011) Transcriptome analysis and functional genomics of sugarcane. *Trop. Plant Biol.* 4, 9–21.

Murad AM, Molinari HBC, Magalhães BS, Franco AC, Takahashi FSC, Franco OL, Quirino BF. (2014) Physiological and proteomic analyses of *Saccharum* spp. grown under salt stress. *PLoS One* 9, e98463.

Mustafa G, Joyia FA, Anwar S, Parvaiz A, Khan MS. (2018) Biotechnological interventions for the improvement of sugarcane crop and sugar production. In: *Sugarcane-Technology and Research;* IntechOpen: London, UK, pp. 113–138.

Ngamhui NO, Akkasaeng C, Zhu YJ, Tantisuwichwong N, Roytrakul S, Sansayawichai T. (2012) Differentially expressed proteins in sugarcane leaves in response to water deficit stress. *Plant Omics* 5, 365.

Nishiyama MYJr, Ferreira SS, Tang P-Z, Becker S, Poertner-Taliana A, Souza GM. (2014) Full-length enriched cDNA libraries and ORFeome analysis of sugarcane hybrid and ancestor genotypes. *PLoS One* 9, e107351.

Oliveira MA, Verma KK, Bhatt R, Oliveira TBA. (2022) Impact of green and organic fertilizers on soil fertility and sugarcane productivity. In: Verma KK, et al. (Eds.), *Agro-Industrial Perspectives on Sugarcane Production Under Environmental Stress*. https://doi.org/10.1007/978-981-19-3955-6_11.

Rodrigues FA, Laia ML, Zingaretti SM. (2009) Analysis of gene expression profiles under water stress in tolerant and sensitive sugarcane plants. *Plant Sci.* 176, 286–302.

Selvarajan D, Mohan C, Dhandapani V, Nerkar G, Jayanarayanan AN, Mohanan MV, Murugan N, Kaur L, Chennappa M, Kumar R. (2018) Differential gene expression profiling through transcriptome approach of *Saccharum spontaneum* L. under low temperature stress reveals genes potentially involved in cold acclimation. *3 Biotech* 8, 195.

Sharma A, Singh RK, Singh P, Vaishnav A, Guo DJ, Verma KK, Li DP, Song XP, Malviya MK, Khan N, Lakshmanan P, Li YR. (2021) Insights into the bacterial and nitric oxide-induced salt tolerance in sugarcane and their growth-promoting abilities. *Microorganisms* 9, 2203. https://doi.org/10.3390/microorganisms9112203.

Sharma A, Song X-P , Singh RK, Vaishnav A, Gupta S, Singh P, Guo D-J, Verma KK, Li Y-R. (2022) Impact of carbendazim on cellular growth, defence system and plant growth promoting traits of *Priestia megaterium* ANCB-12 isolated from sugarcane rhizosphere. *Front. Microbiol.* 13, 1005942. doi:10.3389/fmicb.2022.1005942

Singh NP, Shami V. (2018) Abiotic stress tolerance in sugarcane using genomics and proteomics techniques. In: *Biotechnology to Enhance Sugarcane Productivity and Stress Tolerance*. CRC Press: Boca Raton, FL, USA, pp. 97–114.

Singh P, Singh RK, Li HB, Guo DJ, Sharma A, Verma KK, Solanki MK, Upadhyay SK, Lakshmanan P, Yang LT, Li YR. (2023) Nitrogen fixation and phytohormone stimulation of sugarcane plant through plant growth promoting diazotrophic *Pseudomonas*. *Biotechnol. Genetic Eng. Rev.* 1–21. https://doi.org/10.1080/02648725.2023.2177814

Singh P, Singh RK, Song QQ, Li HB, Guo DJ, Malviya MK, Verma KK, Song XP, Lakshmanan P, Yang LT, Li YR. (2021) Comparative analysis of protein and differential responses of defense-related gene and enzyme activity reveals the long-term molecular responses of sugarcane inoculated with *Sporisorium scitamineum*. *J. Plant Interactions* 16(1), 12–29.

Singh RK, Singh P, Li HB, Song QQ, Guo DJ, Solanki MK, Verma KK, Malviya MK, Song XP, Lakshmanan P, Yang LT, Li YR. (2020) Diversity of nitrogen-fixing rhizobacteria associated with sugarcane: A comprehensive study of plant-microbe interactions for growth enhancement in *Saccharum* spp. *BMC Plant Biol.* 20, 220. https://doi.org/10.1186/s12870-020-02400-9.

Solanki MK, Verma KK, Dastogeer KMG, Mora-Poblete F, Mundra S. (2023) Microbial resilience in plant nutrient management towards sustainable farming. *Front. Microbiol.* 14, 1280811. https://doi.org/10.3389/fmicb.2023.1280811

Song S, Qi T, Huang H, Ren Q, Wu D, Chang C, Peng W, Liu Y, Peng J, Xie D. (2011) The jasmonate-ZIM domain proteins interact with the R2R3-MYB transcription factors MYB21 and MYB24 to affect jasmonate-regulated stamen development in Arabidopsis. *Plant Cell* 23, 1000–1013.

Song X-P, Mo F, Verma KK, Wei J, Zhang X, Yang L, Li YR. (2019) Effect of sugarcane smut (*Ustilago scitaminea* Syd.) on ultrastructure and biochemical indices of sugarcane. *Biomed J. Sci. Tech. Res.* 002950. https://doi.org/10.26717/BJSTR.2019.17.002950.

Song XP, Mo F, Yan M, Zhang X, Zhang B, Huang X, Huang D, Pan Y, Verma KK, Li YR. (2022) Effect of smut infection on the photosynthetic physiological characteristics and related defense enzymes of sugarcane. *Life* 12, 1201. https://doi.org/10.3390/life12081201

Song XP, Verma KK, Tian DD, Zhang XQ, Liang YJ, Huang X, Li CN, Li YR. (2021) Exploration of silicon functions to integrate with biotic stress tolerance and crop improvement. *Biol. Res.* 54, 19. https://doi.org/10.1186/s40659-021-00344-4

Sreenivasulu N, Sopory SK, Kavi Kishor PB. (2007) Deciphering the regulatory mechanisms of abiotic stress tolerance in plants by genomic approaches. *Gene* 388, 1–13.

Tian D, Qin L, Verma KK, Wei L, Li J, Li B, Zhou W, He Z, Wei D, Huang S, Long S, Huang Q, Li C, Wei S. (2023) Transcriptomic and metabolomic differences between banana varieties which are resistant or susceptible to *Fusarium wilt. PeerJ* 11, e16549. https://doi.org/10.7717/peerj.16549

Tripathi P, Chandra A, Prakash J. (2019) Physio-biochemical assessment and expression analysis of genes associated with drought tolerance in sugarcane (*Saccharum* spp. hybrids) exposed to GA_3 at grand growth stage. *Plant Biol.* 21, 45–53.

Velez-Bermúdez IC, Wen T-N, Lan P, Schmidt W. (2016) Isobaric tag for relative and absolute quantitation (iTRAQ)-based protein profiling in plants. In: *Plant Proteostasis*. Springer: New York, NY, USA, pp. 213–221.

Verma KK, Anas M, Chen Z, Rajput VD, Malviya MK, Verma CL, Singh RK, Singh P, Song XP, Li YR. (2020) Silicon supply improves leaf gas exchange, antioxidant defense system and growth in sugarcane responsive to water limitation. *Plants* 9, 1032, https://doi.org/10.3390/plants9081032.

Verma KK, Joshi A, Song X-P, Liang Q, Xu L, Huang H-r, Wu K-C, Seth CS, Arora J and Li Y-R (2024a) Regulatory mechanisms of plant rhizobacteria on plants to the adaptation of adverse agroclimatic variables. *Front. Plant Sci.* 15, 1377793. https://doi.org/10.3389/fpls.2024.1377793

Verma KK, Joshi A, Song X-P, Singh S, Kumari A, Arora J, Singh SK, Solanki MK, Seth CS, Li Y-R. (2024b) Synergistic interactions of nanoparticles and plant growth promoting rhizobacteria enhancing soil-plant systems: A multigenerational perspective. *Front. Plant Sci.* 15, 1376214. https://doi.org/10.3389/fpls.2024.1376214

Verma KK, Li DM, Singh M, Rajput VD, Malviya MK, Minkina T, Singh RK, Singh P, Song XP, Li YR. (2020a) Interactive role of silicon and plant-rhizobacteria mitigating abiotic stresses: A new approach for sustainable agriculture and climate change. *Plants* 9, 1055. https://doi.org/10.3390/plants9091055.

Verma KK, Rajput VD, Singh M, Verma CL, Minkina T. (2022a) Influence of waterlogging on plant physiology and molecular biology. In: *Understanding Abiotic Stresses*. Nova Science Publishers, Inc.: New York.

Verma KK, Siddiqui N, Singh M. (2022b) Understanding the plant physiological and molecular adaptation strategies to soil salinity. In: *Understanding Abiotic Stresses*. Nova Science Publishers, Inc.: New York. https://doi.org/10.52305/UBUJ4024

Verma KK, Singh P, Song X-P, Malviya MK, Singh RK, Chen G-L, Solomon S, Li YR. (2020b) Mitigating climate change for sugarcane improvement: Role of silicon in alleviating abiotic stresses. *Sugar Tech.* 22(5), 741–749. https://doi.org/10.1007/s12355-020-00831-0.

Verma KK, Singh RK, Song QQ, Singh P, Zhang B-Q, Song X-P, Chen G-L, Li YR. (2019a) Silicon alleviates drought stress of sugarcane plants by improving antioxidant responses. *Biomed J. Sci. Tech. Res.* 002957. https://doi.org/10.26717/BJSTR.2019.17.002957

Verma KK, Song X, Degu HD, Guo D, Joshi A, Huang H, Xu L, Singh M, Huang D, Rajput VD, Li YR. (2023a) Recent advances in nitrogen and nano-nitrogen fertilizers

for sustainable crop production: A mini-review. *Chem. Biol. Technol. Agric.* 10, 1–14. https://doi.org/10.1186/s40538-023-00488-3
Verma KK, Song X, Verma CL, Huang H, Singh M, Xu L, Wu J, Khan ZH, Zhang B, Li Y. (2023b) Mathematical modeling of climate and fluoride effects on sugarcane photosynthesis with silicon nanoparticles. *Plant Physiol. Biochem.* 204, 108089. https://doi.org/10.1016/j.plaphy.2023.108089
Verma KK, Song XP, Budeguer F, Nikpay A, Enrique R, Singh M, Zhang B, Wu, JM, Li YR. (2022c) Genetic engineering: An efficient approach to mitigating biotic and abiotic stresses in sugarcane cultivation. *Plant Signal. Behav.* 17(1), e2108253.
Verma KK, Song XP, Joshi A, Rajput VD, Singh M, Sharma A, Singh RK, Li DM, Arora J, Minkina T, Li YR. (2022d) Nanofertilizer possibilities for healthy soil, water and food in future: An overview. *Front. Plant Sci.* 13, 865048. https://doi.org/10.3389/fpls.2022.865048
Verma KK, Song XP, Joshi A, Tian DD, Rajput VD, Singh M, Arora J, Minkina TM, Li YR. (2022e) Recent trends in nano-fertilizer for advancing sustainable agriculture under the era of climate change to ensure food security for future generations. *Nanomaterials* 12, 173. https://doi.org/10.3390/nano12010173.
Verma KK, Song XP, Li DM, Singh M, Wu JM, Singh RK, Sharma A, Zhang BQ, LI YR. (2022f) Silicon and soil microorganisms improves rhizospheric soil health with bacterial community, plant growth, performance and yield. *Plant Signal. Behav.* 17(1), e2104004.
Verma KK, Song XP, Lin B, Guo DJ, Singh M, Rajput VD, Singh RK, Singh P, Sharma A, Malviya MK, Chen GL, Li YR. (2022g) Silicon induced drought tolerance in crop plants: Physiological adaptation strategies. *Silicon* 14(6), 2473–2487. https://doi.org/10.1007/s12633-021-01071-x
Verma KK, Song XP, Rajput VD, Boldyreva V, Zhang BQ, Minkina T, Li YR. (2022h) Morpho-physiological, biochemical and ultrastructural modifications on sugarcane to prolonged water deficit. In: Verma KK, et al. (Eds.), *Agro-Industrial Perspectives on Sugarcane Production Under Environmental Stress*. https://doi.org/10.1007/978-981-19-3955-6_11.
Verma KK, Song XP, Rajput VD, Solomon S, Li YR, Rao GP. (2022i) *Agro-Industrial Perspectives on Sugarcane Production under Environmental Stress*. Springer Nature: Singapore, https://doi.org/10.1007/978-981-19-3955-6
Verma KK, Song XP, Singh M, Tian DD, Rajput VD, Minkina T, Li YR. (2023c) Association of silicon and soil microorganisms induces stress mitigation, increasing plant productivity. In: de Mello Prado R. (ed.), *Benefits of Silicon in the Nutrition of Plants*. pp. 299–328. https://doi.org/10.1007/978-3-031-26673-7_17
Verma KK, Song XP, Tian DD, Guo DJ, Chen ZL, Zhong CS, Nikpay A, Singh M, Rajput VD, Singh RK, Minkina T, Li YR. (2021a) Influence of silicon on biocontrol strategies to manage biotic stress for crop protection, performance and improvement. *Plants* 10, 2163. https://doi.org/10.3390/plants10102163
Verma KK, Song XP, Tian DD, Singh M, Verma CL, Rajput VD, Singh RK, Sharma A, Singh P, Malviya MK, Li YR. (2021b) Investigation of defensive role of silicon during drought stress induced by irrigation capacity in sugarcane: Physiological and biochemical characteristics. *ACS Omega* 6, 19811–19821.
Verma KK, Song XP, Verma CL, Chen ZL, Rajput VD, Wu KC, Liao F, Chen GL, Li YR. (2021c) Functional relationship between photosynthetic leaf gas exchange in response

to silicon application and water stress mitigation in sugarcane. *Biol. Res.* 54, 15. https://doi.org/10.1186/s40659-021-00338-2

Verma KK, Song XP, Verma CL, Malviya MK, Guo DJ, Rajput VD, Sharma A, Wei KJ, Chen GL, Solomon S, Li YR. (2021d) Predication of photosynthetic leaf gas exchange of sugarcane (*Saccharum* spp.) leaves in response to leaf positions to foliar spray of potassium salt of active phosphorus under limited water irrigation. *ACS Omega* 6, 2396–2409.

Verma KK, Song X-P, Xu L, Huang H-R, Liang Q, Seth CS, Li Y-R. (2023d) Nano-microplastic and agro-ecosystems: A mini-review. *Front. Plant Sci.* 14, 1283852. https://doi.org/10.3389/fpls.2023.1283852

Verma KK, Song XP, Yadav G, Degu HD, Parvaiz A, Singh M, Huang HR, Mustafa G, Xu L, Li YR. (2022j) Impact of agroclimatic variables on proteo-genomics in sugarcane (*Saccharum* spp.) plant productivity. *ACS Omega* 7, 22997-23008.

Verma KK, Song XP, Zeng Y, Guo DJ, Sing M, Rajput VD, Malviya MK, Wei KJ, Sharma A, Li DP, Chen GL, Li YR. (2021e) Foliar application of silicon boosts growth, photosynthetic leaf gas exchange, antioxidative response and resistance to limited water irrigation in sugarcane (*Saccharum officinarum* L.). *Plant Physiol. Biochem.* 166, 582–592.

Verma KK, Song XP,Zeng Y, Li DM, Guo DJ, Rajput VD, Chen GL, Barakhov A, Minkina TM, Li YR. (2020c) Characteristics and correlation of leaf stomata and its relationship with photosynthesis on *Saccharum* spp. under different irrigation and silicon application. *ACS Omega* 5, 24145–24153.

Verma KK, Vatsal S, Singh M, Rajput VD, Minkina T. (2022k) Coping with water stress: Effects and adaptive mechanisms to plants. In: *Understanding Abiotic Stresses*. Nova Science Publishers, Inc.: New York.

Verma KK, Wu KC, Verma CL, Li DM, Malviya MK, Singh RK, Singh P, Chen GL, Song XP, Li YR. (2020d) Developing mathematical model for diurnal dynamics on photosynthetic responses in sugarcane responsive to different irrigation and silicon application. *PeerJ.* 8, e10154. https://doi.org/10.7717/peerj.10154.

Verma KK, Liu X-H, Wu K-C, Singh RK, Song QQ, Malviya MK, Song X-P, Singh P, Verma CL, Li YR. (2020e) The impact of silicon on photosynthetic and biochemical responses of sugarcane under different soil moisture levels. *Silicon* 12, 1355–1367.

Verma KK, Wu K-C, Singh P, Malviya MK, Singh RK, Song X-P, Li YR. (2019b) The protective role of silicon in sugarcane under water stress: Photosynthesis and antioxidant enzymes. *Biomed. J. Sci. Tech. Res.* 002685. https://doi.org/0.26717/BJSTR.2019.15.002685

Waclawovsky AJ, Sato PM, Lembke CG, Moore PH, Souza GM. (2010) Sugarcane for bioenergy production: An assessment of yield and regulation of sucrose content. *Plant Biotechnol. J.* 8, 263–276.

Wang J, Roe B, Macmil S, Yu Q, Murray JE, Tang H, Chen C, Najar F, Wiley G, Bowers J. (2010) Microcollinearity between autopolyploid sugarcane and diploid sorghum genomes. *BMC Genom.* 11, 261.

Wu K-C, Huang C-M, Verma KK, Deng Z-N, Huang H-R, Pang T, Cao H-Q, Luo H-B, Jiang S-L, Xu L. (2022) Transcriptomic responses of *Saccharum spontaneum* roots in response to polyethylene glycol–6000 stimulated drought stress. *Front. Plant Sci.* 13, 992755. https://doi.org/10.3389/fpls.2022.992755

Xu L, Deng Z-N, Wu K-C, Malviya MK, Solanki MK, Verma KK, Pang T, Li Y-J, Liu X-Y, Kashyap BK, Dessoky ES, Wang W-Z and Huang H-R. (2022) Transcriptome analysis reveals a gene expression pattern that contributes to sugarcane bud propagation induced by Indole-3-Butyric Acid. *Front. Plant Sci.* 13, 852886. https://doi.org/10.3389/fpls.2022.852886

Xu P, Huecksteadt TP, Hoidal JR (1996) Molecular cloning and characterization of the human xanthine dehydrogenase gene (XDH). *Genomics* 34, 173–180.

Xu S, Wang J, Shang H, Huang Y, Yao W, Chen B, Zhang M. (2018) Transcriptomic characterization and potential marker development of contrasting sugarcane cultivars. *Sci. Rep.* 8, 1683.

Yang Y, Gao S, Su Y, Lin Z, Guo J, Li M, Wang Z, Que Y, Xu L. (2019) Transcripts and low nitrogen tolerance: Regulatory and metabolic pathways in sugarcane under low nitrogen stress. *Environ. Exp. Bot.* 163, 97–111.

Zeng Q, Ling Q, Fan L, Li Y, Hu F, Chen J, Huang Z, Deng H, Li Q, Qi Y. (2015) Transcriptome profiling of sugarcane roots in response to low potassium stress. *PLoS ONE* 10, e0126306.

Zhang J, Zhang X, Tang H, Zhang Q, Hua X, Ma X, Zhu F, Jones T, Zhu X, Bowers J. (2018) Allele-defined genome of the autopolyploid sugarcane *Saccharum spontaneum* L. *Nat. Genet.* 50, 1565.

Zhao D, Li Y-R. (2015) Climate change and sugarcane production: Potential impact and mitigation strategies. *Int. J. Agron.* 2015, 547386.

Index

C

I

J

K

L

M

N

P

www.ingramcontent.com/pod-product-compliance
Ingram Content Group UK Ltd.
Pitfield, Milton Keynes, MK11 3LW, UK
UKHW022159190926
475702UK00003B/115
* 9 7 8 1 7 7 9 6 4 1 1 2 0 *